Bayesian Approach
to Interpreting
Archaeological Data

Bayesian Approach to Interpreting Archaeological Data

Caitlin E. Buck

School of History and Archaeology, University of Wales, Cardiff, U.K.

William G. Cavanagh

Department of Archaeology, University of Nottingham, Nottingham, U.K.

Clifford D. Litton

Department of Mathematics, University of Nottingham, Nottingham, U.K.

JOHN WILEY & SONS

Chichester · New York · Brisbane · Toronto · Singapore

Other Wiley Editorial Offices

John Wiley & Sons, Inc., 605 Third Avenue,
New York, NY 10158-0012, USA

Jacaranda Wiley Ltd, 33 Park Road, Milton,
Queensland 4064, Australia

John Wiley & Sons (Canada) Ltd, 22 Worcester Road,
Rexdale, Ontario M9W 1L1, Canada

John Wiley & Sons (Asia) Pte Ltd, 2 Clementi Loop #02-01,
Jin Xing Distripark, Singapore 0512

British Library Cataloguing in Publication Data

A catalogue record for this book is available from the British Library

ISBN 0 471 96197 3

Produced from camera-ready copy supplied by the authors.
Printed and bound in Great Britain by Bookcraft (Bath) Ltd.
This book is printed on acid-free paper responsibly manufactured from sustainable
forestation, for which at least two trees are planted for each one used for paper production.

Contents

Series Preface

Statistics in Practice is an important international series of texts that provide direct coverage of statistical concepts, methods and worked case studies in specific fields of investigation and study.

With sound motivation and many worked practical examples, the books show in down-to-earth terms how to select and use a specific range of statistical techniques in a particular practical field within each title's special topic area.

The books meet the need for statistical support required by professionals and research workers across a wide range of employment fields and research environments. The series covers a wide variety of subject areas: in medicine and pharmaceutics (e.g. in laboratory testing or clinical trials analysis); in social and administrative fields (e.g. for sample surveys or data analysis); in industry, finance and commerce (e.g. for design or forecasting); in the public services (e.g. in forensic science); in the earth and environmental sciences; and so on.

But the books in the series have an even wider relevance than this. Increasingly, statistical departments in universities and colleges are realizing the need to provide at least a proportion of their course-work in highly specific areas of study to equip their graduates for the work environment. For example, it is common for courses to be given on statistics applied to medicine, industry, social and administrative affairs, and the books in this series provide support for such courses.

It is our aim to present judiciously chosen and well-written workbooks to meet everyday practical needs. Feedback of views from readers will be most valuable to monitor the success of this aim.

Vic Barnett
Series Editor
1996

Preface

This book is intended for undergraduate and postgraduate students of archaeology and cognate disciplines, and for professional archaeologists. We aim to introduce to members of the archaeological community the ideas underlying the Bayesian approach to the statistical analysis of data and their subsequent interpretation. The major advantage of the Bayesian approach is that it allows the incorporation of relevant prior knowledge or beliefs into the analysis. By doing so it provides a logical and coherent way of updating our beliefs—from those held before (prior to) observing the data to those held after (posterior to) taking the data into account. Central to the discussion are how to elicit the prior beliefs, how to combine them with the data, and how to interpret the posterior beliefs.

We explore the processes through which archaeologists handle their data and how these can be made more rigorous and effective by formal mathematical and statistical modelling. A great strength of the Bayesian approach is not only that it permits the inclusion of prior knowledge, but that it can combine into a single analysis the expertise of different specialists. We present in, we hope, an accessible manner the procedures of mathematical and statistical modelling within the Bayesian framework; we illustrate their power and effectiveness through a series of real case studies in areas commonly of interest to archaeologists. Realizing that other readers might find the case studies of interest, we have presented some background to each.

In the course of writing we came to realize the difficulty of the task we had set ourselves. Smooth progression, within the constraints of a reasonably sized book, was a particular problem. We start by assuming relatively little previous knowledge, and progress to some areas of statistics which are at the cutting edge of current research. One of our kindest readers commented on the switchback-like experience of hurtling from straightforward description to very difficult sections and back again. We are conscious that there is a lot more to Bayesian statistical analysis than we have been able to present here, and the references throughout the text should help guide readers to more detailed expositions.

The book consists of two major parts and a concluding chapter. Chapters 1–

8 cover the principles underlying the Bayesian approach and practice, whilst Chapters 9–12 contain a series of case studies. After setting the scene in the first two chapters, mathematical modelling, modelling uncertainty and statistical modelling are considered in Chapters 3–6. These cover a lot of ground rather briefly, and Chapter 6, in particular, deals with some highly technical aspects; although difficult we judged it important to include an explanation of the distributions which underlie the analyses of the case studies. Chapter 7 lies at the core of our argument, as an exposition of the principles of Bayesian inference. We regret that we were not able to proceed more rapidly to this discussion, but thought it wiser to explain the background of mathematical and statistical modelling first. Some of the issues of implementation, which have so revolutionized the application of Bayesian methods in recent years, are discussed in Chapter 8.

The second part of the book is given over to case studies. In Chapters 9 and 10 we consider a number of developed applications in radiocarbon dating and spatial analysis. Chapters 11 and 12 consider potential application areas, such as provenance studies and other dating methods, in some detail. The book closes with an examination of theoretical and practical consequences of Bayesian analysis, and looks to the immediate future of Bayesian archaeology. We have collaborated closely and each author has contributed parts to every chapter; nevertheless responsibility for drafting and subsequent coordination of Chapters 3, 7, 8, 9 and 12 fell to C. D. L., Chapters 1, 2, 4, 5, 6 and 10 to W. G. C. and Chapters 11 and 13 to C. E. B.

It is a particular pleasure to thank all our colleagues and friends who have helped us. We owe a particular debt to Adrian Smith, who, whilst at the University of Nottingham, first stimulated us to follow the Bayesian approach, and has supported our research ever since. Bob Laxton has been unstinting not only with his ideas and advice but also with moral support— the Consumerland example owes not a little to his practical conviviality. Andres Christen has also helped the progress of the book with stimulating and fresh ideas. We are most grateful to those who read earlier drafts, identified misconceptions, inconsistencies and errors, and greatly improvd the text with their suggestions: Box Laxton, Sue Hills, Tony O'Hagan, Vic Barnett, Neil Brodie, Andres Christen and Richard Jones. For their helpful advice and encouragement we also thank Mike Baxter, Alex Bayliss, Paul Budd, Morven Leese, Clive Orton and Mike West. David Taylor and Jane Goodard prepared the illustrations, and we are grateful for their skill and their patience. Data, and help in their interpretation, have been given by Michael Hughes, Adrian Olivier, Stephen Shennan and Jim Zeidler. The University of Nottingham and the University of Wales, Cardiff, have awarded grants to support the research leading to the book. In particular the University of Nottingham awarded C. D. L. and W. G. C. a Support for Key Resarch Staff grant (No. 94 SKRS006) which enabled them to take the sabbatical leave during which the book was written.

During the past ten years the authors have collaborated on a number of research projects into the application of Bayesian statistics to archaeology supported by the Science and Engineering Research Council, and conducted by the Depatments of Mathematics and Archaeology at the University of Nottingham. PhD theses submitted at the University of Nottingham have also furthered the research which led to this book, notably H. J. Zainodin's *Statistical Models and Techniques for Dendrochronology*, A. Christen's *Bayesian Interpretation of 14C Results* and especially C. E. B.'s *Towards Bayesian Archaeology*. Without the impetus and example of the last, this book would not have been written.

We wish to thank our colleagues at Nottingham and elsewhere for many helpful discussions and suggestions over that period. There are very many who have helped with advice, by supplying information and with a patient endurance of our importunities. We are conscious of our debt of gratitude to them all. Our families have, over the last year, put up with the short tempers and delicate sensitivities of the authors, and as a token of our affection we dedicate this book to them.

1

The Bayesian approach to statistical archaeology

1.1 Why use the Bayesian approach?

Bayesian statistics uses probability as a means of measuring one's strength of belief in a particular hypothesis being true. By doing so, it emphasizes that the interpretation of data is conditional on the information available and on an individual's understanding of it at that time. This stance has fundamental implications for the way in which data analysis is carried out, and these will become clearer as this book progresses. We believe that the Bayesian approach has much to offer archaeologists and welcome this opportunity to demonstrate its strengths.

We are not alone in this belief. In his Distinguished Lecture in Archaeology presented at the 91st Annual Meeting of the American Anthropological Association held in San Francisco in 1992, Cowgill (1993, p. 554) has commented:

> One very important aid to improved thinking is . . . the 'Bayesian' approach to statistical inference

whilst Orton (1992, p. 139) also endorses the method by the statement

> I have for many years advocated the . . . Bayesian approach.

and Bowman and Leese (1995) have urged its use in radiocarbon dating:

> Bayesian methodology is particularly suited to [the] calibration of groups of dates.

Encouraged by the views of these and other scholars (see for example Ruggles, 1986) we believe that an exposition of the Bayesian approach and its potential for archaeology will be welcomed. The book's publication is timely, for a formidable obstacle which lay in the path of the Bayesian method has recently been overcome, at least in part. The computational demands of the method were, until recently, so great that they stretched even the most powerful computers. Recent advances in numerical techniques, however, and increases in computer power have now pushed back these limitations.

The book can be considered timely in another respect. As approaches to archaeology have, in recent years, become philosophically more aware, the discipline has become more questioning and self-critical. As a consequence archaeologists are more conscious of the assumptions underlying their inferences, and more critical of the traditional discourse. The result is uncertainty. The Bayesian approach to data analysis is, precisely, a mechanism for handling uncertainty and formalizing the relationship between presupposition and conclusions. We have not, here, pursued the philosophical question of scientific reasoning and the problem of induction. To those who would like to read more about the philosophical background we recommend Howson and Urbach (1993). For the views of a philosopher who is a Bayesian on Mondays, Wednesdays and Fridays, but has his doubts on other days (and rests on Sundays), see Earman (1992).

At the heart of the Bayesian approach is the combination, in a single formal analysis, of our present understanding of a problem together with the data which bear on that problem, in order to make inferences about the problem to hand in a logical and coherent manner. A brief example will help illustrate what we have in mind.

1.1.1 An illustrative example

Dating is vitally important to archaeologists, because without a time-scale there is no history or prehistory. We take as an example the excavations which have been conducted at the Early Bronze Age site of St Veit-Klinglberg in Austria. This key site was selected for excavation because of the light it would shed on the development of metal-working in an area renowned for its rich copper-ore deposits. Work was carried out in the late 1980's by Shennan (1989). The challenge presented to the statisticians was to combine information from radiocarbon estimates (incorporating the correction of those estimates through calibration against the tree-ring curve) with the stratigraphic information available from the excavator. The radiocarbon samples were obtained from short-lived specimens, such as seeds and bone, and were dated using the Oxford accelerator dating facilities. In all, fifteen samples were sent for radiocarbon analysis and ten of them could be related one to another by means of stratigraphic information. The relationships thus established are shown schematically in Figure 1.1, where the numbers given are context numbers representing stratified levels. The radiocarbon determinations are listed in Table 1.1. Readers not familiar with radiocarbon dating should note that the information in Table 1.1 is in a standard form. In the first column we give the archaeological context number which allows

Table 1.1 The radiocarbon determinations from the ten related contexts at St Veit-Klinglberg, Austria.

Context number	Radiocarbon determination	Laboratory identifier
758	3275±75	OxA-3899
814	3270±80	OxA-3897
1235	3400±75	OxA-3900
493	3190±75	OxA-3898
925	3420±65	OxA-3882
925	3370±75	OxA-3883
923	3435±60	OxA-3881
1168	3160±70	OxA-3901
358	3340±80	OxA-3903
813	3270±75	OxA-3902
1210	3200±70	OxA-3904

the archaeologist at St Veit-Klinglberg to identify uniquely the level or archaeological context from which the sample was taken. In the second column we give the radiocarbon determination in the form $x \pm \sigma$ where x is the radiocarbon age and σ reports the laboratory's estimate of the error on x. In the third column we give the unique number assigned to the sample by the radiocarbon dating laboratory.

For example the sample from context 758 was sent to the Oxford accelerator laboratory where it was given the identifier OxA-3899. The determination return by the laboratory was $x = 3275\,\text{BP}$ with $\sigma = 75$. (Note that BP indicates that x is given "before present" and has not been corrected by calibration against the tree-ring curve. In what follows, where we use the term *radiocarbon determination* we quote x and σ, but BP is usually omitted.)

Common sense suggests that if the prior phasing and stratigraphic information are incorporated into the calibration process, better estimates for the dates will result. Of course we need to be confident that the stratigraphy and phasing are reliable, and the archaeologist must advise on this. Bayesian methods permit, indeed demand, that just such information (if it is available) be included. Thus Figure 1.2 shows the calibrated date range for one of the radiocarbon determinations if we ignore the stratigraphy and Figure 1.3 the calibrated range if we include the prior archaeological information. Comparing the histograms side-by-side it is plain that the second date range (Figure 1.3) is much shorter — a marked increase in precision has been achieved by including the prior knowledge. Without using the stratigraphic evidence we can be fairly confident that the date lies in the 430-year period from 3825 to 3395 BP. (BP

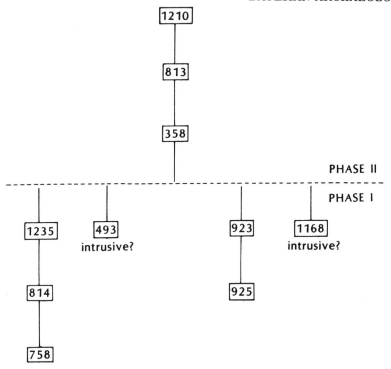

Figure 1.1   Schematic representation of the relationship between the phases and
contexts from the site of St Veit-Klinglberg, Austria. This represents the prior
archaeological information used in the Bayesian analysis of the radiocarbon
determinations associated with the site. (After Buck *et al.*, 1994a, p. 431, fig. 1.)

denotes "Before Present" where conventionally 0 BP is taken as AD 1950.)
Contrast this with the situation when the stratigraphic evidence *is* included
in the analysis: the comparable period is only 150 years long, spanning 3535
to 3385 BP — a dramatic improvement!

Many other improvements were attained by means of this type of analysis;
the study looked at questions such as the duration of the two phases,
incorporating the remaining five dates (for which no relative stratigraphic
information was available), the starting and finishing dates for the site, and
so on (see Buck *et al.*, 1994a; Geake, 1994). Quite how these improvements are
achieved will be examined in Chapter 9. For the moment we wish to emphasize
the straightforward gains from combining various types of information within
the coherent framework which the Bayesian approach imposes. Fundamental
to the approach are mathematical and statistical modelling: what is meant by
these terms will be developed in subsequent chapters. To begin with, however,

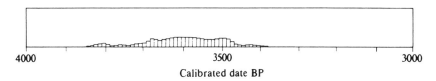

Figure 1.2 Histogram of the calibrated radiocarbon date for context 358 from St Veit-Klinglberg making no allowance for the stratigraphic information. The higher the histogram is for an interval, the more likely the date lies in that interval.

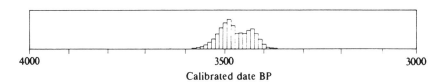

Figure 1.3 Histogram of the calibrated radiocarbon date for context 358 from St Veit-Klinglberg having taken the archaeological prior information into account. The higher the histogram is for an interval, the more likely the date lies in that interval.

we shall draw a distinction between descriptive statistics and model-based statistics.

1.2   Statistics and archaeology

1.2.1   Descriptive statistics in archaeology

Field work and research have always produced vast amounts of both qualitative and quantitative data, the latter comprising counts, measurements, weights and so forth. In most cases, however, analyses have been limited to *descriptive statistics*, that is to say graphs, histograms, or averages and the like. Such methods are in fact inescapable. For without some means of compressing their bulk, the data derived from archaeological research could never be interpreted or published. As a result archaeologists are skilled in using summary statistics such as means, medians and modes, and are familiar with histograms, pie charts and scatter-plots. Most introductory texts on quantifying archaeology have clear explanations of their use (Shennan, 1988). However it may be worthwhile to devote some space to refreshing the memory in order to lay the foundations for some of the ideas and concepts to be introduced later.

Probably the most successful way of communicating numerical results to other researchers is visually, by the appropriate use of pie charts, bar charts,

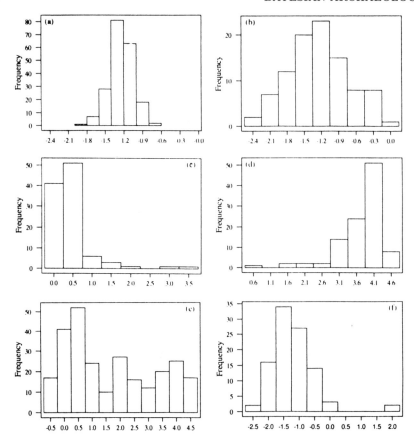

Figure 1.4 (a) Roughly symmetric histogram with the observations close together. (b) Roughly symmetric histogram with the observations widely spread. (c) Asymmetric histogram skewed to the right. (d) Asymmetric histogram skewed to the left. (e) Multi-modal histogram. (f) Roughly symmetric histogram with an outlier.

graphs, histograms and other such diagrams. If we focus our attention on histograms, we might ask which are the salient features that an observer wishes to identify. Questions that spring to mind include:

Where is the centre of the histogram?
Are the observations close together or are they widely spread?
What shape is the histogram?
Is the histogram symmetric/asymmetric?
If it is asymmetric, is it skewed to the left or to the right?
How many modes (or peaks) does the histogram have?

Are there any outliers present?

Answers to these questions and others can provide the observer with some insight into the problem at hand. In Figure 1.4 is a variety of histograms illustrating many of the features just itemized.

Next there are the methods of *summarizing* data by using one or two numerical values, called *summary statistics*. These are useful when comparing the results, say, from several excavations or laboratory projects, although compressing large amounts of data into just one or two values is not without its risks. The purpose of calculating these summary statistics is to describe the whole mass of data as concisely and succinctly as possible. In practice we wish typically to communicate, through appropriate measures, the *location* and *spread* of the data. These are often called, respectively, central tendency and dispersion:

(i) *Measures of central tendency*: (i.e. what is a typical observation?) Summary statistics include the *mean*, the *mode(s)* and the *median*.
(ii) *Measures of dispersion*: (i.e. how spread out are the observations?) Summary statistics include the *variance*, the *standard deviation*, the *range* and the *inter-quartile range*.

Definitions of these statistics and ways to calculate them can be found in any basic text on elementary data analysis (see for example Daly *et al.*, 1995). In particular we refer archaeological readers to chapter 3 of Shennan (1988) where these measures are described, their advantages and disadvantages discussed, and their application to archaeological data illustrated. Suggestions for other methods of data summarization including those based on the techniques of *exploratory data analysis* may be found in Velleman and Hoaghlin (1981). Chambers *et al.* (1983) describe the use of more advanced graphical display methods. Description, however, is not to be confounded with *inference* and for this we need to move up to a new level of analysis.

1.2.2 Beyond descriptive statistics

In order to adopt the Bayesian approach (and this is equally true of other statistical methodologies) it is necessary to move beyond exploratory methods in statistics, and to use a model-based approach. In advocating this, we in no way intend to diminish the importance of exploratory data analysis. In every project, researchers need to get a feel for the problem under investigation, by looking at the data and how they are distributed. Descriptive statistics, plots and histograms, and all the techniques briefly referred to above, have an important role in the very process of model-building. Indeed, as we shall see, they can be vital in the process of formulating prior information, a key stage in the development of the Bayesian approach.

All the same, it is instructive to consider here the changes of approach required when mathematical models are applied in archaeology. Certainly summary statistics, such as the proportion of sheep bones amongst the animal bones recovered from a site, are vital. Nevertheless merely to state these observations reveals their limitations. We are led at once to ask further questions: for example, what do the sheep bones tell us about the operation of early farming? Can we recognize from the ages at which the sheep died whether they were raised mainly for their meat (males slaughtered young, females kept for breeding) or mainly for wool and milk (with a different slaughtering pattern)? These more significant questions need mathematical modelling. By making use of mathematical models archaeologists begin to investigate the underlying processes that gave rise to the data they collect.

The potential of a model-based approach is realized only when archaeologists and mathematicians work in conjunction; and there are benefits on both sides. Interesting and challenging problems are presented to a mathematician working in the subject. They rarely suit standard models and it is common for detailed collaboration to be required. Despite the effort required, those who make such collaborative links are quite clear about the benefits to be had. For example, Fieller and Flenley (1988, p. 79) affirm that "investment in statistical modelling and analytic techniques can yield dividends" and go on to emphasize this in the context of their research (Fieller and Flenley, 1988, p. 81):

> Analysis of particle size data can proceed by either of two distinct routes. The first attempts only to obtain a simple numeric summary and description of the observed data; the second postulates a statistical model for the distribution of the sizes and proceeds to estimate the parameters of the model by statistical techniques. The first approach describes only the actual data obtained, the second attempts to investigate the process underlying the data.

Read (1987) has made a similar point. He uses a case study to demonstrate that archaeological results are more interpretable when applying model-based, rather than just exploratory, data analysis. Although such carefully coordinated projects are still relatively rare, their numbers are growing all the time (see in particular the recent review by Fieller, 1993).

1.2.3 Interdisciplinary collaboration

When setting out to write this book we realized from the outset that it might produce a natural reaction. What we propose is a demanding and challenging methodology. Model-building requires close collaboration between archaeologists, often other subject specialists, and statisticians. It is undoubtedly rewarding and an enriching experience, but equally it requires time and effort. Traditional archaeologists might object that they have computer packages to carry out their statistical analyses, and they can analyse their data themselves. There are dangers here, partly from what Orton (1992, p. 137) has memorably termed "the Audrey Syndrome":

> An ill-favoured thing, sir, but mine own.

Home-made solutions are liable to lag behind advances in statistics, and can fall into traps that an expert would avoid. In a broader sense, moreover, standard statistical recipes will produce standard statistical answers — but will they include *all* the relevant information in the analysis? Rather than look back to the abuse of computer packages, let us now look forward to exploiting the revolution in computer communications. Dispersed experts can communicate and exchange their ideas almost instantaneously.

1.3   Aims of the book

The aims of this book are, therefore,

(i) to explain, in the context of archaeological investigation, the basic rationale of the Bayesian approach to statistical inference,

(ii) to highlight the advantages which the Bayesian approach brings to archaeologists, and

(iii) to illustrate a range of archaeological problems which have successfully been addressed using its methods.

Plainly we have written with an archaeological audience in mind, and take relatively little knowledge of statistics for granted. We hope, nevertheless, that colleagues and students working in other areas will find some of the applications of interest. In the early chapters, many of the examples are rather simplistic and artificial. This is a deliberate ploy in order to illustrate the points that we are making. However, in a few cases some complex statistical argument is developed; we judged it better to include the mathematics even where it is complicated, and leave the readers to decide where they wish to follow and where they wish to skip, or to seek further advice. In later chapters, more realistic problems are tackled using a series of case studies to illustrate how the Bayesian approach has been used to solve actual archaeological problems. The level of mathematics required does vary from case study to case study and at times can be quite demanding.

1.3.1 Further reading

In his recent review of "archaeostatistics" Fieller (1993) stressed the rich variety of applications, which makes the subject so difficult to define. This impression is reinforced by any overview of the publications on mathematical methods and statistics in archaeology. Archaeology has been well served by a number of monographs in recent years. A milestone publication was Doran and Hodson's *Mathematics and Computers in Archaeology* (1975), the first textbook aiming to introduce the subject. This has been followed by Orton's *Mathematics in Archaeology* (1980) and Shennan's *Quantifying Archaeology* (1988), the latter complementing the former in its treatment: Orton introduces, in an accessible way, the processes whereby common archaeological problems can be addressed using mathematical and statistical approaches. Shennan presents, in rather more detail, methods of summary statistics and classical statistical tests on archaeological data sets.

Fletcher and Lock (1991) have also produced a review of simple descriptive techniques, statistical tests and computing for archaeologists. Amongst the more specialized works attention might be drawn to Hodder and Orton's *Spatial Analysis in Archaeology* (1976) and Baxter's *Exploratory Multivariate Analysis in Archaeology* (1994). Whilst texts such as these refer, in some cases, to Bayes' theorem and Bayesian inference, they have not attempted a full presentation.

Papers on more current work can be found in the proceedings of (and by attending) the regular gatherings of those working in the field; particular mention might be made of *Computer Applications and Quantitative Methods in Archaeology* held annually and with an excellent record of prompt publication. Commission 4 of the *International Union of Prehistoric and Protohistoric Sciences* also meets regularly and publishes quickly (see recently Pavuk, 1993; Johnson, 1994).

Numerous periodicals contain papers on advances in mathematical and statistical methodology in archaeology, among them *Antiquity, American Antiquity, Archaeometry, Archeologia e Calcolatori, Journal of Archaeological Science, Science and Archaeology* and *Quantitative Anthropology*. Finally we might mention the contributions in edited papers such as *Quantitative Research in Archaeology: Progress and Prospects* (Aldenderfer, 1987).

Because the Bayesian approach has not been widely applied in archaeology, we have frequently been thrown back on our own resources, and used case studies developed by ourselves. Of course this is in no way intended to diminish the important work carried out by the international community of scholars working in the field. Important new models are constantly being developed which can be adapted as readily to a Bayesian as to a classical approach.

1.4   Plan of the book

In Chapter 2 we describe in a non-technical manner the basic rationale underpinning the Bayesian approach. This will be illustrated by archaeological examples. The chapters which follow can be divided into two parts: Chapters 3–8 consider the theory and basic statistical modelling, including the use of Bayes' theorem in its generalized form. The second part, Chapters 9–12, takes the reader through a series of case studies grouped under four main heads: radiocarbon dating, spatial analysis, sourcing and provenancing, and other dating methods. This opens different paths through the book: some might prefer to turn to a particular application in the later section, and read through the examples, only afterwards seeking clarification in the earlier chapters. Other readers will prefer to clear the theoretical ground first, and then examine the case studies. Either way, in the next chapter it will be useful to introduce some of the key terms which are unavoidable in so technical a subject. We shall pursue this aim by discussing them initially in the context of mathematical modelling and then in the context of statistical modelling; needless to say these will be developed in greater detail later.

2

Outline of the approach

2.1 Structure of the chapter

The aim of this chapter is to give a broad overview of the essential features of the Bayesian approach to data analysis. We will cover some of the important aspects of the approach, namely mathematical modelling, updating information, Bayes' theorem and Bayesian inference. We must emphasize that our style will be informal; we are trying to give the reader an *intuitive* feel of the subject. In places, terms, such as probability, will be used in their everyday sense without being properly defined. This may cause concern to some readers, but later in the book, a more rigorous stance will be adopted.

2.2 Mathematical modelling

2.2.1 Models and archaeology

There was a particular vogue in the archaeology of the 1960's and 1970's to espouse the term model; sometimes it seemed to mean almost anything and almost nothing. This is unfortunate since if we are to get beyond the merely descriptive or summarizing use of statistics, then mathematical and statistical modelling are vital. Thus in writing of an "aid to improved thinking" in support of the Bayesian approach Cowgill (1993, p. 554) had in mind firstly the requirement of prior information, but secondly the formulation of an explicit model. In fact, whatever their field of interest most statisticians stress the importance of model-building and this has been recognized in archaeological applications by amongst others Doran and Hodson (1975, especially p. 26–29), Orton (1980) and Fieller and Flenley (1988, 1993). Archaeologists today would also put emphasis on the primary importance of theory, a view from which we certainly should not wish to differ; a useful mathematical model will be built on the foundation of a well-formulated theory. But this does not detract from the importance of also building good models. Let us first of all examine some examples of models.

2.2.2   A mathematical model for St Veit-Klinglberg

We return to St Veit-Klinglberg. During the excavation each context (archaeological unit or stratigraphic level) was given a unique number. From several of these contexts organic material was sampled for radiocarbon dating. We will consider ten such contexts of specific interest, these being contexts 358, 493, 758, 813, 814, 923, 925, 1168, 1210 and 1235 (see Figure 2.1 of Chapter 1). In order to develop a mathematical model of St Veit-Klinglberg, we need to introduce some mathematical notation. We let θ_i represent the (unknown) calendar date when the organic material found in context i ceased metabolizing. For example θ_{358} denotes the calendar date of context 358, θ_{923} that of context 923, and so on. Thus our mathematical model for these contexts from St Veit-Klinglberg includes the ten mathematical terms: θ_{358}, θ_{493}, θ_{758}, θ_{813}, θ_{814}, θ_{923}, θ_{925}, θ_{1168}, θ_{1210} and θ_{1235}, which represent the unknown calendar dates of the archaeological contexts. These mathematical terms are called the *parameters* of the model.

The relationship between the parameters of the model is derived from the archaeological information, which in this case is the stratigraphy shown schematically in Figure 1.1. Thus we can use

$$\theta_{814} > \theta_{1235}$$

to mean that the unknown true date of context 814 is earlier than (greater than in dates BP) that of context 1235. The radiocarbon determinations listed of Table 2.1 in Chapter 1, together with the information from the radiocarbon calibration curve, supply the data for the model.

2.2.3   The modelling of corbelled structures

The technique of corbelling was used in antiquity (and is still used today) as a means of roofing space with stone. It is effective because it exploits the strength of stone in compression, and because it avoids the weakness of stone in tension. In investigating ancient techniques of corbelling Cavanagh and Laxton (1981) have developed a mathematical model of the structural mechanics of corbelled domes. This model is based on a broader theory of mechanics known as Plastic Theory: this theory explains how structures adapt their shapes to contain changes in stress; how, for example, the wings of aeroplanes move and flex in order to adapt to the changing forces of air pressure to which they are subject. Heyman (1966, 1982) has shown that stone buildings can also be viewed as plastic structures, and it was his insight which inspired the mathematical approach adopted. Thus we note, in passing, that the model of corbelled structures was constructed on the basis of a deeper theoretical foundation.

Given this theoretical foundation, Cavanagh and Laxton (1981) went on to

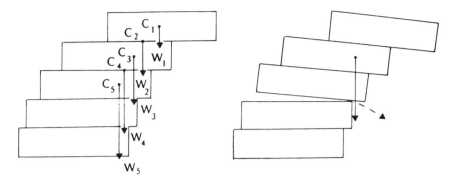

Figure 2.1 Schematic representation of a corbelled vault showing the line of force brought to bear by the weight of the masonry above each course of the wall.

develop a model for the structural stability of corbelled buildings. In brief, we can suppose that as each course of masonry is built up, it must overlap the course below by only so much that, when the building is complete, the vertical forces will act within the wall (Figure 2.1). If the forces act outside the wall, then the whole structure will collapse. The mathematical model proposes how, in theory, a line of force should run from top to bottom. This is found by calculating the moments (the action of the weight of the masonry) of the courses from the top of the vault down to the bottom, as they would come to bear on each supporting course. Imagine first the action of the topmost course on the course below, then the top two courses on the third, the top three on the fourth, and so on down to the bottom course. It was found that the relationship between the depth, represented by x, of any given level below the apex of a corbelled dome, and the radius of the dome at that level, denoted by $r(x)$ (see Figure 2.2) is given by the following expression:

$$r(x) = cx^d \tag{2.1}$$

or

radius at depth x = constant (c) × depth (x) raised to the power d.

Thus it was possible to express our understanding in a generalized mathematical form: a mathematical model has been created which will apply to all corbelled vaults. The parameters of the model are c and d. Their value may vary from building to building but, given suitable data, they can be estimated for each particular building. Moreover these parameters do have a physical interpretation. The parameters c and d tell us about the size and the shape (or curvature) of the vault respectively.

By way of a more detailed interpretation we can say that as you go down from the top of the vault (as x increases), so the radius gets larger. If $d = 1$,

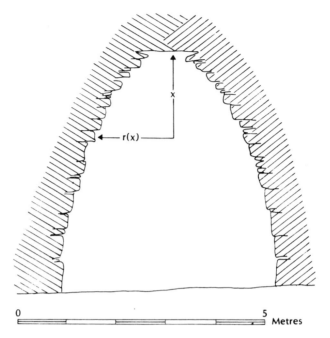

Figure 2.2 Section through the Minoan tholos tomb at Stylos, east Crete, showing the depth, x, below the capstone and the radius, $r(x)$, at that depth. Note that the capstone has cut off the topmost part of the vault, so that the apex of the dome has to be estimated.

the section of the corbelled vault is a straight line (the shape of the vault is a cone). If $d < 1$ the shape of the section is concave (the dome is like an elongated hemisphere), which is the case for most buildings. If $d > 1$ the shape of the section will be convex, and the dome bottle-shaped.

Given this model it is possible to further our insight into the early technology of building, by investigating particular examples: this has been done, for example, for the neolithic megalithic tombs of Brittany, the nuraghi of Sardinia and the tholos tombs of Minoan Crete and Mycenaean Greece. Furthermore it is also possible to study the interaction between different societies and the spread of technological knowledge.

2.3 Updating information

We mentioned in the second paragraph of this chapter the important concept of prior information. We shall not, at this stage, attempt to define this concept in a technical way. A number of related notions can be brought together under the term, such as prior belief, prior knowledge, prior understanding and so forth. For the present purposes we shall retain the term "information", as this is the most concrete expression, though the reader should bear in mind that other terms could be more appropriate in other circumstances.

2.3.1 Prior information

Perhaps the fundamental (though not the only) difference between the classical and Bayesian approaches to statistical inference lies in the use of prior information or belief. To quote Cowgill (1993, p. 554) again:

> ... in the Bayesian approach one's prior knowledge and beliefs can be built explicitly into the equations and can thus be made overt rather than covert. One can formalize the interaction between prior beliefs and new data.

Let us reconsider the example of St Veit-Klinglberg where the archaeologist is insisting that the radiocarbon determinations must be reconciled with the prior stratigraphic information. Some would claim that the results of "hard" science, such as radiocarbon dating, should not be subordinated to the less controlled observations of excavation, which, even in the best of circumstances, is an inexact procedure. Against this the laboratories are the first to acknowledge that a certain degree of imprecision in the dating is inevitable. Others have objected that the two types of information cannot be matched (Reece, 1995). It seems perverse, however, to ignore what is known merely because the knowledge is uncertain, for all knowledge is uncertain. On the other hand the advantages of including that knowledge in the analysis by means of prior probabilities will become clearer as this book progresses.

2.3.2 Comment

Why is prior information so important? In a general sense it is a familiar concept. Let us think for a moment about what archaeologists tend to hold as good practice when they approach a new piece of work. Consider, for example, the excavation of a newly discovered Bronze Age site in Europe. All professional archaeologists would expect colleagues undertaking such work to be experts in Bronze Age European archaeology. This means that the person coordinating the work will have undertaken similar excavations, will have read a great deal in the same and related fields and will have discussed the project with other experts. They will have assimilated a vast range of *a priori* knowledge before they ever lift the first turf. Then, when the excavation commences, any features or artefacts recovered will be interpreted in the light of all this background knowledge.

If the final report does not set the site in its context, with constant references to information from other sites, the excavator would be criticized. In some cases final interpretations will *only* be possible with reference to the prior information provided by previous work. In others the new data will cause the excavator (and colleagues) to update previously held assumptions.

Both these situations are, in some sense, normal outcomes of archaeological excavation. We see the Bayesian approach as offering a formal way of providing a comparable framework in which to undertake statistical analysis of archaeological data. As such it has natural intuitive appeal for those already working in modern, professional archaeology. Moreover, it also appeals to theoreticians since it offers a clearly structured and coherent framework in which to make inferences. Throughout this book we will endeavour to demonstrate how this can be achieved.

Already in the St Veit-Klinglberg example we have stressed the advantages gained through incorporating the prior stratigraphic information into the whole analytical procedure. That said, it must be stressed that the archaeologists need to be reasonably sure of their information. If, for example, there is some uncertainty about the association of a sample with its context, or about the ascription of a context to a phase, then the prior information used must reflect this. As a consequence, the results of the analysis and the posterior information will also reflect this uncertainty.

2.3.3 Data

We have summarized the elements of prior information and the model: how a problem can be formulated mathematically by suggesting a relationship between the various parameters under investigation.

The radiocarbon example modelled the dating of archaeological phases relating levels and periods to the samples. We have also seen a model of the forces acting within a corbelled building. This is a theoretical model, it is not based on any particular corbelled vault. The model indicates how the action of the forces can be summarized in terms of the shape of a corbelled dome. Next we need to introduce the data to the model: the radiocarbon determinations reported by the dating laboratory, or the actual physical measurements from a corbelled building.

2.3.4 Posterior information

When we have a model, prior information about its parameters and some data we are in a position to commence the Bayesian computational process. This process combines the data and the prior information through the model to obtain the *posterior information*. In other words, the handle is turned, all the calculations are carried out, and the results are worked out; these results are referred to as the posterior information. The combining of the data with the prior information about the parameters of the model gives us the updated information about the various parameters under investigation (posterior information about the calendar dates, θ_{358}, θ_{493} and so on, of the samples from the contexts at St Veit-Klinglberg; posterior information about c and d for particular corbelled buildings). It is precisely this process of updating information which gives Bayesian inference an attractively natural feel.

All archaeologists approach a given problem with a set of information, a framework of prior understanding. New data are obtained, and in the light of them, we adapt and change our information to a new state of understanding: our posterior information. As time progresses this then becomes the new prior information, which sometimes (but not always) can be further shaped by new experiments or fresh data, or can be transformed through a new insight or a new model devised under the influence of a more refined theory. Whilst some of the technicalities of Bayesian statistics are complex, the underlying principles are straightforward.

2.4 Bayesian inference

We have so far couched the discussion in terms of information. However for the Bayesian statistician the key concept is that of *probability* and the probability of certain hypotheses being true. It is not our intention to give a precise definition or explanation of probability at this stage in the book. This will be done later. In the mean time we shall rely upon the reader's everyday interpretation of probability or chance.

To be useful in Bayesian data analysis, prior information needs to be converted to prior probability. If that can be done we replace "prior information" by "prior probability" in the preceding sections. The same applies to replacing "posterior information" with "posterior probability". In this section of the chapter, therefore, the same sequence of "prior", model and "posterior" will be followed as before, but with emphasis on their probabilistic meaning.

2.4.1 Bayes' theorem

At this stage it is appropriate to make an initial statement, although an informal one, of Bayes' theorem, the basis on which the prior information (probability), the model and data are combined in a logical and coherent manner to give the posterior information (probability). We must emphasize that we shall not define our terms precisely, as our aim here is simply to convey to the reader in as intuitive a manner as possible the basis of the Bayesian method.

The parameters are the unknowns under investigation, for example the true calendar dates, denoted by θ_{358}, θ_{493} and so on, of the organic samples from St Veit-Klinglberg, or the c and d in the mathematical model of the corbelled vaults.

The prior probability captures the prior information about possible values for the unknowns, before taking any account of the data. For example, we might believe, before knowing the results of the radiocarbon analysis, that there is a 60% chance that contexts 758, 814 and 1235 of the early Bronze Age village at St Veit-Klinglberg were laid down in that order and within the period 4500–3500 BP — see Figure 1 of Chapter 1.

The *prior probability* is a mathematical expression which we represent by

$$Pr(\text{parameters}),$$

where, for the moment, "Pr" stands for probability (in an everyday sense). It is a function of the parameters — the value of $Pr(\text{parameters})$ depends on the value of the parameters. One set of parameter values will give rise to one value for $Pr(\text{parameters})$, another set will give another value. $Pr(\text{parameters})$ is chosen so as to reflect the *a priori* prospect that the parameters take values within a certain specified range. In other words, we can simply interpret this prior probability as a way of representing our prior information about the values of the parameters before we even look at the data.

Next we need to introduce the concept of the *likelihood* which provides the link between the data and the parameters, this link being made by the model. The likelihood is denoted by

$$l(\text{parameters}; \text{data}),$$

the "l" standing for likelihood. The likelihood introduces the data into the analysis, data such as the radiocarbon determinations in the case of St Veit-Klinglberg, or the measurements of the radius at particular depths of the vaults in the investigation of corbelled buildings. It establishes a connection between the data, the radiocarbon determinations of St Veit-Klingberg, and the parameters of the model, the calendar dates denoted by θ_{358}, θ_{1210} and so on. For the corbelled buildings, the likelihood relates the radius and depth measurements to the c and d of the mathematical model given in (2.1).

We are now in a position to express our *posterior probability* as a combination of the data and the prior information. This is denoted by

$$Pr(\text{parameters}|\text{data}),$$

which can be read as the probability that the unknown parameters take particular values, given (or in the light of) the data and the prior information. In other words this represents the results of the Bayesian analysis, for example our updated estimate of the calendar dates of an archaeological context, given the new radiocarbon dates.

These three terms, the *prior probability*, the *likelihood* and the *posterior* probability, are connected via Bayes' theorem which may be expressed succinctly in words as

Posterior probability is proportional to the likelihood times the prior probability,

or slightly more mathematically as

$$Pr(\text{parameters}|\text{data}) \propto l(\text{parameters}; \text{data}) \times Pr(\text{parameters}).$$

(Note that the mathematical symbol "$\propto$" means "proportional to". Suppose that y is proportional to x, written $y \propto x$, then the equation relating y to x will be $y = kx$ where k is a constant — k being called the *constant of proportionality*. The relationship says that if x is doubled (trebled, halved or increases twenty-fold) then y is doubled (trebled, halved or increases twenty-fold) as well.)

Thus the end product of a Bayesian analysis will usually be some probability statement, referred to as the posterior probability or, for short, the posterior. This posterior probability then needs to be interpreted in terms of posterior information about the problem to hand.

2.4.2 Prior information about St Veit-Klinglberg

Recall that the date of context i is represented by θ_i. That is, θ_{758} means the (unknown) calendar date of context 758 and likewise for the other nine contexts. We will use the "greater than" symbol, "$>$", to denote that one context was deposited before another. We have chosen this symbol as the dates are quoted BP so that a larger value of θ_{758} means an *earlier* date. For example, from the stratigraphy there is clear evidence that the organic sample found in context 813 was deposited before that in context 1210 but after that in context 358. Therefore the archæological information tells us that θ_{358} is before θ_{813} which is before θ_{1210}. In terms of our mathematical symbols we have

$$\theta_{358} > \theta_{813} > \theta_{1210}.$$

In fact using the stratigraphic information given about all ten contexts in Figure 1.1 we end up with the following set of constraints:

$$\theta_{758} > \theta_{814} > \theta_{1235} > \theta_{358} > \theta_{813} > \theta_{1210},$$
$$\theta_{493} > \theta_{358},$$
$$\theta_{925} > \theta_{923} > \theta_{358} \text{ and}$$
$$\theta_{1168} > \theta_{358}.$$

The above mathematical relationships summarize the prior archæological information. So long as the dates satisfy the above relationships the

archaeologist will be content. But dates that do not satisfy these constraints
would conflict with the prior information and therefore are impossible on
archaeological grounds. Apart from this information about the partial ordering
of the dates, no other chronological information is available. Quite how this
information is modelled and used in our Bayesian analysis will be explained
in detail in Sections 7.5.3 and 9.4.1.

2.4.3 *Prior information about the corbelled structures*

Suppose we were faced with analyzing a new set of radius and depth data
from a corbelled structure. Before we look at the data and before we carry out
the statistical analysis, what do we know? Firstly, to make the mathematical
model realistic, both parameters, c and d, must be assumed to be positive. (If
c were negative the building would appear inverted, with its top at the bottom
and its bottom at the top; if d were negative, instead of widening from top
to bottom, the dome would narrow in the shape of a funnel.) Mathematically
this is expressed as $c > 0$ and $d > 0$. Depending upon the problem, we may, if
there is no other relevant available knowledge, stop here. On the other hand,
in some situations, we may have some similar structures nearby that have
already been analysed. The results of the previous analyses may provide some
further (limited) prior information on the values of c and d for the current
structure.

2.5   Questions and answers

We have outlined above a mathematical model for the stability of corbelled
domes, just as one was formulated for the analysis of the radiocarbon
determinations from St Veit-Klinglberg. Next we examine what inferences
archaeologists may wish to make.

2.5.1 *Questions about St Veit-Klinglberg*

There are many possible questions that could be posed here. Four that
immediately spring to mind are the following.

(i) *Does context 758 date before or after context 925?* Mathematically, is
$\theta_{758} > \theta_{925}$ or not?

(ii) *What is the date of the earliest context?* This means we have to answer
questions about the largest (or maximum) of θ_{758}, θ_{493}, θ_{925} and θ_{1168}.

(iii) *What is the total time-span of the occupation as represented by the organic
samples?* This involves calculating the difference between the earliest
date, as represented by the largest of θ_{758}, θ_{493}, θ_{925} and θ_{1168}, and the

latest date, as represented by θ_{1210}. In other words we wish to make inferences about $\max(\theta_{758}, \theta_{493}, \theta_{925}, \theta_{1168}) - \theta_{1210}$.

(iv) *What are the lengths of Phase I and Phase II?* Here we need to make statements about the difference between the beginning of Phase I, as represented by $\max(\theta_{758}, \theta_{493}, \theta_{925}, \theta_{1168})$, and the end of Phase I which is given by the smallest (or minimum) of $\theta_{493}, \theta_{923}, \theta_{1168}$ and θ_{1235}. For Phase II, we simply require the difference between the dates of contexts 358 and 1210, that is $\theta_{358} - \theta_{1210}$.

To some extent the questions are almost endless. In practice, only those individuals directly involved in the excavation and subsequent interpretation know which are the important ones to be answered. In fact, the excavator took samples from more contexts than those described above. Some were from contexts that could not be related by stratigraphy to the rest of the main sequence because of soil erosion. Here radiocarbon dating could be used to build up a complete picture of the development of the site. Considering one such sample from context 460, does it date to before Phase I, within Phase I, between the two phases, within Phase II or after Phase II?

2.5.2   Questions about corbelled structures

It is even more difficult to be brief here and readers who wish to follow the full details are recommended to consult the articles referenced above. Suffice it to say that the value of the exponent d in (2.1) represents the shape of the dome, which in turn relates to the method by which the artisans constructed the buildings. The value of d is useful, for example, as a means of seeing whether groups of buildings were erected following the same or different traditions. Thus statisticians might be asked whether two groups of buildings were constructed following essentially a similar method, or whether they could be the product of two different traditions.

Accordingly, the principle of corbelling was investigated for the Bronze Age tholos tombs of Greece. These magnificent buildings, some of them princely tombs, have a long and interesting history. The majority are found on the mainland of Greece, and belong to the Mycenaean culture (see Figure 2.3), but a small group is also known from the island of Crete where the older Minoan culture flourished (see Figure 2.2). The latter group probably date to the fourteenth century BC, an interesting period in the history of the island when it was subject to foreign Mycenaean rule. The Minoan/Mycenaean tombs of the island were plainly imitating the princely fashion of the mainland, but were mainland craftsmen brought over to construct them? Just by looking at the tombs it is difficult to judge.

It was commented above that the exponent d in (2.1) related, in particular, to the shape of the curvature of the vault, and, hence, to the methods the

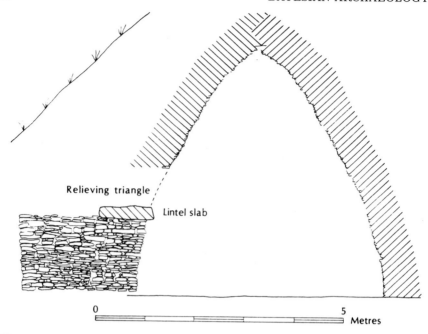

Figure 2.3   Section through the Mycenaean tholos tomb at Kardista, mainland
Greece.

ancient builders used. We can, therefore, set up two models whereby the
mainland tombs might be described by

$$r_1(x) = c_1 x^{d_1}$$

and the Cretan tombs by

$$r_2(x) = c_2 x^{d_2}.$$

If the values of d_1 and d_2 are found to differ, then we have some reason
to argue that whilst the type of tomb is the same, nevertheless two different
groups of builders constructed the two regional variants. Here our point is to
draw out the relationship between the mathematical model and the statistical
analysis. The statistical analysis investigates the question of whether d_1 and
d_2 are approximately the same (expressed mathematically as $d_1 \approx d_2$) thereby
indicating the same architectural tradition; or whether d_1 and d_2 can be
considered to be different (expressed as $d_1 \neq d_2$) which would point towards
differing architectural traditions.

Figure 2.4 Rev. Thomas Bayes FRS.

2.6 The history of the Bayesian approach

Bayesian ideas are not new, indeed, they can be traced back to the Rev. Thomas Bayes FRS (Figure 2.4) who was born about 1702 and died in 1761. His most famous work, in which he formulated the concept of "inverse" probability (posterior probability), was published posthumously (Bayes, 1763). Similar ideas, but without the detailed mathematical foundation, are to be found in the earlier work of Bernoulli (1713).

Only recently have Bayesian ideas become well-established in statistical practice. During the middle of this century, classical approaches were almost uncontested within the statistical community and it was not until about 1960 that a steady revival in the Bayesian approach occurred. Development of the Bayesian statistical framework can be followed in the works of Jeffreys (1961), Lindley (1965), DeGroot (1970) and Box and Tiao (1973); the current state of Bayesian theory can be found in Bernardo and Smith (1994) and O'Hagan (1994).

A useful introductory text is Berry (1995) whilst Gelman *et al.* (1995) provide a full acount of the Bayesian approach to modelling and data analysis with particular emphasis on the practical aspects of the methodology. Gelfand and Smith (forthcoming) will review the development of computational aspects of Bayesian statistics, whilst O'Hagan (1988) and Smith (1988) both provide useful introductions to the Bayesian view of probability. Stimulating reading on the foundations of the Bayesian approach is offered by de Finetti (1974, 1975) while Howson and Urbach (1993) give a Bayesian perspective on scientific reasoning. A less mathematical exposition can be found in Lindley (1985). Barnett (1982) and Stuart and Ord (1991, chapter 31) compare the conventional or classical approach to statistics with the Bayesian framework.

2.7 Tailpiece

In this chapter, we have introduced, albeit informally, some of the key concepts of Bayesian data analysis; namely prior probability, model, likelihood and posterior probability. These will constantly recur throughout this book. The reader should now be in a position to press forward with Chapter 3 and really get to grips with mathematical modelling, although a glance at the case studies in the later chapters will give a flavour of the actual applications of the Bayesian method. We have, thus far, aimed to keep the algebra relatively straightforward, and have made an effort to explain the various expressions. Those readers, in particular, who are tempted to go directly to the later chapters will find themselves plunged into much more complex expressions with summations, integrals, derivatives, products, exponential expressions and so forth. Depending on the mathematical sophistication of the reader, some of this will have to be taken on trust. All the same, the essential pattern of prior probability, likelihood and posterior probability, all linked via Bayes' theorem, should still come through in each application.

Equally importantly, if less insistently, we hope to have conveyed a feel for the processes of Bayesian inference. They demand a particular responsibility on the part of the archaeologist, a responsibility not only of understanding but also for the expertise contributed to the analyses. Their use of prior information together with its incorporation into the modelling process means that the prior views affect the results in an important way. Consequently, the careful elicitation of that prior information and its conversion into a probabilistic form is a field of vital importance to Bayesian statisticians. In the case studies addressed in Chapters 9–12 attention will be paid to the methods of incorporating expert knowledge into the statistical process.

3

Modelling in archaeology

3.1 Introduction

The development and construction of mathematical and statistical models is an essential prerequisite for all Bayesian statistical analyses. In general, however, archaeologists (unlike mathematicians or statisticians) do not have a common understanding of the term *model*. Therefore it is necessary to spend some time in the early part of this chapter comparing and contrasting the use of the term *model* in both archaeology and mathematics (including statistics).

Before we carry out this comparison, we give a simple example to focus ideas. Suppose we were asked to design a cuddly toy elephant suitable for a one-year-old child, perhaps, to be given by a doting grandparent. Which features of a real elephant need to be captured by the toy? Four legs, big body, head with big floppy ears, eyes, trunk and perhaps a small tail? By incorporating these features scaled to roughly the right proportions, one would end up with a model that most grandparents would instantly recognize as an elephant. Details like the colour, mouth, feet and so on, are not essential for this purpose. Of course, if the model were being designed for a different use (say, teaching about Africa in a junior school) then more details would be required: perhaps it should be anatomically more correct.

What is the point of this rather naive example? Well, a model can be thought of as *reality viewed from a distance*. In the toy elephant example, the toy (model) presents an impression of an elephant (reality) when seen from a long way off. It captures the major features that can be seen from a distance. In general, when modelling a situation, one is trying to identify the major factors involved and how they are linked together. In other words, *a model is a simplified representation of some aspect of reality created for a specific purpose*. Note that the model must be influenced by the purpose for which it is to be used: the model of an elephant designed for a one-year-old child might not be suitable for a 10-year-old. Before discussing mathematical and statistical models we turn our attention to what archaeologists might understand by the term.

3.2   Archaeological ideas

From the very beginning archaeologists have used the term model in a broad
and rather vague sense, perhaps closer to "theory" in scientific contexts.
Thus Piggott's famous definition talks of "some working hypothesis or in
scientific language, a model of the past ..." and "Models of the more ancient
past have been numerous — theological and teleological, moral and aesthetic,
racial and cyclical" (Piggott, 1965, p. 6). The first paper in the book *Models
in Archaeology* written a few years later (Clarke, 1972a) proclaims a more
rigorous programme. All the same in addressing the question "What is a
model?" Clarke's definition is still vague (Clarke, 1972b, p. 1–2):

> Models are pieces of machinery that relate observations to
> theoretical ideas, they may be used for many different purposes
> and they vary widely in the form of machinery they employ, the
> class of observations they focus upon and the manner in which they
> relate the observations to the theory or hypothesis. It is therefore
> more appropriate to describe models than to attempt a hopelessly
> broad or a pointlessly narrow definition for them.

Such a general description has its uses as it enables archaeologists to have
different types of model tailored to their own particular field of study. The
contributors to the above volume (Clarke, 1972a) produce many different
types of models. However all the models share the notion of a simplified
representation of reality.

Perhaps the one type of model not done justice in Clarke (1972a) is the
mathematical model. This is not because the general description of Clarke
(1972b) explicitly excludes mathematical models but because that volume was
more concerned with archaeological theory than with specific methodological
considerations. Read, on the other hand, has written "Some Comments on
the Use of Mathematical Models in Anthropology" explicitly on this theme.
He opens (Read 1974, p. 3):

> As is the case for students in other disciplines, anthropologists
> are becoming increasingly aware that the use of mathematical
> models can be a very powerful tool in the formulation of theory.
> Mathematics is useful if for no other reason than that the very
> nature of a mathematical model requires precise definitions, exact
> statements of relationships between variables, and quantification
> of data.

More recent publications discussing and giving examples of the uses of
mathematical models in archaeology are Aldenderfer (1987), Judge and
Sebastian (1988) and Voorrips (1990). In the latter two works, mathematical
models are seen as increasingly important tools in interpreting the past.

3.3   Mathematical models

As we have said before, mathematical modelling is a key activity in science and in archaeology, although applications in the latter are rather limited. Since some archaeologists tend to blur the distinction between model and theory it is essential for what follows that we have a clear understanding of what we mean by model. We will use the term model to mean mathematical model, adopting the definition of Meyer (1984, p. 2):

> A mathematical model is a model whose parts are mathematical concepts such as constants, variables, functions, equations, inequalities etc.

If our knowledge of some of the variables is uncertain but can be modelled by the use of probabilities then we have a *statistical model*. The latter will refer to models that include some stochastic or random effects. In the remainder of this book we use the term model to mean mathematical model or, if there is some probabilistic element involved, statistical model.

3.3.1   A simple example

Suppose I wish to travel from home to the nearest international airport to catch the 09:55 flight to Hamburg in order to attend an important archaeological conference. If I intend to travel by car, at what time should I leave home? This is a very common type of problem that many people have to face.

The travel agent has told me that I need to check in at the airport at least one hour before the flight is due to depart. Moreover I will need to allow about half an hour to park the car and get to the check-in desk. Therefore I ought to arrive at the airport car park by 08:30 at the very latest.

This still leaves me with the problem of how long I should allow for the actual journey. Luckily I live quite close to a major motorway which goes almost directly to the airport. To help me make my decision, I decide to construct a simple mathematical model. To do so we have to introduce some notation. Let D represent the distance in kilometres from my home to the airport, let S represent the speed, in kilometres per hour, that I anticipate travelling at, and let T be the journey time measured in hours. Then a simple model is that

$$\text{time} = \frac{\text{distance}}{\text{speed}}.$$

Mathematically this is expressed by the equation

$$T = \frac{D}{S}.$$

The distance, D, is fixed at about 220 kilometres, but I can to some extent choose the speed, S, at which I travel. Of course I must obey the speed limit of 120 kilometres per hour. If I anticipate driving at a speed of 100 kilometres per hour, that is $S = 100$, then the journey time, T, predicted by the model is given by

$$T = \frac{220}{100} = 2.2 \text{ hours}$$

which is 2 hours 12 minutes. If I could travel at the speed limit then the time would be $220/120$ hours which is just a little (10 minutes) under 2 hours. If $S = 80$, then the journey time would be 2 hours 45 minutes.

Therefore if I could guarantee that there would be no hold-ups or traffic jams and that I would be able to travel at the speed limit, then I could leave home at about 06:30. However my knowledge of the motorway, gleaned from television, radio and news reports, as well as from my own personal experience, suggests that delays are likely particularly as I would be approaching the airport during the morning rush-hour which is between 07:30 and 09:00. A more pessimistic estimate of the speed of 80 kilometres per hour may be more appropriate with a resulting journey time of about 2 hours 45 minutes and a departure time from home of, say, 05:30. Since my airline ticket is non-transferable and I am a born pessimist, I may want to start even earlier just to be sure!

In the example several modelling suppositions have been made that make the real-life problem simpler. For instance, I know full well that my speed will not be constant all the way but will vary according to local traffic conditions. In fact, I cannot say in advance that between junctions 22 and 20 of the motorway my speed will be only 60 kilometres per hour and that between junctions 20 and 17 I will be able to travel at the speed limit. To build this type of information into the model would make it too sophisticated and cumbersome for the purpose at hand. For a similar reason, I choose to ignore the 5 kilometres I have to travel on non-motorway roads before I get to the motorway near home. Obviously my speed on these roads will be less than 120 kilometres per hour.

In other words, the model specified above is very much an approximation. It is capturing the important features of the journey and ignores details that are not so important. Of course it could well be that I have over-simplified the situation but the calculation that I leave home before 05:30 seems sensible, especially as it is in accord with my experience of previous journeys to the airport. If the answer had been, through some miscalculation, "leave at 22:30 the previous evening" I would have been very suspicious and would have checked my model formulation and my arithmetic!

We all have faced similar situations to that described above. We have made similar assumptions and carried out similar calculations in order to plan a journey. It is important to recognize that we have been able to predict our

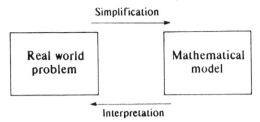

Figure 3.1   The modelling cycle.

arrival time by using quite a simple model that does not capture all the detailed features of a car journey. Using a more realistic model not only may complicate the mathematics but may make a solution virtually impossible to obtain. Sensible, simplifying assumptions *must* be made if we are to provide a solution.

3.4. The modelling cycle

Let us now try to analyse what we have done in the previous section.

  (i) We started with a problem, that of trying to estimate a sensible time to leave home.
 (ii) We developed a simple but hopefully realistic model of the time it takes to drive from home to the airport.
(iii) We used this model and some data, the distance from home to the airport and my estimated speed, to predict the journey time. Hence, we were able to calculate a departure time.
(iv) Finally we looked at the answer to see whether it was sensible in the light of previous experience and other knowledge.

Notice that we have carried out four steps. In general when developing a mathematical model we can break the process down into a number of sequential steps similar to those above.

In constructing a model we carry out what is commonly called the *modelling process* or the *modelling cycle*. We start with a real-life problem and make simplifications to construct a model. We use the model together with some data to provide an answer and interpret it in the light of the purpose of the model. This process may be repeated several times before the model is considered satisfactory. Figure 3.1 shows the process as a cycle and Figure 3.2 shows how it is applied to the previous example of estimating the journey time to the airport.

As in the example of the journey to the airport with its four main stages, so the modelling cycle in general can be thought of as consisting of four main

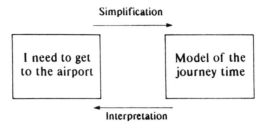

Figure 3.2   The modelling cycle for the journey to the airport.

stages, namely *purpose, simplification, model* and *interpretation*.

In examining mathematical modelling it is enlightening to look at each stage carefully. If we return to the airport example we can see what happens at each stage. Then we can generalize to other situations.

(a) I first discussed the situation and clarified my purpose by specifying the real-life problem. *So the first step is to specify the problem.*

(b) Next I set up a simple model for the journey. *So the second step is to set up a model.*

(c) I then addressed a slightly different question of "finding the time to travel from home to the airport". *So the third step is to formulate the mathematical question to be answered.*

(d) I then did a calculation to obtain an estimated journey time of 2 to 3 hours depending upon traffic conditions. *The fourth step was to use the mathematical model to answer the mathematical question.*

(e) Then the solution of the mathematical problem was interpreted in the light of the original purpose, i.e. I should leave at 05:30. *The fifth step was to interpret the solution.*

(f) I then questioned the solution: was it realistic, did it accord with my previous experience? *The sixth step was to compare with reality.*

(g) Depending upon how the model and its solution compared with reality, I might have gone round the modelling cycle again until I was satisfied. Eventually I would have to travel to the airport. *So the seventh and final step is to use the result.*

These modelling steps can be incorporated into the modelling cycle given in Figure 3.3. This type of formulation is useful in many mathematical modelling situations. In principle the modeller can go round the cycle as many times as they like, but in practice two or three times are usually sufficient. The first time round the cycle often results in an over-simplified model which may be used to give a "feel" for the problem and its solution. Refinements can be added to improve the model by making it more realistic, and therefore probably more complex, until it is considered to be satisfactory. Finally, if the model proved completely unsatisfactory, it might have to be abandoned and

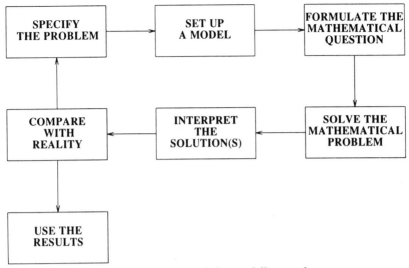

Figure 3.3   Steps of the modelling cycle.

replaced by another.

Each step of the modelling cycle can be thought of as an instruction to help us to decide what to do next. It also helps when trying to analyse what other researchers have done. Unfortunately not all modelling falls neatly into the seven steps or follows such a logical order. Sometimes it may be necessary to back-track, particularly if a solution to the mathematical problem cannot be obtained. It could be realized at Step (e) that there was a "better" Step (b). Therefore the diagram in Figure 3.3 is very much a guide which suggests a logical basis for a modelling sequence. But there are no hard and fast rules, everyone approaches modelling in a slightly different manner.

3.5   An illustrative example

In 1955 Alexander Thom published an article on geometry of the British megalithic monuments (Thom, 1955) which was the forerunner of his investigations into the design of stone circles (see Thom, 1962, 1964, 1967). By surveying megalithic monuments in England, Wales and Scotland (see Figure 3.4) and studying their dimensions, he postulated the existence of a megalithic unit of measurement: the "megalithic yard". Since that time more research has been carried out, but the basic model has remained unchanged. It is the methods used to test the model which have developed through time.

In its simplest form, Thom's model states that if the megalithic yard exists then the basic dimension (for example the diameter) of monuments in which

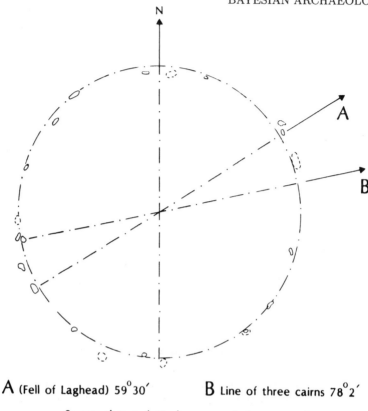

A (Fell of Laghead) $59^{\circ}30'$ B Line of three cairns $78^{\circ}2'$

Stones shown dotted are now below ground

Figure 3.4 Plan of the megalithic stone circle at Cauldside, Scotland (after
Thom, 1967, p. 59, fig. 6.3); one of the stone circles analysed in attempts to
establish whether a megalithic yard was used in antiquity. (Reproduced by
permission of Oxford University Press.)

the unit was used must be composed of whole number (integer) multiples
of the megalithic yard. Let Q denote the megalithic yard (assuming that it
exists). If the diameter of a stone circle is denoted by D then the model is
that D is a whole number multiple of Q. However it is difficult to imagine
that any prehistoric constructors had the technology and the ability to work
to high accuracy. Hence the need to allow for some error. Mathematically this
is expressed as

$$D = MQ + \epsilon$$

where M is a positive integer constant and ϵ is an error term. The latter,
which recognizes that the model is only an approximation to the real world,
may arise from a variety of sources including: errors made by the constructors,

errors made by the modern surveyors, errors associated with the finite sizes of the stones themselves, errors associated with the re-erection of fallen stones and so on.

To illustrate the use of the model let us consider a particular Scottish circle at the site coded M2/14 by Thom (1967, p. 36). (We will use Thom's original measurements which were recorded in *feet*, abbreviated to *ft*.) Its diameter was measured to be 21.8 ft. Thus we have for this circle $D = 21.8$. By analysing many circles, Thom estimated the megalithic yard to be 2.72 ft, that is $Q = 2.72$. If we take the diameter to be 8 megalithic yards, that is $M = 8$, we get $MQ = 8 \times 2.72 = 21.76$. Thus the error term, ϵ, for this circle is equal to $D - MQ = 21.8 - 21.76 = 0.04$ ft.

According to Thom's theory, circles built at this and other sites in Britain were constructed using the same basic unit of measurement but to different dimensions. For example, another circle at the same site had a diameter of 44.1 ft and by taking $M = 16$, its error term is equal to 0.58 ft. A circle at a different site had a diameter of 64.0 ft, so setting $M = 24$ we have an error term of -1.28 ft. We note that the error terms differ from circle to circle.

However the value of the megalithic yard is unknown and so for any stone circle the error term is also unknown. To make any progress, we need to model how the error may vary. This leads to considering what values of the error, represented by ϵ, are likely. Generally small errors, positive or negative, would be thought to be more likely than large ones. In other words large errors are believed to be unlikely whereas smaller ones should be more common. This raises the issue of how best to model this random variation in the error term from circle to circle by using a *statistical* model.

In this example we have a simple illustration of a mathematical model used in archaeology. Of course the original question was how, given Thom's collection of data from a wide range of British circles, one can assess whether the megalithic yard does exist and, if so, how it can be estimated.

3.6 Examples of mathematical models in archaeology

In the remainder of this chapter, we will give some examples of mathematical models used in archaeology, which are intended to give the reader an idea of the wide range of problems tackled and the rationale underlying the models. For each example we will give some background to the problem and discuss its archaeological importance. We then describe how a suitable mathematical model may be formulated for the problem.

If appropriate we will indicate how the model may be developed further to incorporate any uncertainties, thus becoming a statistical model.

3.6.1   Analysis of field survey data

Recent increases in the application of scientific methods to the surface survey of archaeological sites have resulted in a plethora of spatial data. Many techniques are used including relatively simple field walking, where the numbers of objects of a particular type or types seen on the surface are recorded, and the scientifically more sophisticated such as soil resistivity, ground and soil magnetic susceptibility and the chemical analysis of soil. Here we concentrate on the analysis of soil phosphate data. Soil samples are usually collected over a transect or a regular grid. Soil phosphate analysis provides evidence not about structural remains such as buildings or roads, but about areas in which organic material was deposited. At most sites macroscopic remains of such depositions have decayed but soil phosphate analysis provides a simple means of locating them.

Cavanagh *et al.* (1988) describe data where soil phosphate analysis was used to determine the area of sites previously found by field walking. The specific purpose of the phosphate survey was to attempt to define the limits of human activity associated with the artefact scatters already located.

Given their specific aim, the authors required a model which reflected only a relatively crude distinction in the data. They defined the phosphate concentrations of interest to be simply *on-site* and *off-site*. The *on-site* zones were those with a high concentration of phosphate whereas the *off-site* zones were those with a low concentration. They were not interested in the more subtle *within-site* variations that might be present.

In terms of a mathematical model, we could use a grid system and define a variable $C_{x,y}$ to indicate whether the position (where x and y are the co-ordinates of each position sampled) was allocated to an *on-site* zone or to an *off-site* zone. That is we define $C_{x,y}$ as follows:

$$C_{x,y} = \begin{cases} 1 & \text{if position } (x,y) \text{ is deemed to be on-site} \\ 0 & \text{if position } (x,y) \text{ is deemed to be off-site.} \end{cases}$$

To illustrate this we give in Figure 3.5 phosphate data collected from site G165 of the Laconia Survey. Certainly there are high phosphate and low phosphate concentrations, but it is difficult to get a clear picture of where they are.

In order to define these zones more precisely we need to develop a model to divide the grid into *on-site* and *off-site* zones. A very simple model could be to say that any cell with a value of 100 or more is defined as *on-site*. Figure 3.6 shows the result of a more elaborate model for partitioning the site into zones of high and low phosphate concentrations.

77	57	55	31	37	45	59	64	64	55	73	32	17	62	108	121
59	55	34	45	41	47	38	28	45	53	33	48	52	80	101	112
45	66	62	34	38	48	32	44	62	80	60	27	60	50	75	108
55	40	41	66	36	68	55	45	21	80	66	88	91	88	83	91
59	57	80	71	19	80	60	60	60	62	+	+	166	77	+	68
60	68	75	85	57	44	30	62	38	91	+	+	68	77	+	59
48	73	101	80	47	64	41	34	47	71	62	116	60	73	52	294
68	80	50	121	131	64	59	47	77	68	143	66	32	50	55	50
71	71	71	91	80	68	57	75	73	77	60	34	47	50	50	101
57	125	91	136	83	68	71	83	62	104	62	62	45	59	41	27
60	83	94	108	80	88	66	71	27	75	80	77	34	57	30	71
55	66	94	+	88	116	83	77	44	41	59	41	57	55	47	48
53	77	91	108	73	108	85	83	53	33	75	23	60	57	47	36
57	71	75	80	73	85	85	73	53	131	57	38	64	38	55	71
62	47	68	80	97	91	77	77	52	41	27	68	68	71	75	66
64	55	59	73	62	73	83	59	36	37	57	68	+	+	108	83

Figure 3.5 Phosphate data collected over a grid at Laconia Survey site LS G165. Raw phosphate concentrations in mg phosphate per 100 g soil; + indicates that no data were available from that cell.

3.6.2 Sourcing

To help focus ideas we will concentrate on one type of material, namely obsidian, although a similar basic model can be adapted for many others such as jade, marble or other raw materials. Obsidian is a volcanic glass which was used for the production of blades, scrapers, projectile points and, on occasion, carvings and jewellery. Unlike flint, which is used in much the same way and is extremely abundant, obsidian is a relatively rare geological material and occurs only at a limited number of locations. Due to its rarity, effectiveness and attractive glassy appearance, obsidian is thought by many experts to have played an important role as a trade and prestige item.

Geologists have demonstrated that obsidian can be characterized by its chemical composition. The problem, therefore, lies in identifying the geological source of the raw material used in the manufacture of obsidian artefacts. Once these sources are identified it is possible to draw conclusions about the patterns of trade in the past.

The chemical composition of the raw material from a potential source can be compared with that of the archaeological artefacts themselves. Up to 30 different chemical elements present in the sample might be measured. Those

KEY

	Number	Mean	S.D.
■	86	4.45	0.26
▨	161	3.94	0.32
□	Missing Data Value		

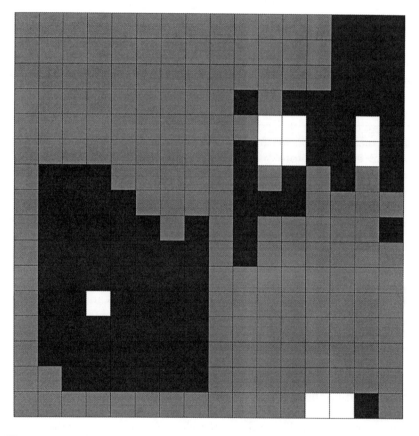

Figure 3.6   Grid of soil phosphate data partitioned into on-site (black) and off-site (grey) areas. (The white cells indicate missing data.)

Table 3.1 The chemical composition (in parts per million) of obsidian from three different sources for five trace elements.

Source	Trace element				
	1	2	3	4	5
1	10	12	15	10	20
2	15	10	27	22	10
3	20	17	12	17	8

elements forming concentrations of less than 1%, known as trace elements, have been found particularly useful in distinguishing sources one from another. Due to the limited number of geological sources of obsidian the model required is straightforward. If each of the geological sources of obsidian is chemically homogeneous, and if it has a chemical composition sufficiently different to be distinguished from the others under consideration, then the source of an obsidian artefact should be identifiable by comparing its chemical composition with those of a range of possible source materials.

To formulate this in mathematical terms, suppose that there are k likely geological sources of obsidian for the artefact under study and that the chemical composition of obsidian from each source has been estimated. The chemical composition of source i can be represented as a series of numbers, one for each of the q elements analysed. (Typically q could be 15 or even more depending upon the laboratory's practice.) Let $x_{i,1}$ be the amount of element 1 present in obsidian from source i. Likewise let $x_{i,2}, x_{i,3}, \ldots, x_{i,q}$ respectively be the amounts of elements $2, 3, \ldots, q$ at source i. Thus for source 1, the chemical composition can be represented by q numbers denoted by $x_{1,1}, x_{1,2}, x_{1,3}, \ldots, x_{1,q}$, source 2 by the numbers $x_{2,1}, x_{2,2}, x_{2,3}, \ldots, x_{2,q}$ and so on for all the remaining sources up to the kth which is represented by $x_{k,1}, x_{k,2}, x_{k,3}, \ldots, x_{k,q}$.

As an illustrative example, consider having $k = 3$ possible sources for a particular specimen of unknown origin. For simplicity, suppose we have measurements in parts per million on $q = 5$ trace elements. These measurements are given in Table 3.1. For source 1, we have $x_{1,1} = 10$, $x_{1,2} = 12$, $x_{1,3} = 15$, $x_{1,4} = 10$ and $x_{1,5} = 20$. For source 2, we have $x_{2,1} = 15$, $x_{2,2} = 10$, $x_{2,3} = 27$, $x_{2,4} = 22$ and $x_{2,5} = 10$. Finally for source 3, we have $x_{3,1} = 20$, $x_{3,2} = 17$, $x_{3,3} = 12$, $x_{3,4} = 17$ and $x_{3,5} = 8$.

Let $y_1, y_2, y_3, \ldots, y_q$ be the chemical compositional data for a specimen of obsidian under study whose origin is unknown. Because of some naturally occurring random variation in the composition of obsidian at one site and because measurements are not totally accurate, even if the specimen were from site i, y_1 would not be exactly equal to $x_{i,1}$ and the same would be true for the other elements. That is, even for two specimens of obsidian from the same source, the proportions of a particular element will vary, albeit slightly, from specimen to specimen. Therefore a model is needed to assign the specimen to source i if the chemical compositions are "close". How to define "close" is crucially important.

Returning to our simple example, note that $y_1 = 10$, $y_2 = 15$, $y_3 = 20$, $y_4 = 15$ and $y_5 = 10$. From which source did the specimen originate? Or did it come from another, as yet undiscovered, source? That is, should k be 4 rather than 3?

3.6.3   Seriation

On excavating archaeological levels, or graves, or other contexts, it is not uncommon to find, from one context to another, that similar types are present, but in different proportions. Thus a series of mesolithic sites might produce stone tools distinguished typologically as scrapers, burins, tranchet axes and so forth. The predominance of one tool type over another will vary over time. In cases where there is no associated stratigraphic sequence for these contexts, archaeologists will often use the finds to produce a chronological order. In other words, they wish to "seriate" the contexts.

The basic principles of seriation can be traced back to Sir Flinders Petrie's work on the pre-Dynastic cemeteries of Egypt. The model was first formally made explicit by Robinson (1951) when his archaeological colleague (Brainerd) asked him to provide a numerical method for seriating pottery assemblages.

Robinson states that

> over the course of time pottery types come into and go out of general use by a given group of people.

That is, it is assumed that artefact types come into use, go out again, but may never come back again into usage. The general form of the model can be represented by a diagram first given by Robinson (1951), a version of which is shown here as Figure 3.7.

To put this in mathematical terms, we usually speak of the data being held in an $m \times n$ matrix, A, where m represents the number of archaeological units to be ordered (for example, graves) and n represents the proportion of different attributes or artefacts associated with those units (for example, grave goods). In this example, a matrix can be thought of as a table of numbers where the rows represent graves and the columns artefact types. A typical entry in the table is denoted by $a_{i,j}$, which is the jth entry in the ith row. In this case, it records the number of artefacts of type j found in the ith grave.

As an example we consider some simple artificial data given in Table 3.2 concerning five types of artefact, labelled A, B, C, D and E, found in six different tombs, labelled 1, 2, ..., 6. Here we have six tombs so that $m = 6$, and five artefact types, hence $n = 5$.

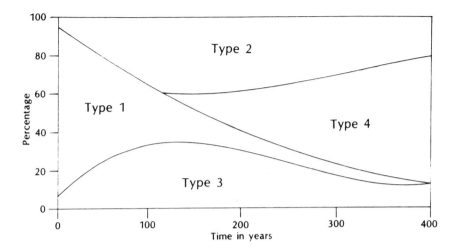

Figure 3.7 The underlying model of seriation.

Table 3.2 Data for the seriation of six tombs based upon five artefact types.

Tomb	Artefact type				
	A	B	C	D	E
1	0	10	60	30	0
2	14	10	5	70	1
3	0	2	0	0	98
4	30	18	2	30	20
5	20	20	0	0	60
6	0	10	40	49	1

The matrix A is said to be *seriated* if, as we look down a particular column, after the numbers have decreased, *then they never increase again*. This is the model proposed by Robinson. It is easily seen that if we reorder the rows (tombs) of Table 3.2 then the tombs seriate. The reordered data are given in Table 3.3. Notice that in this table, the entries in a column never increase after they have decreased. That is, based on the model stated above, the chronological order is either 3, 5, 4, 2, 6 and then 1, or the reverse. The reader might like to confirm that there is no other order of the six tombs which satisfies the model.

Table 3.3   Seriated data for the six tombs based upon five artefact types.

Tomb	A	B	C	D	E
			Artefact type		
3	0	2	0	0	98
5	20	20	0	0	60
4	30	18	2	30	20
2	14	10	5	70	1
6	0	10	40	49	1
1	0	10	60	30	0

3.6.4   Tree-ring dating

Oak timber has been used in the U.K. and elsewhere as a structural building material since prehistoric times. By taking samples of the wood from such a building it is often possible to provide a highly reliable estimate of its construction date. Here we give a brief outline of tree-ring dating.

It has been observed that oak trees add exactly one growth ring for each year of their life and that the width of the ring is affected by climatic variation, often referred to as a *climatic signal*. In a year poor for tree growth the ring is narrow; in a year good for tree growth a wider ring results. Since climatic variation is local, it is often possible to match the ring pattern on one oak tree to that of another grown in the same area during the same time-span. That is to say oaks respond in similar ways to the common climatic signal. In fact in some cases this matching can be done by eye, either by comparing the wood samples or more usually by comparing graphs of the tree-ring widths for the two samples concerned.

A long fixed chronology, called a *master chronology*, is made by matching the growth rings of one oak tree with another, and that with another and so on through time up to a tree whose felling date is known. By matching samples of unknown felling date to the master chronology we then have a means of dating samples from archaeological and historical contexts. In Figure 3.8 we give plots of several tree-ring width sequences taken from samples from oak timbers from Bede House Farm, North Luffenham, Leicestershire, U.K. (Litton, forthcoming). Notice that the sequences are not exactly the same but do demonstrate some similarities. It is these similar patterns of wide and narrow rings that were used to match the samples. Also notice the similarity between the average sequence (composed from all the Bede House Farm samples) and the East Midlands master tree-ring chronology for oak which is used to date it.

As we have seen in this example, there is considerable potential for ring-width variation, from tree to tree, between rings growing in the same year.

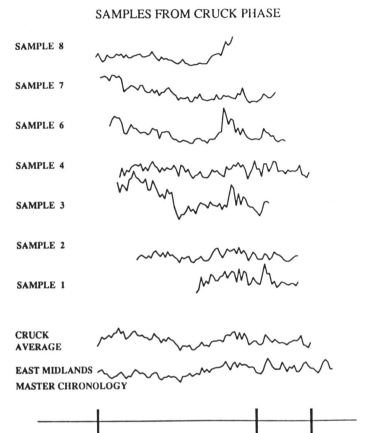

SAMPLES FROM CRUCK PHASE

SAMPLE 8

SAMPLE 7

SAMPLE 6

SAMPLE 4

SAMPLE 3

SAMPLE 2

SAMPLE 1

CRUCK
AVERAGE

EAST MIDLANDS
MASTER CHRONOLOGY

1291   1381

Figure 3.8   Plot of tree-ring sequences, taken from samples of the oak timbers
from the cruck phase of Bede House Farm, North Luffenham, Leicestershire, U.K.
At the bottom is the East Midlands Master Chronology and just above that is the
"average" chronology for Bede House Farm.

Thus there is a natural variation within the data. In order to allow for this
several workers have established more rigorous statistical methods for dating
samples. The variation may be caused by a number of different factors such as
micro-environmental conditions, insect attack, pollarding or other interference
by man, fire and so forth. These non-climatic factors might affect one tree
growing at the same time in the same region, but not another.

   In general oak trees put on wider rings in their early years of growth
and narrower rings later on. This could obviously make matching ring
widths between trees extremely difficult and so mathematical techniques are

employed to remove or "filter" out this effect (but to keep in the climatic signal) before attempts are made to match samples. The result of such filtering is that each sequence of tree-ring widths is represented by a sequence of indices rather than by the original measurements of the widths.

In mathematical terms, given the tree-ring data in their index form, we assume that there is an underlying climatic signal which results in idealized and underlying indices $\theta_1, \theta_2, \ldots, \theta_n$ for years $1, 2, \ldots, n$. If there are J trees with tree j ($j = 1, 2, \ldots, J$) having ring indices denoted by $y_{1,j}, y_{2,j}, \ldots, y_{n,j}$ (corresponding to years $1, 2, \ldots, n$) then these indices are not identical to the underlying indices, but are a *noisy* version of them. As mentioned above, this noise is caused by local environmental conditions and other such micro-effects which affect each tree in slightly different ways and must be allowed for in the model. The basic model adopted is

$$y_{i,j} = \theta_i + Z_{i,j} \quad i = 1, 2, \ldots, n; \; j = 1, 2, \ldots, J$$

where $Z_{i,j}$ is the noise component for the jth tree in year i. An important question here is how best to statistically model the noise component.

3.6.5 Radiocarbon dating

Radiocarbon dating is one of the most important scientific dating techniques used by archaeologists. Here we will only give a brief overview of the principles involved and we refer the reader to Bowman (1990) or Aitken (1990) for more detailed accounts.

Radioisotope dating depends upon the basic principle that radioactive isotopes are inherently unstable so that after some time and without external influence, they undergo transitions to more stable isotopes. Associated with such transitions is the emission of ionizing radiation.

Based upon this fundamental assumption, Libby *et al.* (1949) developed the theory of radiocarbon dating. This states that cosmic rays bombarding the earth's atmosphere produce neutrons which interact with nitrogen, ^{14}N, in the atmosphere to produce radioactive isotopes of carbon, ^{14}C. In the form of carbon dioxide, ^{14}C isotopes (and the stable isotopes ^{12}C and ^{13}C) enter plants by photosynthesis. When animals eat plants the carbon isotopes are then taken into the food chain. Throughout the life of the plant or animal carbon is taken up from and exhaled/excreted into the environment as part of a process known as the carbon cycle. When the plant or animal dies its carbon atoms are released to the environment only when tissue decomposes.

If tissue is preserved, the carbon in the cells is no longer being replaced as part of the carbon cycle so that the carbon present at the time of death remains "locked" into them. Any ^{14}C isotopes in the cells decay radioactively back to ^{14}N and are not replaced. The ratio of ^{14}C to the stable ^{12}C and ^{13}C gradually diminishes. It is possible, therefore, by measuring the quantities of various carbon isotopes to determine the sample's age.

Physicists have postulated the following mathematical model of radioactive decay which has been verified by experimental observations. Consider an organic sample that died at a particular time which we will, for convenience, call time zero. (We do so in order that time will be measured relative to the time of death.) Let t represent an arbitrary time later, measured from the time of death. Let $N(t)$ denote the number of ^{14}C atoms in the sample at time t. The law of radioactive decay says that the number of ^{14}C atoms decaying in a very small period of time is proportional to the number of ^{14}C atoms present at that time. Mathematicians denote the rate of decay by the symbols

$$\frac{dN(t)}{dt}.$$

The next step is to say that, since the rate of decay is proportional to the number of atoms present, we have

$$\frac{dN(t)}{dt} = -\lambda N(t)$$

where λ is the decay constant which has been estimated. (This type of equation is known as a "differential equation".) The number of ^{14}C atoms present when the organic matter died (that is, when $t = 0$) is given by $N(0)$. By using mathematics outside of the scope of this book (the interested reader should perhaps consult Burghes and Borrie, 1981, p. 24–33), the above differential equation can be "solved" to obtain another equation which directly relates $N(t)$ to t and $N(0)$. The equation is

$$N(t) = N(0)e^{-\lambda t}.$$

By rearranging this we can calculate the age, t, by the formula

$$t = -\frac{1}{\lambda} \ln\left(\frac{N(t)}{N(0)}\right).$$

For example, suppose we have a sample with initially 1 million atoms of ^{14}C and we observe it again when it only has two hundred thousand left. In mathematical terms, $N(0) = 1\,000\,000$ and $N(t) = 200\,000$. Noting that the decay rate, λ, for ^{14}C has been estimated experimentally to be about 1/8033, we can estimate the age of the sample as

$$t = -8033 \ln\left(\frac{200\,000}{1\,000\,000}\right) = 12\,929.$$

In other words the sample is about 13 000 years old.

The model given above is the original model used by the early workers in radiocarbon dating. Unfortunately there are a number of complications that have become apparent since those early days and the model used currently is somewhat more complex. In particular, it has been shown that the levels of ^{14}C in the earth's atmosphere have not been constant over time. As a result there is now an internationally recognized radiocarbon calibration curve, based upon measurements of tree rings of known date, which laboratories use to convert the radiocarbon date to the calendar time-scale.

Here is a clear-cut *archaeological* example of the modelling cycle being applied in practice. The initial model was shown to be inadequate and so a more complex model was developed. No doubt as our knowledge and understanding improves further refinements will be made.

3.7   Summing-up

In this chapter we have endeavoured to convey the essential features of mathematical and statistical modelling. Firstly we have recognized that a model is a simplified representation of some aspects of reality, created for a specific purpose, and that the purpose will often influence the complexity or simplicity of the model. Secondly we have introduced the idea of the modelling cycle and the steps involved. Of course, it is itself a simplification of what happens in practice, which is not always carried out in quite so logical a manner! As we have seen, the modelling cycle is so-called because initially a very simple model may be continually refined by modifying it in the light of experience until it meets the required specification. Lastly we have illustrated some of the wide variety of mathematical and statistical models that are in use in archaeology, a field that is not noted for its mathematical sophistication. Moreover we hope that we have conveyed some of the underlying rationale of the models.

In future chapters, these models and others will be used to analyse and interpret archaeological data arising from real problems. But first we need to spend some time explaining the basic ideas of *probability* and *statistical modelling*.

4

Quantifying uncertainty: the probability concept

4.1 Introduction

In Chapter 1, we discussed the use of statistics to summarize possibly large and complex data sets using measures of central tendency, such as the mean, mode or median, and measures of dispersion, such as the standard deviation or the inter-quartile range. It was stressed there that such descriptive statistics get one only so far, they merely describe a set of data and are rarely sufficient to further our understanding of the underlying problem. In the light of this we proceeded, in Chapter 3, to discuss the mathematical modelling of archaeological problems, whereby we attempt to capture the essential structure of a problem and express it in terms of mathematical relationships, such as functions and equations relating the parameters of interest to each other in some specified way. Next we need, as it were, to introduce the data to the mathematical model, in the hope of a happy marriage of the two. In order to achieve this it is necessary to call upon the skills of the statistical modeller, the subject of this and the next two chapters.

Recall that in Chapter 3 a mathematical model for investigating the megalithic yard was proposed, where D represents the diameter of the circle in megalithic yards and is a whole number multiple, M, of Q which represents the megalithic yard itself:

$$D = MQ + \epsilon.$$

In this expression we have included an error term, denoted by ϵ, which accounts for various inaccuracies. On the basis of this model and data collected from many stone circles, we hope to be able to judge how probable it is that there was a megalithic yard and what its value was. To do so we need to identify the nature of these inaccuracies in order to include them in the error term. This will involve using probabilistic and statistical techniques to model as best as we can the error term. In a nutshell, we will be using *statistical modelling* which is concerned with measuring degrees of uncertainty and with how uncertainty can be incorporated into a model.

How then is uncertainty to be measured? The statistician uses probabilities, and in what follows we shall introduce the notion of probability from an intuitive point of view. We do not propose to adopt a rigorous axiomatic or mathematical approach here, as it would take us well beyond the scope of the present book; what can be stressed is the fact that there is a firm theoretical underpinning to probability (see Kolmogorov, 1933). Readers who wish to pursue the theory of probability from the Bayesian point of view are referred to de Finetti (1974), Lindley (1965), Bernardo and Smith (1994) and O'Hagan (1988).

Rather we shall proceed directly to a discussion of the quantification of probability in a consistent and logical manner, and the rules of probability including Bayes' theorem (which is central to our philosophy). In the next chapter we shall extend our study to statistical modelling, and consider how uncertainty may be modelled by means of probability distributions. These principles will be illustrated by means of examples, and the conventions and notation used will be explained as we go along.

4.2   Quantifying uncertainty

4.2.1   Knowledge and ignorance

In a much quoted definition, de Finetti (de Finetti, 1974, p. xi; quoted by Bernardo and Smith, 1994) has written:

> The only relevant thing is uncertainty — the extent of our own knowledge and ignorance. The actual fact of whether or not the events considered are in some sense *determined*, or known by other people, and so on, is of no consequence . . .

What this definition entails is that uncertainty, and its measurement through probability, is *conditional upon our knowledge at the current time.* The importance to the Bayesian approach of the principle that probability is conditional cannot be understated. Other notions of the nature of probability look to arguments on the basis of the outcome of a long series of identical trials, but we adopt the view that probability is no more and no less than a measure of belief: a measure of one's own personal belief that an event will come about, given that one has some background knowledge. *Uncertainty, probability and subjectivity are bound up together.*

The language of probability is a familiar part of everyday discourse: weather charts today will frequently show probabilities, and most people have no problem in interpreting a 60% chance of thunderstorms in the locality tomorrow as meaning that there is a 3 in 5 chance that there will be torrential rain, and a 2 in 5 chance that there will not. The probability is based on the weather forecaster's understanding at the time, on the basis of satellite information, reports from weather stations and so forth. And such probabilities are used to decide actions: if the chances are high of rain so heavy that any attempt at excavation is likely to cause more harm than good, then the director of a project might plan to process finds rather than proceed with digging. This is an intuitive application of probability which is entirely compatible with the Bayesian approach.

The use of odds in gambling is a clear case of the estimation of probabilities: odds of 4 to 1 against Dobbin winning the Derby means that the bookmaker estimates that Dobbin has no more than a 1 in 5 chance (or a probability of not more than 0.2) of winning; the punter who puts his shirt on Dobbin must think it has a better chance.

In quantifying the odds, bookmakers use their prior knowledge (Dobbin's form, the track-record of the other animals in the race, the jockeys' weights and so forth) and their judgement. Likewise informed gamblers use *their* prior knowledge and make the same calculation using the evidence available to them.

The same process can be used by any expert in any field: in interpreting an aerial photograph an archaeologist might submit odds that a particular mark, say a circular crop-mark, represents a particular type of monument, say a henge monument (Figure 4.1). Certainty is not possible and all experts have stories of misleading marks, for example the potential henge monument that turned out to be the site where the local gymkhana (horse-riding competition) had taken place a little while before the photograph was taken (see Wilson, 1982, chapter 4). Moreover the uncertainty can be quantified and a probability ascribed.

4.2.2 How to assess uncertainty

What does the probability really mean? If an expert in the interpretation of aerial photographs says there is a 0.7 probability that a particular photograph shows a henge monument, what does this mean? How was the figure of 0.7 reached? What does it signify in contrast to a value of 0.9 or 0.668 or some other value? Is it possible that the experts state higher probability when they are feeling good, and lower probabilities when they are feeling depressed? There are two issues intertwined, the issue of quantification and that of subjectivity.

There are various ways of arriving at probability assessments, and theory does not prescribe a necessarily right way (though there are necessarily wrong ways). One method is measurement by reference to a standard. Here the standard often considered is an urn containing 100 balls. (It is interesting to comment here that many textbooks on probability and statistics use examples

Figure 4.1 Aerial photograph of a henge monument (?) at Libberton, Lancashire. (By courtesy of the University of Cambridge Committee for Aerial Photography.) (Reproduced by permission of Cambridge University Collection of Air Photographs: copyright reserved.)

based on drawing different-coloured balls from urns to illustrate the rules of probability. How many of the authors of these books realize that the ancient Athenians had a special device, called a *kleroterion*, which held balls and was used for selecting jurors and other officials by lot? See Figure 4.2.) The balls are identical except that some are coloured black and the remainder white; matters are so arranged that any one of the 100 balls has an equal chance of being drawn. What is the probability that a black ball will be drawn? This clearly depends on the number of black balls in the urn. If there are b black balls (and $100 - b$ white balls) then the probability equals $b/100$; thus if we have 80 black balls, that is $b = 80$, then the probability is 0.8.

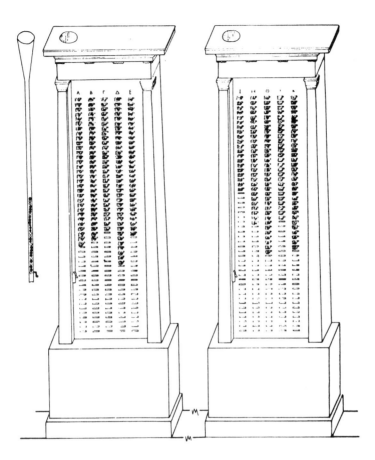

Figure 4.2   Kleroterion from the Agora excavations, Athens. (After Thompson and Wycherley, 1972, p. 54, fig. 15.) (Reproduced by permission of American School of Classical Studies at Athens: Agora Excavations.)

We return to the expert in the interpretation of aerial photographs. Recall that the expert's probability that the mark on the photograph is a henge is 0.7. Let us compare this situation with that of drawing a ball from an urn. Suppose that the urn initially contains 100 black balls (all the balls are black). Then, one by one, a black ball is replaced with a white one (always maintaining a total of 100); as the number of black balls decreases, the probability that a ball drawn at random from the urn is black approaches the expert's probability that the mark is a henge. And when there are 70 black balls and 30 white, the two situations are the same — and only then.

Given a choice between the two situations which would the expert select: to gamble on drawing a black ball, or to gamble on the crop-mark being a henge? A rational choice would depend on the number of black balls in the urn: if just one, the gamble on the ball being black would seem not to be worthwhile; if 99, then it would seem a good bet. When the number of black balls is 70, the two situations are the same and the expert will be indifferent between them.

We can now apply this to another situation. Let us say the expert is asked to pass judgement on a different photograph, under the same conditions. How probable does he or she think that the site is, indeed, a henge monument? In order to help the expert to come to a conclusion they are offered a wager. Either gamble on drawing a black ball from the urn or gamble on the correct identification of the monument. As the proportion of black balls in the urn decreases, the more the balance of preference tilts towards gambling on the identification of the crop-mark as a henge. Eventually there comes a point when, with b black balls in the urn, the expert is indifferent between the situation of drawing a black ball and identifying the mark as a henge. At this point, we can say that the probability it is a henge monument is $b/100$. By this means we have assessed the expert's uncertainty concerning the presence or not of the henge. Notice we have not assessed anyone else's uncertainty, it is a personal probability.

4.3   Mathematical notation

In order to model uncertainty we need to introduce some mathematical concepts and notation. The conventions we use include the following.

(i) Consider a situation about which we have imperfect information, so that there is some uncertainty or doubt about it. The situation might be the result of a future horse race, the completion year of a medieval building or the provenance of a pot. In all three cases we are uncertain because we have incomplete information. For the horse race, our information is incomplete because the race is still to take place. For the building, it was finished in one and only one year, but our information, based on documentary evidence, typology and similar evidence, is not sufficient to give a precise date. Likewise for the pot, it came from one and only one provenance, but we do not have sufficient evidence to say which.

(ii) For each situation, we can list the possible *elementary outcomes*. For a horse race with 11 runners, ignoring deadheats and other unlikely occurrences, there are 11 possible winners of the race and so we have 11 distinct elementary outcomes. For the medieval building, we can be sure that the period of interest is from AD 500 to AD 1600, so that we have 1101 distinct elementary outcomes. For the pot we may have evidence for say 20 provenances and the pot may have come from any one of these, so we have 20 distinct outcomes. But in this situation there is always the possibility of it coming from an as yet unknown (to us) provenance. Hence we have a total of 21 elementary outcomes.

(iii) We let Ω represent the collection or *set* of all the possible elementary outcomes of a situation. In the case of the medieval building, Ω consists of 1101 different *elements*, one for each possible year of construction.

(iv) An *event* is a collection or set of one or more elementary outcomes from Ω. We let E denote an *event*; this might be the outcome of a race, the date of a medieval building, the identification of a provenance, whatever happens to be the subject of the enquiry. Alternatively it may refer to a collection or set of elementary outcomes. For example an event could be that the completion date of the building lies between AD 1560 and AD 1590; that the pot came from the south-western United States without being precise about the exact whereabouts in that area.

In what follows we shall often have occasion to refer to several possible outcomes within the event space. We will do so either by using different letters to represent different events or by using subscript, to distinguish them. Thus E_1 might represent the event that a building dates to the thirteenth century, that is E_1 consists of the set of elementary outcomes $1200, 1201, \ldots, 1299$. E_2 might represent that it dates to the fourteenth, E_3 that it dates to the fifteenth and so on.

(v) The class of all events of interest associated with a given situation is called the *event space*.

(vi) We use H to denote the relevant background information: the bookmaker's information about the previous performance of the horse and its competitors, the architectural historian's knowledge about the date of a building, or the archaeologist's beliefs about the provenance of the pot found on a site, the information acquired through earlier research about the function and nature of the archaeological sites under investigation, and so forth.

(vii) The letter P denotes probability. Thus $P(E|H)$ is read as the probability (P) of an event (E) given $(|)$ the currently available information, knowledge or history (H). This emphasizes the point that assessments of probability are *conditional*, that they are made in the light of one's prior information and understanding. For the sake of simplicity of presentation, probabilities may be written $P(E)$, but even with this notation there is always a tacit implication that this probability is conditional on the information available at the time.

The view adopted in this book is that assessments of probability are subjective and made in the light of experience: there is no difference in kind between the bookmaker's estimate of odds, the architectural historian's view of a date for a medieval building, the doctor's diagnosis, the archaeologist's opinion about the provenance of a pot, or the uncertainty in a scientist's estimate of the distance of the sun from the earth. People will have different opinions of the philosophical appeal or the philanthropic benefit of the one or the other, but in terms of the understanding of probability they are equivalent. To say that assessments of probability are subjective, however, is not to imply that the processes of statistical inference are arbitrary. On the contrary there is a system of rules and laws which make sure that there is no inconsistency in manipulating probabilities.

4.3.1   The rule of coherence

So far we have discussed the question of putting numbers to chances, the problem of quantification; the discussion of this will be developed further in later case studies when we come to examine the elicitation of prior information. Once quantified, however, there is need of a logical framework for dealing with preferences or probabilities, and there are certain general laws or axioms which establish the rules of operation. Thus bookmakers are faced with the problem that not only must they think of the probability of Dobbin's winning, but they must combine that with the odds of each of the other animals in the race; they also need to combine that probability with a margin for their own profit, and maybe other eventualities, such as the chance that the meeting might be abandoned altogether. Once all the probabilities have been estimated there are certain rules and laws for combining them in such a way that inconsistencies will not arise.

The first among the axioms we wish to mention has been referred to as the "transitivity of preferences" and establishes that the events (such as the event of a horse winning the Derby) and preferences (or probabilities) between them should fit together, or *cohere*. Let E_1, E_2 and E_3 represent three possible outcomes to a race (E_n standing here for an *element* of the event "winning the Derby" and the subscript distinguishing which of several possible such elements is being considered). Thus in our case E_1 stands for the elementary event "Dobbin wins the Derby". The law states that

$$E_1 < E_2 \quad \text{and} \quad E_2 < E_3 \quad \text{implies} \quad E_1 < E_3.$$

The less than sign "$<$" in this case means I believe that the probability of one event is less than the other. If you will, this could stand for a statement along the lines: if you prefer Maria-my-love (E_2) to Dobbin (E_1) as winner, and if you prefer Shergar (E_3) to Maria-my-love (E_2), then these two statements imply that you prefer Shergar (E_3) to Dobbin (E_1).

Coherence applies to preferences between judgements about events. Suppose we are interested in the sequence of three graves; we have some stratigraphic evidence, and two graves contain coins. Grave 3 holds a coin of the Roman Emperor Augustus, Grave 2 a coin of Tiberius; Grave 1 cuts Grave 2 (Figure 4.3). It is reasonable to place the graves in the order in which the emperors ruled, Grave 3 earlier than Grave 2 since Augustus was succeeded by Tiberius. Here we must admit that the evidence is less than certain: coins can remain in circulation for some time; thus in the U.K. even in the 1960's it was not unusual to find a Victorian penny in one's change, more than sixty years after it had been minted. Based on the coin evidence, therefore, we believe that it is fairly *likely* that Grave 3 is earlier than Grave 2. Thus we might define Event 1 (E_1) as the event that Grave 3 is earlier than Grave 2, with a reasonable, but not very high probability.

Let us next consider, as E_2, the event that Grave 3 is earlier than Grave 1. Here we have the evidence that Grave 1 cuts Grave 2, and so is later than Grave 2, over and above the coin evidence to place Grave 3 earlier than Grave 2. So we can be a little more confident that E_2 is true than that E_1 is true, that is

$$E_1 < E_2.$$

Next we consider the event (E_3) that Grave 2 is earlier than Grave 1. The probability of this event being true is rather high, because it is based on stratigraphy. We might suggest, nonetheless, that the event is not absolutely certain, because our belief is based on observations made in the course of excavation, and there is a remote chance that the stratigraphic relations were not properly observed. All the same it seems reasonable to say that E_2 is less likely than E_3 or

$$E_2 < E_3$$

because E_3 is based directly on the stratigraphy, and does not, like E_2, rely (in part) on coin circulation. It follows from this that

$$E_1 < E_3,$$

a conclusion which common sense confirms.

These relationships are established on the principle of coherence, which demands that we combine the two types of evidence in a consistent way. Indeed in most real-life problems there is a need to bring together many different types of information: thus in developing the chronology of a site there can be stratigraphic information, artefact typologies, perhaps tree-ring dates, coin evidence and so on. Coherence demands that these are combined in a consistent manner, so that inferences concerning the stratigraphy relate with inferences concerning the pottery typology and so forth. This requirement also leads us to the laws for combining and manipulating probabilities, which are the concern of the next section.

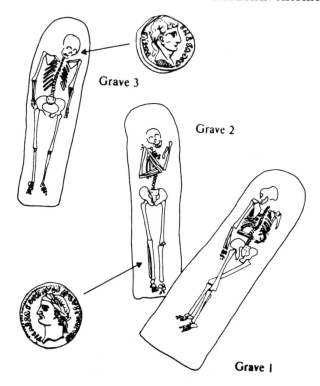

Figure 4.3   Schematic diagram of Graves 1, 2 and 3 together with the coins
found in them.

4.4   Some rules of probability

There are many rules governing the manipulation of probabilities. These rules
can be deduced from the axioms of probability — the basic building blocks of
probability theory. As we have stated above, in this context we see no call to
provide an axiomatic approach to probability. Instead we state the rules that
we feel are essential for the development of our exposition of the Bayesian
approach. Readers who wish to follow a detailed axiomatic approach might
consult Bernardo and Smith (1994).

4.4.1   Rule 1

The first rule states:

$$0 \le P(E|H) \le 1,$$

that is to say the probability of an event lies between 0 and 1 inclusive. This is conventional and, for example, percentages could be used, but it indicates that probabilities when measured are assessed as proportions.

If we assess that, on the basis of our information H, an event, E, has a probability of 1 we are saying that we believe that it is absolutely certain to occur. In contrast a probability of 0 expresses our belief that the event is impossible.

4.4.2   Rule 2, the Addition Rule

Recall that an event E is a set of outcomes. We talk of the *union* of the two events $(E_1$ or $E_2)$ if either (or both) occur, and only then. The union consists of all the outcomes in E_1 and all the outcomes in E_2. The Addition Rule states that the probability of the union of two *exclusive* events, that is to say two events which have no outcomes in common, is equal to the sum of the two probabilities:

$$P(E_1 \text{ or } E_2|H) = P(E_1|H) + P(E_2|H). \tag{4.1}$$

Example

Suppose that a historian of vernacular architecture was interested in dating a simple, unaltered, timber-framed building (Figure 4.4). They could argue that it was either fourteenth century or fifteenth century, but plainly it could not have been completed in both. In this case, E_1 is taken to represent the event that the building was built in the fourteenth century, and E_2 the event that the building was constructed in the fifteenth century. If there is thought to be a 0.2 chance that the building is fourteenth century, and a 0.7 chance that it is fifteenth century, then the Addition Rule tells us that there is a 0.9 chance that it is either fifteenth or fourteenth century (by implication there is a 0.1 chance that it was built in neither).

Of course, not all events are exclusive. Continuing the above example, let D_1 and D_2 be the events that the building was constructed in the periods 1300–1450 and 1380–1500 respectively. Suppose that the historian assesses that

$$P(D_1|H) = 0.6 \text{ and } P(D_2|H) = 0.75.$$

Note that D_1 and D_2 overlap for the period 1380–1450 and so are not exclusive, they have some years in common.

Clearly we cannot say that

$$P(D_1 \text{ or } D_2|H) = P(D_1|H) + P(D_2|H) = 1.35$$

because their sum is greater than 1, which violates Rule 1! For events that

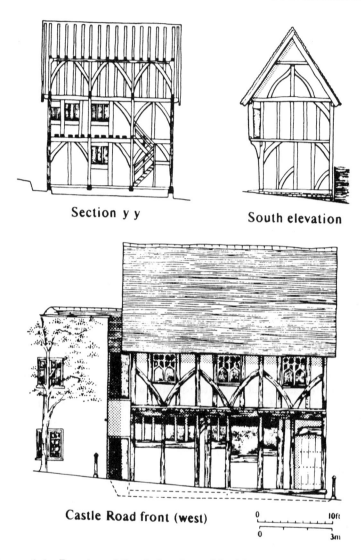

Section y y

South elevation

Castle Road front (west)

Figure 4.4 Drawing of the timber-framed building known as "Severns",
Nottingham, thought on stylistic grounds to be fifteenth century. Tree-ring dating
has now established a construction date of mid-fourteenth century.

are *not* exclusive the Addition Rule becomes

$$P(D_1 \text{ or } D_2|H) = P(D_1|H) + P(D_2|H) - P(D_1 \text{ and } D_2|H).$$

That is to say, it is the sum of the two probabilities minus the probability of where the two "overlap", that is where they have elements (in this case a number of years) in common. "D_1 and D_2" represents the event that the building was constructed at some time in the period 1380–1450. If $P(D_1 \text{ and } D_2|H)$ were not subtracted there would be double counting: the probability of the event "D_1 and D_2" would enter into the equation twice.

If we suppose that the probability of the building being constructed in the period 1380–1450 was assessed as 0.45 then we have

$$P(D_1 \text{ or } D_2|H) = 0.6 + 0.75 - 0.45 = 0.9.$$

Thus, based on current information, the probability that the building was constructed in the period 1300–1500 is 0.9.

Example

These rules have many other consequences. For example, suppose we believe that, based on the information currently in our possession, Dobbin has a 0.2 probability of winning (counting a dead heat as a win) the Donkey Derby. Let E_1 be the event that Dobbin wins, so that we have $P(E_1) = 0.2$. Let E_2 be the event that Dobbin does *not* win. Plainly either Dobbin wins or it does not, so that the event "E_1 or E_2" is certain to happen. Therefore we have $P(E_1 \text{ or } E_2) = 1$ and hence using the Addition Rule (4.1) we have

$$1 = 0.2 + P(E_2|H).$$

Consequently the probability that Dobbin does not win is 0.8 ($= 1 - 0.2$).

4.4.3 Rule 3, the Multiplication Rule

We talk of the *intersection* of two events E_1 and E_2 if both of the events occur. The Multiplication Rule says that for any events E_1, E_2 and information H,

$$P(E_1 \text{ and } E_2|H) = P(E_2|E_1 \text{ and } H)P(E_1|H). \tag{4.2}$$

For convenience we remove explicit reference to the prior knowledge and history H, though always bearing in mind its silent presence, and so we have the equation

$$P(E_1 \text{ and } E_2) = P(E_2|E_1)P(E_1). \tag{4.3}$$

Example

Let us take the example of an archaeologist who is investigating the character of sites in a region on the basis of aerial photographs. "Cult" sites (cemeteries, sanctuaries and the like, see Figure 4.5) might be distinguished from habitation sites (farms, settlements, hill-forts and so forth). Let E_1 represent the event that a site spotted on the photograph dates to the Bronze Age, and E_2 the event that a site is a cult site. Suppose that, on the basis of many years of experience of interpreting aerial photographs, the probability a site in the region dates to the Bronze Age is assessed to be 0.3 (that is $P(E_1|H) = P(E_1) = 0.3$). Furthermore the probability of a Bronze Age site being a cult site is assessed to be 0.8 (that is $P(E_2|E_1$ and $H) = P(E_2|E_1) = 0.8$). The probability that a particular site is both a cult site *and* dates to the Bronze Age can be calculated using the Multiplication Rule in the following way:

$$P(E_1 \text{ and } E_2) = P(E_2|E_1)P(E_1) = 0.8 \times 0.3 = 0.24. \qquad (4.4)$$

Note that the probability of an Iron Age cult site would be much lower (such sites do not show up well on aerial photographs), hence the importance of bringing out the point that E_2 is conditional on E_1. Both probabilities are conditional on the same background information and the researcher's understanding of the archaeology of the region.

Conditional probability

We note that provided $P(E_1) > 0$, (4.3) may be rearranged as

$$P(E_2|E_1) = \frac{P(E_1 \text{ and } E_2)}{P(E_1)}$$

which is the conventional definition of the conditional probability of event E_2 occurring given E_1 has already occurred (and we have knowledge H).

Independence

Two events E_1 and E_2 are said to be *independent* given our information H, if

$$P(E_1 \text{ and } E_2|H) = P(E_1|H)P(E_2|H). \qquad (4.5)$$

Expressed in words, this means that for two events E_1 and E_2, if the probability that both occur is equal to the product of the individual probabilities, then E_1 and E_2 are independent. A simple example of this is the throwing of a fair die. The probability of throwing a six is $\frac{1}{6}$, the probability of throwing a five is also $\frac{1}{6}$. As we believe that the result of one particular throw does not affect the result of a subsequent throw, it is reasonable to suppose

Figure 4.5   Aerial photograph of a long barrow, Bronze Age burial mound (?),
and Iron Age "Celtic fields" and enclosure, south of Wonston, Hampshire, U.K.
The burial mounds are difficult to distinguish and interpret because of the complex
field systems also showing up in the photograph (see St Joseph and Wilson, 1976).
(By courtesy of the University of Cambridge Committee for Aerial Photography.)
(Reproduced by permission of Cambridge University Collection of Air
Photographs: copyright reserved.)

that the probability of throwing a six followed by a five is $\frac{1}{6} \times \frac{1}{6} = \frac{1}{36}$. That
is, we believe the two events are independent.

If E_1 and E_2 are independent then using (4.2) and (4.5) we have

$$P(E_2|E_1 \text{ and } H)P(E_1|H) = P(E_1 \text{ and } E_2|H) = P(E_2|H)P(E_1|H).$$

Hence, provided that $P(E_1|H) > 0$, if E_1 and E_2 are independent we have

$$P(E_2|H) = P(E_2|E_1 \text{ and } H). \tag{4.6}$$

This means that if $\overline{}$ nd E_2 are independent, the extra information provided by E_1 does not alter the probability of E_2 occurring. In other words the occurrence of event E_1 provides us with no additional information (additional, that is, to our original information, H) about E_2.

Example

Consider the dates of the contexts from St Veit-Klinglberg. For contexts for which we have stratigraphic information, evidence about the date of one context does have some effect on our beliefs about the date of another. To be more specific, consider the dates of contexts 814 and 1235; these are represented by θ_{814} and θ_{1235} respectively. The stratigraphic information is that context 814 was laid down before context 1235 and so we have $\theta_{814} > \theta_{1235}$. Therefore if evidence comes to light that context 1235 dates to before 1500 BC (3450 BP) then this implies that context 814 must also be before that date. In other words evidence about one date influences our beliefs about another and so the two dates cannot be considered to be independent.

On the other hand, consider contexts 1308 and 1319 which are unrelated by the stratigraphy. In broad terms we know that the dates of these contexts fall in the same wide time interval, perhaps from 2500 BC to 500 BC, which covers the period of occupation of the site. If, however, knowing that the date of context 1308 lies in the interval 1300 BC to 1200 BC does not provide any additional information about the date of context 1319, then the two dates can be considered to be independent. If the date of context 1308 does affect our beliefs about that of context 1319 then the two dates are not independent.

4.4.4   The theorem of total probability

We have already met the notation of conditional probability $P(E_1|E_2)$, where we are assessing the probability that event E_1 occurs, on the condition of event E_2 having occurred. Likewise the probability of event E_1 occurring given that event E_2 has *not* occurred is indicated by $P(E_1|\bar{E}_2)$. The notation $\bar{E}_2$, spoken as "E_2 bar", is used to indicate that event E_2 does not happen; it can be referred to as the complement or the negation of E_2.

Now either E_2 happens or it does not, and it cannot, on the same occasion, both happen and not happen. In other words E_2 and $\bar{E}_2$ are *exhaustive* (they cover all the relevant possibilities) and *exclusive* (they cannot both happen at the same time). Thus we can write

$$P(\bar{E}_2) = 1 - P(E_2).$$

Moreover, the probability of E_1 occurring can be expressed as follows:

$$
\begin{aligned}
P(E_1) &= P([E_1 \text{ and } E_2] \text{ or } [E_1 \text{ and } \bar{E}_2]) \\
&= P(E_1 \text{ and } E_2) + P(E_1 \text{ and } \bar{E}_2)
\end{aligned}
\tag{4.7}
$$

since $[E_1$ and $E_2]$ and $[E_1$ and $\bar{E}_2]$ are exclusive events. Using the Multiplication Rule we note that

$$P(E_1 \text{ and } E_2) = P(E_1|E_2)P(E_2)$$

and

$$P(E_1 \text{ and } \bar{E}_2) = P(E_1|\bar{E}_2)P(\bar{E}_2).$$

Hence we obtain

$$P(E_1) = P(E_1|E_2)P(E_2) + P(E_1|\bar{E}_2)P(\bar{E}_2). \tag{4.8}$$

This is a particular case of a more general result that we state later in this section in (4.9).

Example

By way of illustration imagine that, from experience, a site director believes that on days when the temperature exceeds $30°C$ the diggers notice and record the positions of only 60% of the finds *in situ*, and the remaining 40% are missed and turn up in the course of sieving. In contrast when the temperature does not exceed $30°C$, the diggers are more vigilant, then 85% are found *in situ* and only 15% turn up in sieving. The weather forecasters predict that on 5 out of 25 working-days available in August the temperature will exceed $30°C$. The site director could be interested in questions such as "on any day, what is the recovery rate for a dig." We interpret this as meaning, "what is the probability that an artefact is found *in situ* on a randomly chosen day in August?"

We can set up the notation by letting R represent the event that an artefact is recorded *in situ*. Let W and $\bar{W}$ represent that the temperature is above or below $30°C$ respectively. The above information can be expressed in terms of probabilities as follows:

$$P(W) = \tfrac{5}{25} = 0.2, \quad P(\bar{W}) = 1 - P(W) = 0.8,$$
$$P(R|W) = 0.6, \qquad P(R|\bar{W}) = 0.85.$$

Then using (4.8) we have

$$
\begin{aligned}
P(R) &= P(R|W)P(W) &+& \quad P(R|\bar{W})P(\bar{W}) \\
&= 0.6 \times 0.2 &+& \quad 0.85 \times 0.8 \\
&= 0.12 &+& \quad 0.68 \\
&= 0.80.
\end{aligned}
$$

Thus when digging in August the site director believes that about four-fifths of the finds will be recognized in excavating, and about one-fifth will come from the sieve. This conclusion might help in organizing the logistics of the operation.

The idea behind (4.7) and (4.8) can easily be extended to more complex situations. Suppose we have n events denoted by $A_1, A_2, \ldots, A_n$ where the A_i are exclusive (two or more of the events cannot occur simultaneously) and exhaustive (the union of the A_i represents all the possible outcomes). Then we have the following result:

$$P(E) = P(E|A_1)P(A_1) + P(E|A_2)P(A_2) + \cdots + P(E|A_n)P(A_n). \quad (4.9)$$

This result, which is a generalization of (4.8), is called *the theorem of total probability*. Its use will be illustrated by the next example.

4.4.5   Getting to a dig on time

An archaeologist can get to a dig by foot, bicycle or public transport or by hitching a lift; we refer to these events as A_1, A_2, A_3 and A_4 respectively. These are the only possible methods of transport available and the archaeologist uses one and only one form each day, hence the events are exclusive and cover all possibilities. The director assesses that the methods of travel are used with probabilities 0.1, 0.2, 0.3 and 0.4 respectively. Moreover, the director of the excavation assesses that the archaeologist has a probability of 0.35 of being late if he or she walks to the dig. The corresponding probabilities of being late when travelling by bicycle, by public transport or by hitching are 0.15, 0.40 and 0.10 respectively.

Let L be the event "arriving late", then this information is interpreted mathematically as

$$P(L|A_1) = 0.35, \quad P(L|A_2) = 0.15,$$
$$P(L|A_3) = 0.40, \quad P(L|A_4) = 0.10.$$

What is the probability that the archaeologist arrives late on a particular day?

We need to evaluate $P(L)$. Using the Multiplication Rule the probability of both walking and arriving late is given by

$$P(L \text{ and } A_1) = P(L|A_1)P(A_1) = 0.35 \times 0.1 = 0.035.$$

Likewise for the other forms of transport

$$P(L \text{ and } A_2) = P(L|A_2)P(A_2) = 0.15 \times 0.2 = 0.03,$$
$$P(L \text{ and } A_3) = P(L|A_3)P(A_3) = 0.40 \times 0.3 = 0.12,$$
$$P(L \text{ and } A_4) = P(L|A_4)P(A_4) = 0.10 \times 0.4 = 0.04.$$

Using the Addition Rule, we add the probabilities of being late for each mode of transport (walking, bicycle, bus, hitching) to get the overall chance of being late, as follows:

$$P(L) = P(L \text{ and } A_1) + P(L \text{ and } A_2) + P(L \text{ and } A_3) + P(L \text{ and } A_4)$$
$$= \quad 0.035 \quad + \quad 0.03 \quad + \quad 0.12 \quad + \quad 0.04$$
$$= \quad 0.225.$$

The interpretation is that the director, based on his or her knowledge, believes that the archaeologist has about a 1 in 4 chance of arriving late.

4.5 Bayes' theorem

4.5.1 The basic theorem

The cornerstone of Bayesian statistics is Bayes' theorem which can be derived as follows. We know from (4.3) that

$$P(E_1 \text{ and } E_2) = P(E_2|E_1)P(E_1)$$

and by interchanging E_1 and E_2 we have

$$P(E_2 \text{ and } E_1) = P(E_1|E_2)P(E_2).$$

But

$$P(E_1 \text{ and } E_2) = P(E_2 \text{ and } E_1)$$

and hence we can deduce that

$$P(E_1|E_2)P(E_2) = P(E_2|E_1)P(E_1).$$

This leads to a result of sufficient importance to be given a name:

Bayes' theorem Let E_1 and E_2 be two events with $P(E_2) > 0$, then

$$P(E_1|E_2) = \frac{P(E_2|E_1)}{P(E_2)} P(E_1). \tag{4.10}$$

This is a result of central importance to our approach.

How is it to be understood? We wish to improve our knowledge about an event by using a *new* piece of information, denoted by E_2. Thus on the left-hand side of the equation we are interested in the probability of E_1 *given* E_2; this is known as the *posterior* probability of E_1.

How do we arrive at the posterior probability? On the right-hand side of the equation we have our starting position, $P(E_1)$: that is the *prior* probability of E_1, based upon our original information. Next we multiply that by the fraction $P(E_2|E_1)/P(E_2)$, which updates our original information about E_1 in the light of the new information E_2. In a nutshell Bayes' theorem is a formal way of expressing how probability changes in the light of new information: how we update our knowledge. If $P(E_2|E_1) > P(E_2)$ then the posterior probability of E_1 will be greater than the prior probability of E_1 and as a result of the new information we have greater confidence in E_1.

4.5.2   A simple example of using Bayes' theorem — selecting the right candidate

The director of an archaeological project is interviewing candidates to carry out post-excavation work on the finds from a dig. The successful appointee will need to be able to recognize and catalogue pottery according to the established typology. As part of the interview, therefore, the candidates are presented with a sherd and asked to ascribe it to the correct type out of five possible categories. The director will need to know if a candidate is merely guessing or does in fact have sufficient experience to be appointed.

Firstly we set up an appropriate notation as follows. Let E_1 be the event that they make the correct assignment and E_2 be the event that the candidate does have the necessary experience. In the director's opinion, if the candidate has the necessary experience then they are certain to get the correct answer. Therefore

$$P(E_1|E_2) = 1.$$

On the other hand if they do not possess the necessary experience, $\bar{E}_2$, then they will guess and therefore have a 1 in 5 chance of being correct. In this case we will have

$$P(E_1|\bar{E}_2) = 0.2.$$

Let q be the prior probability that a candidate has the relevant experience, that is

$$P(E_2) = q.$$

As a consequence, the prior probability that the candidate does not possess the relevant experience is

$$P(\bar{E}_2) = 1 - q.$$

The project director requires $P(E_2|E_1)$: the posterior probability of a candidate having the necessary expertise given that they have made the correct assignment. By Bayes' theorem, we have that the posterior probability of E_2 is given by

$$P(E_2|E_1) = \frac{P(E_1|E_2)}{P(E_1)} P(E_2).$$

To calculate $P(E_2)$ we note that from (4.8) we have

$$P(E_1) = P(E_1|E_2)P(E_2) + P(E_1|\bar{E}_2)P(\bar{E}_2).$$

Therefore

$$P(E_1) = (1 \times q) + \left(\frac{1}{5} \times (1 - q)\right) = \frac{4q + 1}{5}$$

and so

$$P(E_2|E_1) = \frac{1 \times q}{(4q + 1)/5} = \frac{5q}{4q + 1}.$$

For example if $q = 0.2$ and the candidate answers correctly then the posterior probability is 0.56. If $q = 0.8$, it rises to 0.95.

Obviously in real life many other considerations need to be taken into account over and above the answering of one question. More likely the candidates will be presented with a variety of sherds to be classified and perhaps with other comparable tasks. Nevertheless the principle remains the same.

4.5.3 A more general form of Bayes' theorem

Suppose we have n events denoted by $A_1, A_2, \ldots, A_n$ where the A_i are exclusive and exhaustive. Let E be another event such that $P(E) > 0$. We replace E_1 by A_i and E_2 by E in (4.10) to get

$$P(A_i|E) = \frac{P(E|A_i)}{P(E)} P(A_i).$$

Using (4.9) to express $P(E)$ in terms of the $P(E|A_i)$ and the $P(A_i)$ we obtain

$$P(A_i|E) = \frac{P(E|A_i)}{\sum_{i=1}^{n} P(E|A_i)P(A_i)} P(A_i). \tag{4.11}$$

This equation is one of several ways of stating Bayes' theorem. Another and probably one of the best ways of remembering it is

$$P(A_i|E) \propto P(E|A_i)P(A_i). \tag{4.12}$$

Bayes' theorem expressed as either (4.11) or (4.12) provides an extremely powerful tool for making inferences. We can think of the A_i as forming a set of hypotheses, of which one and only one is true. If event E is observed, then the prior probability of A_i, $P(A_i)$, changes to the posterior probability of A_i, $P(A_i|E)$. The term $\sum_{i=1}^{n} P(E|A_i)P(A_i)$ is the weighted average of how likely E is to occur if A_i were true, with the weight given by the prior probability of A_i being true. If E is observed, the probability of A_i increases if $P(E|A_i)$ is larger than this average. The terms $P(E|A_i)$ are known as the *likelihoods*; $P(E|A_i)$ is the likelihood given to event A_i by event E.

4.5.4   The director makes inferences

Let us return to the "getting to the dig on time" example introduced in Section 4.4.5. Without directly asking the archaeologist, the director wishes to make inferences about the form of transport used.

If the archaeologist is late, what is the probability that they walked?

We interpret this as finding the probability that they walked (event A_1) *given* that they were late (event L). Thus we need to calculate $P(A_1|L)$. By Bayes' theorem we have

$$P(A_1|L) \; = \; \frac{P(L|A_1)}{P(L)} P(A_1) = \frac{0.35}{0.225} \times 0.1 = 0.156.$$

The director interprets this to mean that when the archaeologist is late there is about a 1 in 7 chance that he or she walked.

If the archaeologist is late, what is the most likely form of transport used?

Carrying out the above calculation for each of the other forms of transport, we find that

$$P(A_2|L) = \frac{P(L|A_2)P(A_2)}{P(L)} = \frac{0.15 \times 0.2}{0.225} = 0.133$$

$$P(A_3|L) = \frac{P(L|A_3)P(A_3)}{P(L)} = \frac{0.40 \times 0.3}{0.225} = 0.533$$

$$P(A_4|L) = \frac{P(L|A_4)P(A_4)}{P(L)} = \frac{0.10 \times 0.4}{0.225} = 0.178.$$

Note that

$$P(A_1|L) + P(A_2|L) + P(A_3|L) + P(A_4|L) = 1,$$

as it should do since one form of transport must be used. This illustrates the consistency of the rules of probability.

So how are these results to be interpreted? We can see that if someone is late the chances of their having walked, cycled or hitched are not too different from one another: somewhere between 1 in 8 (=0.125) and around a sixth (0.167); but the bus has a 50:50 chance of being the mode of travel if someone is late. Hence public transport wins (or rather loses) hands down!

4.5.5   Burial rites and Bayes' theorem

We shall explore increasingly complex examples of the application of Bayes' theorem as the book progresses. By way of preparation we have developed an application which will require close attention by the reader. We believe, for those not familiar with this sort of argument, that it has some useful points to make, and we recommend that it is followed through step by step.

Let us suppose that within a cemetery there is a tendency in period I for female burials to receive a type of vase (call it a pyxis) more commonly than male burials. But in the later period II, fashion changes towards a more unisex distribution. The data are given in Table 4.1.

As is often the case with information from archaeologists this needs a little disentangling. First of all let us establish some notation: A will be used to denote the event that a pyxis is found in a grave; M the event that a male and F that a female was placed in a grave; D_1 the event that the burial belongs to period I and D_2 that it belongs to period II. How exactly are the data in the table to be interpreted? It appears to mean that in period I, of the male burials identified, 30% had pyxides so that we can say

$$P(A|M \text{ and } D_1) = 0.3. \tag{4.13}$$

Similarly we have

$$\begin{aligned} P(A|F \text{ and } D_1) &= 0.9, \\ P(A|M \text{ and } D_2) &= 0.4, \\ P(A|F \text{ and } D_2) &= 0.6. \end{aligned} \tag{4.14}$$

The archaeologist has chosen to present the data simply as presence or absence; the number of vases in the graves is judged as of no consequence. But this gets us only so far: these probability statements are marked by circumscribed conditions, or dependences, which arise directly from the way the data are conveyed in Table 4.1. If we look more closely at this it will be plain that the archaeologist is not giving us all the information we need. Perhaps the first thing to observe is that the table of data is not a question. What is the archaeologist interested in finding out? In a broad sense the way the data are presented indicates that we are concerned with how the roles of men and women are signalled in terms of funerary offerings. But this is not good enough, let us pose a particular enquiry.

Suppose, then, we find a grave containing a pyxis and a skeleton whose sex cannot be recognized as the bones are too decayed. The grave is of particular interest because of its wealthy offerings, indicating high status; an informed judgement of the sex of the person interred would aid the archaeologist's understanding of the social structure. Based on other artefacts found in the

Table 4.1 Information from an imaginary cemetery site about the association of finds and sex of the person buried.

Period	% Males with pyxides	% Females with pyxides
I	30	90
II	40	60

grave, we assess the probability that it dates to period I as 0.3, and the probability that it dates to period II as 0.7. That is we have

$$P(D_1) = 0.3 \text{ and } P(D_2) = 0.7.$$

What is the probability that it is a male burial from period I?
We need to calculate $P(M \text{ and } D_1|A)$. Using Bayes' theorem we have

$$P(M \text{ and } D_1|A) = \frac{P(A|M \text{ and } D_1)P(M \text{ and } D_1)}{P(A)}.$$

From Table 4.1 we know that $P(A|M \text{ and } D_1) = 0.3$ but what is known about $P(M \text{ and } D_1)$ and $P(A)$?
We need to evaluate $P(M \text{ and } D_1)$ which can be expressed, using the Multiplication Rule, as

$$P(M \text{ and } D_1) = P(M|D_1)P(D_1). \tag{4.15}$$

The archaeologist has given us no information regarding the probability of finding a male burial in period I, $P(M|D_1)$; we need an answer to the question. We take the answer to be 0.5 (our belief in the light of no evidence to the contrary): there were as many males as females buried in the cemetery at that time. Armed with this we can say

$$P(M \text{ and } D_1) = 0.5 \times 0.3 = 0.15.$$

What of period II? Again, in the light of no other information, we take $P(M|D_2) = 0.5$ also.
What then of $P(A)$? By the theorem of total probability, see (4.9), we have:

$$P(A) = \quad P(A|F \text{ and } D_1)P(F \text{ and } D_1) + P(A|M \text{ and } D_1)P(M \text{ and } D_1) \\ +P(A|F \text{ and } D_2)P(F \text{ and } D_2) + P(A|M \text{ and } D_2)P(M \text{ and } D_2).$$

For this we need to calculate probabilities of the type $P(A|F \text{ and } D_1)$ (the probability of finding a pyxis given that the burial is a woman dating to period I) and $P(F \text{ and } D_1)$ (the probability of a female burial dating to period I). The figures for this can be calculated in a similar fashion to (4.14) and (4.15):

$$\begin{aligned} P(A) &= \quad (0.9 \times 0.15) + (0.3 \times 0.15) + (0.6 \times 0.35) + (0.4 \times 0.35) \\ &= \quad 0.135 + 0.045 + 0.21 + 0.14 \\ &= \quad 0.53. \end{aligned}$$

The interpretation of this result is that just over 50% graves have at least one pyxis.

We are now in a position to calculate $P(M \text{ and } D_1|A)$ using Bayes' theorem (4.11) with events $E_1 = (M \text{ and } D_1)$ and $E_2 = A$:

$$P(M \text{ and } D_1|A) = \frac{P(A|M \text{ and } D_1)P(M \text{ and } D_1)}{P(A)} = \frac{0.3 \times 0.15}{0.53} = 0.085.$$

This can be read as: there is a posterior probability of just under 1 in 10 that the rich burial with the pyxis is that of a man who was buried in period I. This is the answer to the question posed above. The readers may wish to work out for themselves $P(M \text{ and } D_2|A)$, $P(F \text{ and } D_1|A)$ and $P(F \text{ and } D_2|A)$.

What is the probability that it is a male burial?

Here we need to evaluate $P(M|A)$, bearing in mind that the other possibilities are $P(F|A)$, $P(M|\bar{A})$ and $P(F|\bar{A})$. Using the Multiplication Rule (4.2) we note that

$$P(M|A) = \frac{P(A \text{ and } M)}{P(A)}.$$

We have already calculated $P(A)$ to be 0.53. Therefore we need to evaluate $P(A \text{ and } M)$ which we do by using the theorem of total probability (4.9):

$$\begin{aligned}
P(A \text{ and } M) &= P(A|M \text{ and } D_1)P(M \text{ and } D_1) \\
&\quad + P(A|M \text{ and } D_2)P(M \text{ and } D_2) \\
&= (0.3 \times 0.15) + (0.4 \times 0.35) \\
&= 0.185.
\end{aligned}$$

So the probability that it is a male burial becomes

$$P(M|A) = \frac{0.185}{0.53} = 0.35.$$

Interpretation

Having gone through the process of clarifying the question in hand, and eliciting the information required in order to carry out the analysis, it has been possible to calculate the chances that the rich burial is of a male: 0.35. By a similar argument the probability of a female burial is 0.65, almost twice as good a chance. But these calculations are all contingent on the decisions made concerning, for example, the proportions of male and female burials. If, on the contrary, the excavator had evidence that during the given period females were under-represented, and formed only 25% of those buried in the cemetery, then the inferences would alter, in the light of the new evidence. In any analysis, assumptions must be clearly stated so that subsequent researchers can understand the rationale used and, if necessary, reanalyse the same data but with different assumptions.

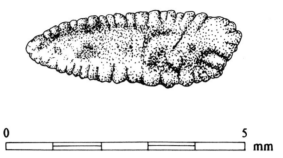

Figure 4.6 Drawing of an otolith, the ear-bone of a cod-fish. The bones are used by archaeologists to identify the time of year when temporary seasonally occupied sites were inhabited.

4.6   Exchangeability

An important but rather difficult concept in Bayesian statistical thinking is that of *exchangeability*. In general terms suppose we have a *population* consisting of *units* which exhibit some characteristic. Then the units are said to be *exchangeable* if our beliefs about the value of a characteristic of one unit are the same as our beliefs about the same characteristic of another.

4.6.1   A fishy example

As a concrete example, our population could be all the 16 month-old cod-fish that have lived in the seas around the islands of Orkney. In studying mesolithic sites on the islands archaeologists recovered large numbers of the ear-bones, or otoliths (see Figure 4.6). The length of the ear-bones is related to the age of the fish — roughly the longer the bone, the older the fish. As the fish spawn at the same time each year, the age of the fish can be used to indicate the time of the year at which the fish were caught. The archaeologists turned, therefore, to the statisticians (see English and Freeman, 1981) to use the measurements of otoliths found in the middens to estimate at what time of year particular sites were occupied.

The characteristics concerning the ear-bone length of a cohort of fish can be estimated on the basis of modern fish caught in the area at a particular time of year. To say that the fish are exchangeable implies that measurements of fish caught in one shoal are believed to be equivalent to those of fish from another shoal caught at the same time of year; the shoals might have been caught at the same time of year in AD 1980 or AD 1970 or 4000 BC.

In Figure 4.7 we display histograms of the lengths of otoliths excavated from various levels in midden sites on the island of Oronsay. On the basis of

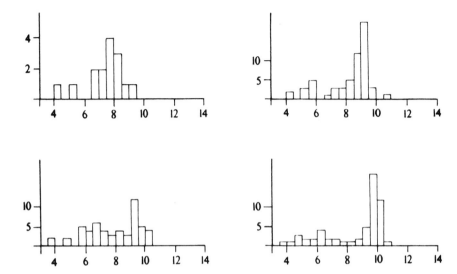

Figure 4.7   Histograms of the otoliths found in various levels and from various sites on Oronsay, Orkney. (After English and Freeman, 1981, p. 131, fig. 1.)

these, inferences were made about the periods of the year when the sites were occupied.

4.6.2   Implications of exchangeability

The implications of exchangeability are important and we illustrate them by a simple example. Suppose an archaeologist wishes to study, through chemical compositional analysis, the provenance of a collection of, say 121, sherds of a particular type. Because of financial constraints only a limited number, say 20, can be analysed. The traditional or classical statistical approach would involve some form of *random sampling*. That is each sherd would be given a unique number and using random numbers, either from tables or from a computer, the sherds to be chosen for the sample would be determined. In this way a representative sample will have been examined.

   In the Bayesian framework the problem would be tackled from a different viewpoint. One would ask whether *a priori* there is any evidence to suspect that the chemical composition of a particular sherd is different from any of the others. Of course by virtue of carrying out an analysis of this kind, one does suspect that the sherds may have different provenances but one is unable to pinpoint particular sherds that are likely or unlikely to have the same provenance. In this case, the sherds can be thought of as exchangeable. In other words, we have no *a priori* evidence that suggests that any two sherds have the same provenance or not.

What is the practical implication of this? It means that, provided the sherds (units) are believed to be exchangeable, *any* 20 can be chosen for the analysis. We could use the first 20 that come to hand! This could be extremely important if the collection were housed in several different museums across the world. Suppose there are 28 in London, 45 in Canberra and 48 in New York, then, provided that the sherds in these collections are exchangeable with each other, we could take the first 20 from London. In contrast, random sampling could well involve using sherds from all three museums. Of course if those sherds held at the Canberra were known to be different in some way from those held at the other two museums then the sherds would no longer be exchangeable and so the above sampling procedure would no longer be valid.

To be more specific, in the nineteenth and early twentieth centuries it was not unusual for several museums to acquire material from the same excavation, to the extent that sherds even from the same vase, for example, might end up in Warsaw, Paris and London. In these circumstances, the sherds are plainly exchangeable. If, on the other hand, sherds come from different excavations, even different areas within the same site, there might be some special (but unknown) factor affecting one group and not the others. This raises some doubts about the exchangeability of all the sherds and therefore, as a precaution, samples should be taken from all the collections in the different museums.

If the units cannot be thought of as exchangeable, then it may be possible to divide them into *subpopulations*, with units within each subpopulation being exchangeable. This could arise in the above example if the 121 sherds could be divided on the basis of thin-section analysis into four groups. Since there may be some relationship between the results of thin-sectioning and chemical compositional analysis, the sherds can no longer be thought of as being exchangeable. However within the subpopulations, the sherds are exchangeable. In this case we would sample from each subpopulation.

4.6.3 Consumerland

As a final example to illustrate the use of Bayes' theorem, consider the following situation. We suppose that archaeologists of the future have excavated a building belonging to the Consumer Culture. Based on their experience they have developed certain theories about this culture and in particular its rich cult traditions. Ritual consumption of food and drink was central to the beliefs of the people of the time. Across the globe many regional variants could be recognized, but two broad cycles of observance are distinguished. Shrines devoted to the dry cycle were known as "restaurants", whereas temples given over to the wet cycle were called "bars".

Figure 4.8 A beer mug. The event of finding a beer mug is denoted by M.

Figure 4.9 A wine goblet. The event of finding a wine goblet is denoted by G.

The question is: is the newly excavated building a "bar" or a "restaurant"? Our data consist of an assemblage of drinking vessels belonging to two basic types: "beer mugs" (Figure 4.8) and "wine goblets" (Figure 4.9). Archaeologists expert in the arcane rituals of the period believed that in a bar one would expect 80% mugs and 20% goblets, whereas in a restaurant one would expect 40% mugs and 60% goblets.

We introduce some notation by letting B and R be the events that the excavated building is a "bar" and "restaurant" respectively. Let M be the event that a glass is a "mug" and G that it is a "goblet". Then using the information in the previous paragraph we have

$$P(M|B) = 0.8, \quad P(G|B) = 0.2,$$
$$P(M|R) = 0.4, \quad P(G|R) = 0.6.$$

Let us suppose that in this particular futuristic excavation of a building, a total of five glasses are found, of which two come under the mugs category and three are classified as goblets. If the building is a bar then the probability that we have two mugs and three goblets is equal to

$$10 \times (0.8)^2(0.2)^3.$$

Where does this expression come from? Firstly consider all the possible orders in which they could have been found by the excavator: they can be listed as follows (G = goblet, M = mug):

$$GGGMM, \quad MGGGM, \quad MMGGG, \quad GMGGM, \quad GGMMG,$$
$$GGMGM, \quad GMMGG, \quad MGMGG, \quad GMGMG, \quad MGGMG.$$

Thus there are ten possible combinations.

Consider just one of these combinations, say, $GMMGG$. Suppose we can consider that the glasses in the original site (whether it was a bar or a restaurant) were exchangeable and this applies to the sample found in the excavation. That is we have no prior belief to suggest that the sample is unrepresentative of glasses in either a bar or a restaurant. If this is the case, given that we are considering a bar, whether a glass is a mug or a goblet is independent of the forms of the other glasses. In which case

$$\begin{aligned} P(GMMGG|B) &= P(G|B)P(M|B)P(M|B)P(G|B)P(G|B) \\ &= (0.2)(0.8)(0.8)(0.2)(0.2) \\ &= (0.8)^2(0.2)^3. \end{aligned}$$

By a similar argument we find that the probability of each of the other orders has the same probability. Hence the probability of finding two mugs and three goblets, given that it is a bar, is

$$10 \times (0.8)^2(0.2)^3 = 0.0512.$$

If it were a restaurant then the probability of two mugs and three goblets would be equal to
$$10 \times (0.4)^2(0.6)^3 = 0.3456.$$

The experts, in fact, have even more information. From a sacred text known as *The Book of the Golden Pages* they know that in the area where the excavation took place, there were roughly twice as many bars as restaurants.

This provided the evidence for assessing that the prior probability of the building being a bar is 2/3, and of it being a restaurant 1/3. That is, *a priori*,

$$P(B) = \frac{2}{3} \text{ and } P(R) = \frac{1}{3}.$$

To make inferences about the use of the building based upon *both* the prior knowledge *and* the data, we need to evaluate $P(B|Z)$ and $P(R|Z)$ where Z represents the event that "two mugs and three goblets" were found. By Bayes' theorem we have

$$\begin{aligned} P(B|Z) &\propto P(Z|B)P(B) \\ &\propto 10 \times (0.8)^2 (0.2)^3 \times 2/3 \\ &\propto 0.0341, \end{aligned}$$

and

$$\begin{aligned} P(R|Z) &\propto P(Z|R)P(R) \\ &\propto 10 \times (0.4)^2 (0.6)^3 \times 1/3 \\ &\propto 0.1152. \end{aligned}$$

To evaluate these probabilities exactly, we need to normalize the above results by dividing each expression by $0.0341 + 0.1152 = 0.1493$, since we know that $P(B|Z) + P(R|Z) = 1$. When the arithmetic is done (and the readers may wish to confirm this by doing the calculations themselves), the posterior probability that the building is a bar is $0.0341/0.1493 = 0.23$, and that it is a restaurant is $0.1152/0.1493 = 0.77$. The archaeological evidence of the two mugs and three goblets has considerably lowered the original expectation that this is a bar in this part of town (i.e. $P(B) = 0.67$ whereas $P(B|Z) = 0.23$!). Based on *all* the evidence it is much more likely to have been a restaurant than a bar.

4.7   Discussion

We have, in this chapter, considered uncertainty and the basic nature of probability. The main concern was to stress the Bayesian perception that assessments of probability are subjective — informed, of course, by experience, knowledge and understanding, but at root subjective. The assessment of probabilities is, therefore, closely bound up with the interpretation and quantification of prior knowledge, which underlies the whole Bayesian approach. The processes of quantification, discussed in the opening sections above, will be further elaborated in future chapters. Here we wish to emphasize that the elicitation and evaluation of prior probabilities presents a particular challenge for those following Bayesian methodology.

The manipulation of probabilities, once quantified, has formed the main part of the discussion in this chapter. The rules of probability have been illustrated by over-simplified examples with a rather limited number of outcomes. This has given them an artificial feel. For a more realistic treatment we need to model uncertainty when the range of possible outcomes is infinite. So in the next chapter we move on to consider statistical distributions. Indeed, the selection of appropriate distributions is a vital aspect of statistical modelling, of linking the model, as an expression of our understanding of a problem, to the data. Moreover statistical distributions are needed to model our prior knowledge about the parameters of the problem.

5

Statistical modelling

5.1 Introduction

The building of mathematical models of archaeological situations was discussed in Chapter 3. This led us, in the last chapter, to consider uncertainty: the way in which statisticians deal with uncertainty, through its quantification, by using probabilities and the manipulation of probabilities. The main thrust of this chapter will continue to be the modelling of uncertainty, and particular emphasis will be placed on statistical distributions.

When discussing the quantification of probability in Chapter 4, we tended to quote a single figure like the odds (such as 4 to 1, or $P = 0.2$) on a horse winning, or a 60% ($P = 0.6$) chance of a thunderstorm. Now we need to recognize that there are more complex situations where a single figure does not adequately cover the problem under investigation. Thus there are cases where there are many possible outcomes, such as the incidence of pots in a grave, where there might be no pots, or one, or two, or three, or more; or the length of time over which a site was occupied, where the answer might lie in a range of possible values. Hence there is a need to investigate how uncertainty varies according to the number of pots recovered, or with changes in the duration of occupation. *Statistical distributions* allow just such generality in describing probabilities. The selection and manipulation of statistical distributions in the modelling process is a difficult and at times very technical discipline. There is much more to the subject then we are able to convey in this book. Therefore, our aim here is to introduce those distributions which are vital for the case studies discussed later. By doing so we hope to give the reader a feel for how uncertainty can be modelled by statistical distributions and how this fits into the overall modelling process.

We shall see that there are very many different types of distribution and much skill lies in recognizing which distribution best models the situation under study. Already two broad classes of data have been referred to: *discrete* data, like the counts of pots per grave or the numbers of coins of different denominations in a hoard, and *continuous* data, such as time or length or chemical concentration. In the sections that follow we will describe firstly the modelling of discrete data and secondly the modelling of continuous data. Here we model only one variable, but later, in Chapter 6, we will look at modelling two or more variables that are related.

5.2   Populations and samples

As an archaeological example we list the floor areas of chamber tombs (see Figure 5.1) dating to the Late Bronze Age (*ca.* 1600–1100 BC) in Greece; the data taken from 134 ($n = 134$) tombs are given in Table 5.1. The mean of this sample of tombs is $7.88\,\mathrm{m}^2$. In fact thousands of such tombs were constructed in Greece during the period. If a different sample of 134 tombs were measured we may be reasonably sure that another average would result, not too different from the first, but not exactly the same. As more and more tombs from the totality are measured and added to the sample the average will settle down. That is to say as n increases the difference between the average area for n tombs and for $n + 1$ tombs from the totality will diminish.

This is one way of approaching the notion that the sample is drawn from some idealized *underlying population*, and the data from the real world will approximate to, but will never be precisely the same as the ideal. The very notion of an ideal population with statistical characteristics is an artifice: thus being able to measure all Greek Late Bronze Age chamber tombs is not merely vastly improbable but absolutely impossible because many of them have been destroyed. In reality, and all too often in a subject like archaeology, it is not possible to get a very large sample. The Bayesian principle that all probabilities are conditional, and the emphasis on prior assessments of probabilities plainly fit in here: population is a way of talking about our model. The statistical idea of an underlying population, it would seem, inevitably implies a model from which it is intended to draw inferences about the real world.

5.2.1   Sample mean and population mean

In the context of statistical modelling, therefore, it is important to distinguish between the *sample mean*, which is normally denoted $\bar{x}$, and the *population mean* which is often represented by μ. The sample mean can be expressed as

$$\bar{x} = \frac{1}{n}(x_1 + x_2 + \cdots + x_n) = \frac{1}{n}\sum_{i=1}^{n} x_i$$

where x_i denotes the value of the ith observation and the mathematical symbol, $\sum$, denotes the *summation* of $x_1, x_2, \ldots, x_n$. The sample mean can be thought of as an approximation to or an estimate of the ideal or population

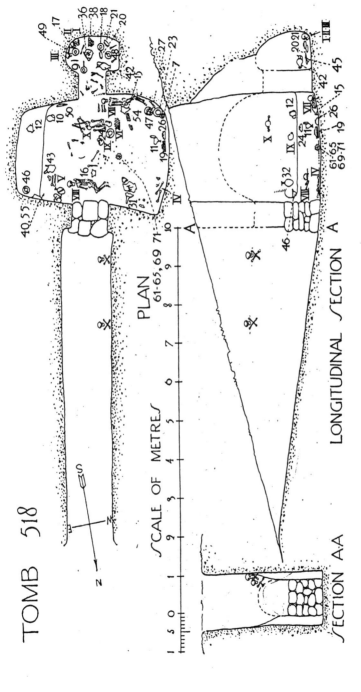

Figure 5.1 Plan of a Greek Late Bronze Age (or Mycenaean) chamber tomb. (After Wace, 1932, p. 75, fig. 29.)

Table 5.1   The floor areas (in square metres) of 134 Greek Late Bronze Age rock-cut chamber tombs.

23.05	3.881	2.802	4.468	0.304	4.964	11.476
4.236	4.725	6.406	5.687	10.154	6.036	1.100
5.161	5.810	3.374	8.014	7.878	25.343	22.555
12.62	19.973	25.277	29.625	9.206	5.943	13.767
0.817	34.599	7.327	7.556	13.240	8.916	13.025
5.506	15.185	13.114	12.627	13.150	7.624	7.784
9.239	0.848	5.556	3.904	14.751	6.470	3.834
5.588	7.964	6.460	5.110	1.840	5.343	4.573
5.972	2.378	8.736	6.508	5.623	4.237	14.091
6.729	10.328	22.763	8.655	9.452	14.337	13.547
2.635	5.562	8.290	7.114	6.300	4.302	8.535
6.442	11.463	18.208	2.734	3.147	23.291	28.606
9.793	7.818	8.173	14.987	17.171	5.905	8.027
9.595	4.945	4.571	21.464	4.483	5.664	4.889
5.709	20.893	4.411	2.283	3.648	2.723	1.882
4.658	2.868	0.572	2.751	1.408	1.216	6.609
7.207	1.201	0.143	2.656	2.336	1.833	2.776
10.35	1.205	5.992	2.100	3.177	4.165	2.164
2.112	1.005	0.737	3.454	4.631	4.781	2.679
4.458						

mean. Sometimes the average will be too high, sometimes it will be too low, and the variation of a series of sample means either side of the population mean is often referred to as the *sampling distribution*.

Other terms that are met in this context are as follows. *Sample frequency* is the number of times an outcome occurs: the number of chamber tombs in the sample with areas between 7 and $8\,\mathrm{m}^2$, and so forth. *Sample relative frequency* is the proportion of the sample frequency to the whole sample, that is the proportion of tombs, compared with the whole sample, which fall within the range indicated.

5.3   Random variables

To introduce the term *random variable*, consider the Consumerland example of Section 4.6.3 in which five glasses were found. Let the random variable X represent the number out of the five that were beer mugs. The variable might take the value 0 or 1 or 2 or 3 or 4 or 5. This is an example of a discrete random variable. Other examples of random variables are the number of pollen grains of a particular species in a section of a pollen column, again discrete, the concentration of phosphate in an archaeological soil sample, a continuous random variable, and the rim diameter of a pot, again continuous. In each case these might take on any one of a whole set of values, discrete or continuous as the case may be. According to this notation, X (upper case) is taken to represent the random variable and x (lower case) to represent the value observed in any particular case. Sometimes, to clarify further, the values in particular cases might be distinguished by a subscript, such as x_1, x_2, x_3. Thus these may be used to talk about the number of mugs found in level 1 or level 2 or level 3.

In other words, a random variable is a rule for assigning a particular number to a particular event. Thus the event might be: one or more flints are found in a level (score 1), or no flints are found in a level (score 0). The rule can be made more complex: thus for each pair of levels we might have a rule that $X = (score\ of\ level\ A\ +\ score\ of\ level\ B) \times 2$, in which case X can be any one of the values 0, 2 and 4; no other value is possible. In this case the *range* (see below) of the random variable X is defined ($x = 0, 2, 4$). Thus for every possible event there is a corresponding number. Expressed more formally we can say that a random variable is a numerically valued function defined on the set of all outcomes of a situation, Ω.

We have, in the last two paragraphs, used some terminology which needs at least a brief explanation.

(i) The *range* of a random variable consists of all the possible values that a random variable might take. Thus if the random variable X is the number of sherds found in a square then it can take the value of any positive integer $0, 1, 2, \ldots$

(ii) We have already met the concept of the set of all outcomes, Ω, in Section 4.3. In the Consumerland example where five glasses were found, the event space consists of all possible combinations of mugs and goblets. Finding the combination "MMGGM" is just one of many possible outcomes. The random variable X (finding a certain number of mugs in a sample of five glasses) assigns numbers to the outcomes. In this case $X = 3$ for the event MMGGM. In Table 5.2 we show how the random variable X is related to all the possible outcomes that could occur.

5.4   Modelling discrete random variables

5.4.1   Probability distributions: introduction

In the statistical modelling of a problem we are concerned with events: the event that a flint artefact is or is not found in a particular archaeological

Table 5.2 Table showing the relationship between the random variable X representing the number of mugs found in a sample of five glasses and the outcomes in Ω. M denotes a mug and G a goblet.

X	Event				
0	GGGGG				
1	MGGGG	GMGGG	GGMGG	GGGMG	GGGGM
2	MMGGG	MGMGG	MGGMG	MGGGM	GMMGG
	GMGMG	GMGGM	GGMMG	GGMGM	GGGMM
3	MMMGG	MMGMG	MMGGM	MGMMG	MGMGM
	MGGMM	GMMMG	GMMGM	GMGMM	GGMMM
4	MMMMG	MMMGM	MMGMM	MGMMM	GMMMM
5	MMMMM				

context, the event that a radiocarbon atom decays, the event that when the floor area of a tomb is measured it is found to be of a particular size. And we are interested in assigning probabilities to the events. The notation $P(\cdot)$ has been used to stand for this; $P(E)$ is shorthand for the probability that an event, denoted by E, occurs. We might let E represent the event that a tomb is greater than 7.00 and less than 8.00   m^2. Then $P(E) = 0.1$ would be interpreted as our belief that one in ten tombs lies within that range; among every ten measured tombs we anticipate that on average there will occur one of the size indicated. We now move on from this to consider a range of possible different outcomes, where we shall need to express the probabilities as a *function*.

5.4.2   Probability mass function

In the case of discrete random variables we have the *probability mass function* (often abbreviated to p.m.f.) which is denoted by

$$p(x) = P(X = x)$$

for all x in the range of X. Note that $p(x)$ is a probability so we have $0 \le p(x) \le 1$. Moreover, if the range of X is $x_1, \ldots, x_n$, then

$$p(x_1) + p(x_2) + \ldots + p(x_n) = 1,$$

or for short

$$\sum_x p(x) = 1.$$

In either case the probabilities are summed over the whole range of X, and the probability and the resulting total must be 1.

Example

As an example let us invent a case of ancient *corrales* in Peru: walled enclosures used to pen animals. In a grid of squares over the grazing lands where they occur, any square might have none, one, two, three, four, but never more than five; obviously any square must have one of these six possible outcomes (it is not difficult to refine the rule to cope with *corrales* which fall only partly into a grid square). We let X be the random variable representing the number of *corrales* in a square. Note that X has the range 0, 1, 2, 3, 4, 5. The probability of these occurring might be assessed, on the basis of current information, as

$$p(0) = 0.5, \quad p(1) = 0.4, \quad p(2) = 0.05,$$
$$p(3) = 0.02, \quad p(4) = 0.02, \quad p(5) = 0.01.$$

We can interpret this to mean, for example, that the probability of no *corrales* is 0.5, the probability of exactly one is 0.4 and the probability of two or more is 0.1. Note that

$$p(0) + p(1) + p(2) + p(3) + p(4) + p(5)$$
$$= 0.5 + 0.4 + 0.05 + 0.02 + 0.02 + 0.01 = 1.$$

That is the probabilities sum to 1.

5.4.3   Cumulative distribution function

The *cumulative distribution function* of a random variable X, denoted by $F(x)$, is defined as the probability that X takes values up to and including x. That is

$$F(x) = P(X \leq x).$$

For the *corrales* example we have

$$F(2) \quad = P(X \leq 2) = p(0) + p(1) + p(2)$$
$$= 0.5 + 0.4 + 0.05$$
$$= 0.95.$$

5.5   Summarizing discrete random variables

In the process of statistical modelling a parallel can be drawn with summary statistics — when data are displayed in bar charts or histograms they will take a particular form: symmetric with one peak, or multi-modal (with several peaks), or skewed to the left or to the right, or some other shape (see Figure 1.4). Since we have moved on from mere summaries of data, to consider the modelling of uncertainty by means of theoretical probability distributions, there is need to communicate their important features to other researchers.

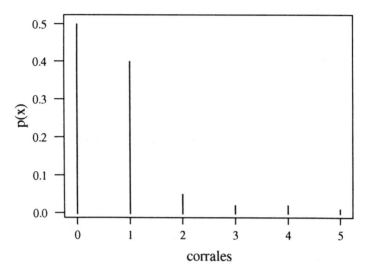

Figure 5.2 Plot of the probability mass function of the (fictional) Peruvian *corrales*.

Undoubtedly the most effective means of communication is visual: a plot of the probability mass function by means of what we shall call *probability histograms*. In terms of (numerical) summaries we require the probability equivalents of sample statistics such as the sample mode, sample median, sample mean and sample variance, which are so useful in describing statistical data. We talk of the mode, the median, the mean and the variance of a random variable or of a distribution. In Table 5.3 we give details of the discrete distributions that we use in this book.

5.5.1 Probability histograms

A visual method for presenting a probability mass function is by means of a probability histogram (we must stress that this is not a histogram of data obtained from a sample of data, but represents the probabilities associated with a random variable taking different values). In Figure 5.2 we plot the probability mass function for the Peruvian *corrales*. The number of *corrales* is recorded along the horizontal axis, and the height of the bars is proportional to their probability. The mode of the distribution (the most probable value) is at $X = 0$.

We now return to the Consumerland example of Section 4.6.3 in order to explore probability histograms more closely. Let X be the random variable representing the number of mugs in a sample of five glasses. There we saw that the probability of two mugs and three goblets in a sample of five glasses from a bar was

$$10 \times (0.8)^2(0.2)^3 = 0.051\,20.$$

Therefore the probability that $X = 2$ given that the building is a bar (denoted by B) is

$$P(X = 2|B) = 10 \times (0.8)^2(0.2)^3 = 0.051\,20.$$

If it were a restaurant then the probability would be

$$P(X = 2|R) = 10 \times (0.4)^2(0.6)^3 = 0.34560.$$

In a similar manner we can calculate the probability of finding 0 or 1 or 2 or 3 or 4 or 5 mugs in a sample of 5 glasses.

Let $p(x|B)$ and $p(x|R)$ be the probability mass functions of X for a bar (B) and restaurant (R), respectively. For example, from above we have

$$p(2|B) = P(X = 2|B) = 0.05120$$

and

$$p(2|R) = P(X = 2|R) = 0.34560.$$

For a bar the complete probability mass function is

$$p(0|B) = 0.00032, \quad p(1|B) = 0.00640, \quad p(2|B) = 0.05120,$$
$$p(3|B) = 0.20480, \quad p(4|B) = 0.40960, \quad p(5|B) = 0.32768.$$

This probability mass function is plotted in Figure 5.3. The mode of the distribution is at $X = 4$. Notice that the probability of three or more mugs is just over 0.94.

For a restaurant the probability mass function is

$$p(0|R) = 0.07776, \quad p(1|R) = 0.25920, \quad p(2|R) = 0.34560,$$
$$p(3|R) = 0.23040, \quad p(4|R) = 0.07680, \quad p(5|R) = 0.01024.$$

and is shown in Figure 5.4. The mode of this distribution is at $X = 2$. Also the probability of three or more mugs is much lower at about 0.31.

5.5.2 Expected value of a discrete random variable

The *expected value* of a random variable, X, or the *expectation of X*, denoted by $E(X)$, is exactly the same as the population mean (the mean of the underlying theoretical population or model). In the case of a discrete random variable X, its expectation is given by

$$E(X) = \sum_x xp(x) \tag{5.1}$$

where the sum is taken over the whole range of X.

Returning to the *corrales* in Peru, we saw that the random variable X, representing the number of *corrales* in a square, had the *range* 0, 1, 2, 3, 4,

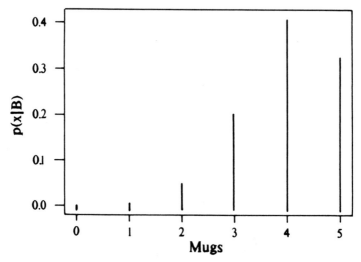

Figure 5.3   Plot of the probability mass function, denoted by $p(x|B)$, of the number of mugs (X) in a sample of $n = 5$ glasses when the site is a bar.

5, and its probability mass function was $p(0) = 0.5$, $p(1) = 0.4$, $p(2) = 0.05$, $p(3) = 0.02$, $p(4) = 0.02$, $p(5) = 0.01$. Therefore using (5.1) the expectation of X can be calculated as

$$
\begin{aligned}
E(X) &= (0 \times 0.5) + (1 \times 0.4) + (2 \times 0.05) + (3 \times 0.02) \\
&\quad + (4 \times 0.02) + (5 \times 0.01) \\
&= 0.69.
\end{aligned}
$$

Perhaps it is worth noting here that the expected value of X does not necessarily equal values that X itself takes, namely, 0, 1, 2, 3, 4 or 5.

If we carry out similar calculations for the Consumerland example the expected number of mugs in a sample of five glasses is four for a bar and two for restaurant.

The *variance* (often denoted by σ^2) of a random variable is a measure of "how spread out" the distribution is. It is defined to be

$$
Var(X) = E[(X - \mu)^2] = \sum_x (X - \mu)^2 p(x) \tag{5.2}
$$

where $\mu = E(X)$ is the mean. In the case of the Peruvian *corrales* we have $\mu = 0.69$ and so the variance of X is

$$
\begin{aligned}
Var(X) &= [(0 - 0.69)^2 \times 0.5] + [(1 - 0.69)^2 \times 0.4] + [(2 - 0.69)^2 \times 0.05] \\
&\quad + [(3 - 0.69)^2 \times 0.02] + [(4 - 0.69)^2 \times 0.02] \\
&\quad + [(5 - 0.69)^2 \times 0.01] \\
&= 0.8739.
\end{aligned}
$$

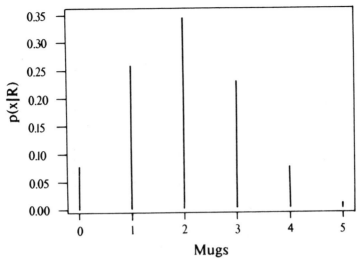

Figure 5.4 Plot of the probability mass function, denoted by $p(x|R)$, of the number of mugs (X) in a sample of $n = 5$ glasses when the site is a restaurant.

We note that there is another way of expressing the variance of a random variable which is

$$Var(X) = E(X^2) - \mu^2. \qquad (5.3)$$

Often it is easier to calculate the variance using (5.3) rather than (5.2). For the *corrales* example we have

$$
\begin{aligned}
E(X^2) = \ & (0^2 \times 0.5) + (1^2 \times 0.4) + (2^2 \times 0.05) \\
& + (3^2 \times 0.02) + (4^2 \times 0.02) + (5^2 \times 0.01) \\
= \ & 1.35.
\end{aligned}
$$

Hence we have

$$Var(X) = E(X^2) - \mu^2 = 1.35 - (0.69)^2 = 1.35 - 0.4761 = 0.8739.$$

Just as the sample mean $\bar{x}$ is distinguished from the population mean μ, so conventionally s^2 is used to denote the sample variance and σ^2 the population variance.

The *standard deviation* is the square root of the variance and converts the measure of how spread out the distribution is back into the same units as the variable under discussion. Thus in the *corrales* example the variance equals 0.8739, and therefore the standard deviation is given by $\sqrt{0.8739} = 0.9348$.

5.6   Examples of discrete distributions

In this section we provide a list of generic distributions that are governed by parameters, together with examples demonstrating typical situations in which they might be used. In introducing each type of probability distribution we follow (if applicable) a standard notation of the form

$$X|n, \theta \sim B(n, \theta)$$

where X represents the random variable given (|) the parameters n and θ. The symbol "$\sim$" is read as X *is distributed as* a particular probability distribution, or more simply X *is* (in this case) Binomial, or whatever is the name of the particular probability function. The symbols in brackets after the name of the distribution denote the parameters which characterize the distribution. Thus in the case of the Binomial distribution, as will be explained in Section 5.6.3, there is the parameter n which stands for the number of events (for example the number of times a coin is tossed), and the parameter θ which stands for the probability of an individual unknown event (for example the probability of a coin landing heads). There are numerous probability distributions, applicable to all sorts of different cases, but we shall introduce here only four discrete random variables. For discussion of others we refer the reader to Daly *et al.* (1995) and Johnson and Kotz (1970–72), or, indeed, any standard statistical text on distributions.

5.6.1   *Uniform distribution*

X is said to have a *discrete Uniform* distribution with parameters m and n, denoted by

$$X|m, n \sim Ud(m, n),$$

if its probability mass function is

$$p(x|m, n) = \frac{1}{n - m + 1} \quad \text{for } x = m, m + 1, \ldots, n.$$

As all values of X are equally likely there is no unique mode. From Table 5.3 we see that the expected value and variance of X are

$$E(X) = \frac{m + n}{2} \quad \text{and} \quad Var(X) = \frac{n^2 - m^2}{12}$$

respectively.

Example

In a study of the later neolithic village site of Skara Brae, Orkney (Buck *et al.*, 1991), fourteen radiocarbon dates were available, and could be ascribed to various phases defined by the stratigraphy. Let X represent the date of a radiocarbon sample taken from the site. Even before the new determinations were taken into account, the archaeologist was prepared to make an estimate as to the bounds of the later neolithic at Skara Brae. Given an understanding of the cultural material, and given all previous research on the subject including previous radiocarbon dates, it was suggested that the site should be no earlier than 5400 BP and no later than 4200 BP. However, no preference was given to any particular year within that range: 5400, 5399, 5398, ..., 4200 were all considered equally probable. That is X was believed to have a discrete Uniform distribution. The two parameters (m and n) correspond, in this case, to the latest date 4200 and to the earliest date 5200 BP respectively. This is an example of a discrete Uniform distribution, with $m = 4200$ and $n = 5400$. The probability that the date of the sample is, say, 4300 is given by

$$p(4300|4200,\ 5400) = \frac{1}{5400 - 4200 + 1} = \frac{1}{1201} \approx 0.00083.$$

The expectation of X, the mean date, is $(4200 + 5400)/2 = 4800$. Its variance is $(5400^2 - 4200^2)/12 = 960\,000$ and its standard deviation is 979.8 years.

5.6.2   Bernoulli distribution

X is said to have a *Bernoulli* distribution with parameter θ, denoted by

$$X|\theta \sim Br(\theta),$$

if its probability mass function is given by

$$p(x|\theta) = \theta^x (1 - \theta)^{1-x} \quad \text{for } x = 0, 1.$$

From Table 5.3 we see that the expected value and variance of X are $E(X) = \theta$ and $Var(X) = \theta(1 - \theta)$ respectively.

Example

In a bar in Consumerland the probability that an individual glass is a mug is 0.8 and that it is a goblet is 0.2. If we let $X = 1$ if the glass is a mug and $X = 0$ if a goblet, then the random variable X has a Bernoulli distribution with $\theta = 0.8$.

5.6.3   Binomial distribution

X is said to have a *Binomial* distribution with parameters n and θ, denoted by

$$X|n, \theta \sim B(n, \theta),$$

if its probability mass function is given by

$$p(x|n, \theta) = \frac{n!}{x!(n-x)!}\theta^x(1-\theta)^{n-x} \quad \text{for } x = 0, 1, \ldots, n,$$

where $n! = n \times (n-1) \times \cdots \times 3 \times 2 \times 1$. A Binomial random variable is used to model a series of n independent trials, each of which can have only one of two possible outcomes, that is a series of independent Bernoulli trials where the parameter θ remains constant from trial to trial. In order to use the Binomial distribution one needs to know the number, n, of trials and the probability of one of the two possible outcomes, θ. From Table 5.3 we see that the mean and variance of X are

$$E(X) = n\theta \text{ and } Var(X) = n\theta(1 - \theta)$$

respectively.

Example

Let us suppose we are investigating the domestic architecture of a particular culture; the houses can be roofed either with a flat roof or with a pitched roof. Normally the distinction is thought to be tied to the amount of rainfall, with pitched roofs in areas with higher rainfall and flat roofs in dry regions. Let us define X as the event that we find a house with a pitched roof. In a given region the probability of a house having a pitched roof might be 0.3; given ten houses in a village what is the chance of finding two with pitched roofs? Here we have $n = 10$ and $\theta = 0.3$ and the probability we wish to evaluate is $P(X = 2|n = 10, \theta = 0.3)$. This can be calculated as

$$p(2|10, 0.3) = \frac{10!}{2!(10-2)!}0.3^2(1 - 0.3)^{10-2} = 0.23$$

and likewise we can find the probability of finding 1 or 4 or 8 houses with pitched roofs (Figure 5.5). The expected number of pitched roofed houses in ten is $10 \times 0.3 = 3$. The variance is $10 \times 0.3 \times (1 - 0.3) = 2.1$ and so the standard deviation is $\sqrt{2.1} = 1.45$.

5.6.4   Poisson distribution

X is said to have a Poisson distribution with parameter λ, denoted by

$$X|\lambda \sim Po(\lambda),$$

if its probability mass function is

$$p(x|\lambda) = \frac{\lambda^x e^{-\lambda}}{x!} \quad \text{for } x = 0, 1, 2, \ldots \tag{5.4}$$

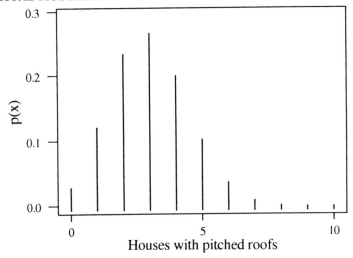

Figure 5.5   Plot of the probability mass function of finding X houses with a pitched roof in a village of ten houses with the probability of a pitched roof set at 0.3; we model the situation using a Binomial distribution with $\theta = 0.3$.

where e is a constant which is 2.7183 approximately. Here (5.4) means that x can take any positive integer value. Thus the probability of no occurrences is

$$p(0|\lambda) = \frac{\lambda^0 e^{-\lambda}}{0!} = e^{-\lambda}$$

where 0! is defined to be 1. The probability of three occurrences is

$$p(3|\lambda) = \frac{\lambda^3 e^{-\lambda}}{3!} = \frac{\lambda^3 e^{-\lambda}}{6}.$$

From Table 5.3 the mean and variance of X are

$$E(X) = \lambda \text{ and } Var(X) = \lambda$$

respectively.

Example

Radioactivity is of value in archaeology, among other reasons, because radioactive isotopes can be used as natural clocks, for example in potassium-argon dating and radiocarbon dating. In the history of the development of nuclear physics it was recognized by Egon Ritter von Schweidler, that radioactive decay was a fundamentally random effect. This was confirmed by experiments where it was found that the decay of radioactive elements did not occur regularly, but randomly over an interval of, say, one minute; sometimes

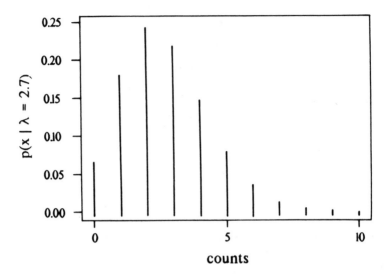

Figure 5.6 Plot of the probability mass function of a Poisson random variable
with mean equal to 2.7.

there would be no decay-events, sometimes one, sometimes two and so forth.
The decay rate for any given isotope is represented by λ decay-events per unit
time.

A consequence of the random nature of radioactive decay is that the number
of decays in a time period of length t time units can be modelled by a Poisson
distribution with parameter λt. (The derivation of this result is beyond the
scope of the book and we refer the reader to Feller, 1968, p. 446–448, or Ross,
1988, p. 133–135.) Thus if we let X represent the number of decay-events in
time t, then

$$X|\lambda \sim Po(\lambda t)$$

and the probability mass function of X is given by

$$p(x|\lambda) = \frac{(\lambda t)^x e^{-\lambda t}}{x!} \quad \text{for } x = 0, 1, 2, \ldots$$

Suppose that for a particular size and type of sample, the average number
of decays of a specific element is 7 per minute. Then, for this specific sample
the probability of 3 decays in one minute, that is $X = 3$ and $t = 1$, is given
by

$$p(3|7) = \frac{7^3 \times 2.7183^{-7}}{3!} = 0.052,$$

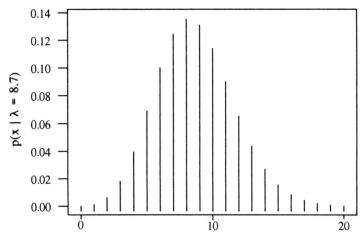

Figure 5.7 Plot of the probability mass function of a Poisson random variable with mean equal to 8.7.

Table 5.3 Table of discrete probability distributions; here $\binom{n}{x} = \frac{n!}{x!(n-x)!}$.

Name	Abbreviation	Probability mass function	Range	Mean	Variance
Discrete Uniform	$Ud(m,n)$	$\frac{1}{n-m+1}$	$m, m+1, \ldots, n$	$\frac{m+n}{2}$	$\frac{n^2-m^2}{12}$
Bernoulli	$Br(\theta)$	$\theta^x(1-\theta)^{1-x}$	$0,1$	θ	$\theta(1-\theta)$
Binomial	$B(n,\theta)$	$\binom{n}{x}\theta^x(1-\theta)^{n-x}$	$0,1,\ldots,n$	$n\theta$	$n\theta(1-\theta)$
Poisson	$Po(\lambda)$	$\frac{\lambda^x e^{-\lambda}}{x!}$	$0,1,2,\ldots$	λ	λ

while

$$p(8|7) = \frac{7^8 \times 2.7183^{-7}}{8!} = 0.1304.$$

Often it is helpful to plot the probability mass function. As we alter λ, the graph takes on different shapes (Figures 5.6 and 5.7). Thus different radioactive elements decay at different rates (measured in terms of their half-lives), and so give different forms to the graph, for example that for ^{14}C is different from that for ^{40}K; it is on the basis of the different half-lives that the age of various samples is calculated.

5.7   Consumerland revisited

An important feature of statistical distributions lies in their power to generalize. To illustrate this point let us return to the Consumerland example of Section 4.6.3. Recall that we were interested in identifying whether a building was a "restaurant" (R) or a "bar" (B), and the decision was made on the basis of the relative numbers of the two types of glasses, mugs and goblets, found at the site. We looked at the specific case where five glasses were recovered from a site, three of which were goblets and two were mugs. Using our Bayesian methodology we calculated that the posterior probabilities of it being a bar or a restaurant are 0.23 and 0.77 respectively. We might have found a total of ten glasses of which six were goblets. Redoing the calculations we would find that the posterior probabilities are now 0.61 and 0.39 respectively.

We now try to generalize the procedure. Suppose n denotes the total number of glasses found at the site. Let X be the random variable representing the number of mugs. On the basis of the finds and our prior information or beliefs, we are interested in whether the site is a bar or a restaurant. That is we want to calculate the posterior probability that it is a bar or a restaurant, namely $P(B|X = x)$ and $P(R|X = x)$ respectively.

Using Bayes' theorem we have

$$P(B|X = x) = \frac{P(X = x|B)P(B)}{P(X = x)} \tag{5.5}$$

and

$$P(R|X = x) = \frac{P(X = x|R)P(R)}{P(X = x)}.$$

Let θ_B denote the probability that a glass is a mug conditional on the building being a bar. Then X has a Binomial distribution with parameters n and θ_B. That is $X|n, \theta_B \sim B(n, \theta_B)$. Therefore we have

$$P(X = x|B) = \binom{n}{x}\theta_B^x(1 - \theta_B)^{n-x}. \tag{5.6}$$

(Note that $\binom{n}{x}$ is another way of writing the formula $\frac{n!}{x!(n-x)!}$.) Likewise if θ_R is the probability of a mug conditional on the building being a restaurant then $X|n, \theta_R \sim B(n, \theta_R)$ so that

$$P(X = x|R) = \binom{n}{x}\theta_R^x(1 - \theta_R)^{n-x}. \tag{5.7}$$

Here, as before, the prior probability that the building is a restaurant is $P(R)$ and that it is a bar is $P(B)$. We note that by the theorem of total probability (4.9)

$$P(X = x) = P(X = x|R)P(R) + P(X = x|B)P(B) \tag{5.8}$$

and therefore using this we can express (5.5) as

$$P(B|X = x) = \frac{P(X = x|B)P(B)}{P(X = x|R)P(R) + P(X = x|B)P(B)}. \qquad (5.9)$$

Hence using (5.6) and (5.7) we have that, conditional on finding n glasses, x of which are mugs, the posterior probability that the building is a bar is

$$
\begin{aligned}
P(B|X = x) &= \frac{\binom{n}{x}\theta_B^x(1 - \theta_B)^{n-x}P(B)}{\binom{n}{x}\theta_R^x(1 - \theta_R)^{n-x}P(R) + \binom{n}{x}\theta_B^x(1 - \theta_B)^{n-x}P(B)} \\
&= \frac{\theta_B^x(1 - \theta_B)^{n-x}P(B)}{\theta_R^x(1 - \theta_R)^{n-x}P(R) + \theta_B^x(1 - \theta_B)^{n-x}P(B)}.
\end{aligned}
$$

The posterior probability that it is a restaurant is of course

$$P(R|X = x) = 1 - P(B|X = x).$$

By using the Binomial distribution, it is possible to generalize our original analysis of the Consumerland problem. Where initially each case was examined separately according to the number of glasses recovered, the use of the statistical distribution enables us to model the whole spectrum of possibilities.

5.8 Modelling continuous random variables

Up to now we have considered random variables modelling situations which are discrete, dealing with entities such as counts of artefacts, the number of radioactive decay-events in a certain time period, separate years and the like. Models of other random variables, such as those representing chemical concentrations or dimensions such as length or areas are measured on a *continuous* scale, and their probability functions take a different mathematical form.

5.8.1 *Probability density function*

In the case of a *continuous* random variable, the equivalent to the probability mass function of a discrete random variable is the *probability density function* (often abbreviated as p.d.f.). If the probability density function of a random variable X is denoted by $p(x)$, then the probability that the random variable X takes values between x_1 and x_2, where $x_1 < x_2$, is the *area* under the curve $p(x)$ but above the horizontal axis between $X = x_1$ and $X = x_2$ (see

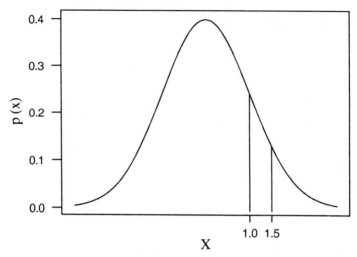

Figure 5.8   Plot of a probability density function. The area below the curve equals 1. The probability $P(1 \leq X \leq 1.5)$ is equal to the area below the curve $p(x)$ and between 1.0 and 1.5.

Figure 5.8). This is written in mathematical notation as the integral

$$P(x_1 \leq X \leq x_2) = \int_{x_1}^{x_2} p(x)\,dx.$$

The probability density function is defined for all values of x in the range of X.

Moreover we have seen in Section 5.4.2 for a discrete random variable the sum of the probabilities over the whole range of X must add up to 1. In the case of a continuous random variable, the probability density function is defined in such a way that the total area under the curve but above the horizontal axis equals 1.

Recall from Section 5.4.3 that the cumulative distribution function of a random variable X is given by

$$F(x_0) = P(X \leq x_0).$$

For a continuous random variable

$$P(X \leq x_0) = \int_{-\infty}^{x_0} p(x)\,dx$$

and therefore its cumulative distribution function is

$$F(x) = \int_{-\infty}^{x_0} p(x)\,dx.$$

When considering discrete data we saw how a probability histogram could be used to illustrate a probability mass function. As far as continuous data are concerned a graph of the probability density function fulfils the equivalent purpose.

5.9 Summarizing continuous random variables

As in the case of discrete random variables there is the need to summarize continuous random variables. This can be done in more or less the same way by plotting the probability density function, and by reporting where most of the probability is concentrated, by calculating the mode, mean, variance and similar measures of the distribution.

5.9.1 Expected value of a continuous random variable

When dealing with discrete random variables we saw, in Section 5.5.2, that the expected value of X, $E(X)$, was expressed as

$$E(X) = \sum_x xp(x).$$

For continuous random variables the expected value of X is given by

$$E(X) = \int_{-\infty}^{\infty} xp(x)\,dx.$$

For a continuous random variable the variance of X is given by

$$Var(X) = E[(X - \mu)^2] = \int_{-\infty}^{\infty} (x - \mu)^2 p(x)\,dx$$

where μ denotes the expected value of X. That is $\mu = E(X)$. Just as for discrete random variables, the variance of X may be expressed as

$$Var(X) = E(X^2) - \mu^2$$

where

$$E(X^2) = \int_{-\infty}^{\infty} x^2 p(x)\,dx.$$

In the following section we give several examples of continuous random variables. A summary of these distributions and their properties is given in Table 5.4.

5.10   Examples of continuous distributions

5.10.1   *Uniform distribution*

X is said to have a continuous *Uniform* distribution with parameters a and b, denoted by

$$X|a, b \sim U(a, b),$$

if it has the probability density function

$$p(x|a, b) = \frac{1}{b - a} \quad \text{for } a \leq x \leq b.$$

In this notation a represents the lower limit of the range of X and b represents its upper limit. Every value between the two limits a and b is equally likely. As a result there is no unique mode. The mean and variance of X are given (see Table 5.4) by

$$E(X) = \frac{a + b}{2} \quad \text{and } Var(X) = \frac{(b - a)^2}{12}$$

respectively.

A special case which is very important is when $a = 0$ and $b = 1$, in which case X is said to have a Uniform distribution on the interval $(0, 1)$.

Example

We can take the same example as we used to illustrate the discrete uniform distribution in Section 5.6.1. There the duration of the later neolithic period was thought of as consisting of 1201 separate years: $5400, 5399, 5398, \ldots, 4200 \, \text{BP}$. It is possible to think of time not as discrete years but as a *continuum*, infinitely divisible into smaller units; if the model is couched in these terms, then the continuous uniform distribution is more appropriate with parameters 5400.0 for the earliest boundary, and 4200.0 as the latest. Thus we have $b = 5400.0$ and $a = 4200.0$. In this case the probability density function of X is

$$p(x|4200, \; 5400) = \frac{1}{5400 - 4200} = \frac{1}{1200} \approx 0.00083 \quad \text{for } 4200 \leq x \leq 5400.$$

The uniform distribution with these parameters is plotted in Figure 5.9. It is, then, possible to calculate, for example, the probability of a date between 4500 and 5000 BP as the area under the curve $p(x)$ between 4500 and 5000 or from

$$P(4500 \leq x \leq 5000) = \int_{4500}^{5000} p(x) \, dx = 0.417,$$

so that the probability is about 0.42.

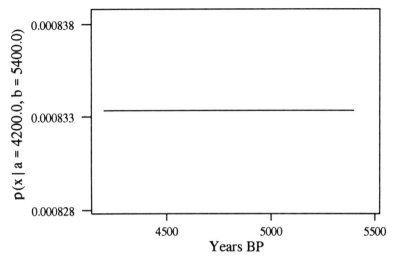

Figure 5.9   Plot of the probability density function of a continuous Uniform random variable on the interval 4200 to 5400 BP.

5.10.2   Beta distribution

X is said to have a *Beta* distribution with parameters α and β, denoted by

$$X|\alpha, \beta \sim Be(\alpha, \beta),$$

if it has the probability density function

$$p(x|\alpha, \beta) = \frac{\Gamma(\alpha + \beta)}{\Gamma(\alpha)\Gamma(\beta)} x^{\alpha-1}(1-x)^{\beta-1} \quad \text{for } 0 \le x \le 1. \tag{5.10}$$

The mathematical term $\Gamma(\alpha)\Gamma(\beta)/\Gamma(\alpha+\beta)$ is needed to ensure that the area under the curve, over the total range between 0 and 1, is equal to unity. $\Gamma(\alpha)$ can be evaluated numerically as it satisfies the relationship $\Gamma(\alpha) = (\alpha-1)\Gamma(\alpha-1)$. In fact if α is a positive integer then $\Gamma(\alpha) = (\alpha-1)!$. For example, if α equals 4, then $\Gamma(4) = 3! = 6$.

The mode, mean and variance of X are given by

$$\frac{\alpha-1}{\alpha+\beta-2}, \quad \frac{\alpha}{\alpha+\beta} \quad \text{and} \quad \frac{\alpha\beta}{(\alpha+\beta)^2(\alpha+\beta+1)}$$

respectively.

Example

In our Consumerland example, see Section 4.6.3, we have fixed the probability that a glass is a mug, given that we are in a bar, at 0.8. But it would be

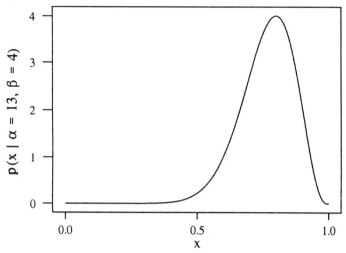

Figure 5.10   Plot of the probability density function of a Beta random variable
with parameters $\alpha = 13$ and $\beta = 4$.

more realistic to suppose that this probability varies from bar to bar. There
might, for example, be regional differences from area to area, according to
local custom. A Beta distribution could be used to model this variation in the
following manner.

Let θ denote the proportion of mugs in a bar, then we know that θ must lie
between 0 and 1 inclusive so that a Beta distribution would be appropriate.
We might believe that 0.8 was the most likely proportion and that either side
of 0.8 the probability dies away. In Figure 5.10 we plot the probability density
function of a Beta random variable with $\alpha = 13$ and $\beta = 4$. (Note that the
mode of the distribution is at $(\alpha - 1)/(\alpha + \beta - 2) = (13 - 1)/(13 + 4 - 2) = 0.8$.)

From this distribution we see that it is unlikely that a bar would have a
proportion less than 0.65. Is this a good model? We leave it to the readers to
judge in the light of their own experience whether a Beta distribution using the
above values properly corresponds to reality. Is a bar in your region at all likely
to have fewer than 65% mugs? Note that by using Beta distributions to model
the uncertainty, we have made the model more realistic. As a consequence of
using an improved model, the posterior probabilities of the site being a bar
or a restaurant will change from those calculated in Section 4.6.3.

5.10.3   Normal or Gaussian distribution

X is said to have a *Normal* distribution with parameters μ (the mean) and σ (the standard deviation), denoted by

$$X|\mu,\sigma \sim N(\mu,\sigma^2),$$

if it has the probability density function

$$p(x|\mu,\sigma) = \frac{1}{\sigma\sqrt{2\pi}} e^{-\frac{(x-\mu)^2}{2\sigma^2}} \quad \text{for } -\infty < x < \infty$$

where e can be taken as 2.7183 approximately. The term $1/\sigma\sqrt{2\pi}$ ensures that the area under the curve between $-\infty$ and ∞ is equal to 1. Note that for convenience this density function can be written as

$$p(x|\mu,\sigma) = \frac{1}{\sigma\sqrt{2\pi}} \exp\left\{ -\frac{(x-\mu)^2}{2\sigma^2} \right\} \quad \text{for } -\infty < x < \infty$$

where $\exp\{z\}$ means e^z. The mean and variance of the distribution are μ and σ^2 respectively. The distribution is symmetric about the mean μ; the mode is at $x = \mu$, as is the median. The Normal distribution is used to model statistically many situations involving a continuous, uni-modal, symmetric random variable. An important and useful property of the normal distribution is that about 68% of the distribution lies between $\mu - \sigma$ and $\mu + \sigma$, about 95% between $\mu - 2\sigma$ and $\mu + 2\sigma$, and about 99% between $\mu - 3\sigma$ and $\mu + 3\sigma$.

A special case is when $\mu = 0$ and $\sigma = 1$, then $X \sim N(0,1)$, which is called the *standard Normal* distribution (see Figure 5.11). Its probability density function is

$$p(x) = \frac{1}{\sqrt{2\pi}} \exp\left\{ -\frac{x^2}{2} \right\}.$$

Its mean, median and mode are all at $x = 0$.

Example

Clay pipes were commonly used in Europe and elsewhere in the seventeenth and eighteenth centuries for the smoking of tobacco. The pipes were manufactured from moulds so that two pipes made from the same mould would have almost identical measurements in terms of the stem width, bowl size and so on. Because of slight differences in manufacture, however, and slight inaccuracies in measuring them, their dimensions would vary slightly from pipe to pipe. To be specific, let us consider the bowl size which we will represent by X. The mould would be designed to give a theoretical size which we denote by μ. Because of the random or stochastic errors that come with any manufacturing process, this underlying value would not be exactly achieved in practice. Some pipes would have a bowl size smaller than the desired value,

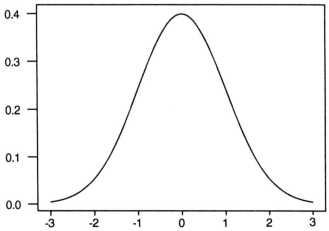

Figure 5.11 Plot of the probability density function of a Normal random variable with mean equal to 0 and variance equal to 1. This is known as the probability density function of the standard Normal distribution.

others would be larger. That is there would be some error term involved. This error or noise term is taken to be unbiased in the sense that if we consider the difference between an actual bowl size and the desired value, μ, negative or positive values of the difference of the same magnitude are believed to be equally likely. However, large errors are considered to be more unlikely than smaller errors, with the probabilities falling away either side of the true value μ. This type of variation can be modelled using a Normal distribution with mean μ and variance σ^2 (see Figure 5.11).

It is known that as time progressed from the seventeenth and through the eighteenth century, pipe bowls in England and the colonies became generally larger, as the duty on tobacco declined and smoking became cheaper. If we look at the products of pipe makers who lived at different times, we can see that the overall size differed (the mean was larger) but the essential shape of the distribution remains the same (Figures 5.12 and 5.13). If, instead, the manufacturing process had become more reliable thus resulting in smaller pipe-to-pipe variation, the variance of the error term would have decreased through time (see Figure 5.14).

5.10.4 *Exponential distribution*

X is said to have an *Exponential* distribution with parameter λ, denoted by

$$X|\lambda \sim M(\lambda),$$

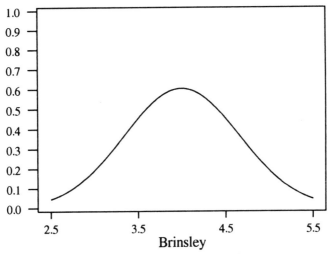

Figure 5.12 Plot of the probability density function of the volume of pipe bowls manufactured by Richard Brinsley, a Nottingham clay-pipe maker who died in 1729, as modelled by a Normal distribution with $\mu = 4.0$ and $\sigma = 0.66$.

if its probability density function is given by

$$p(x|\lambda) = \lambda e^{-\lambda x} \quad \text{for } 0 < x < \infty.$$

For the Exponential distribution with parameter λ, its mean and variance are

$$E(X) = \frac{1}{\lambda} \text{ and } Var(X) = \frac{1}{\lambda^2}$$

respectively.

Example

In countries like Britain it is not uncommon for archaeological surface surveys to be carried out in the following manner. The field walkers, following a transect across a field, mark the position of each artefact seen with a cane. (In areas such as the Mediterranean and the Near East artefact concentrations are normally so dense that such a methodology is impractical.) If the artefacts can be considered to be randomly scattered about the survey area then, just as the number of radioactive decay events occurring in a particular time period may be modelled by a Poisson distribution (see Section 5.6.4), so the number of sherds in a specified area can be modelled by a Poisson distribution. Furthermore the number of sherds found along a transect will also have a Poisson distribution.

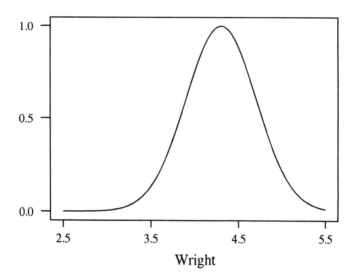

Figure 5.13 Plot of the probability density function of the volume of pipe bowls manufactured by John Wright, a Nottingham clay-pipe maker, who was active *ca.* 1729, as modelled by a Normal distribution with $\mu = 4.3$ and $\sigma = 0.4$.

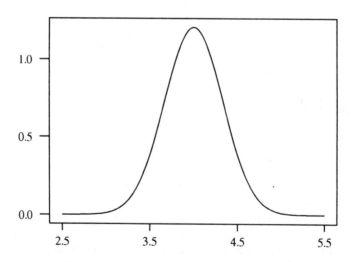

Figure 5.14 Normal distribution with $\mu = 4.0$ and $\sigma = 0.33$. Compare this distribution with that for the pipe bowls made by Richard Brinsley in Figure 5.12.

To be more specific, let λ denote the rate per metre at which sherds are found along a transect. Let M be the random variable representing the number of sherds found in x metres, then M has a Poisson distribution with parameter λx so that

$$p(m|\lambda) = \frac{(\lambda x)^m e^{-\lambda x}}{m!} \quad \text{for } m = 0, 1, 2, \ldots$$

Let X represent the *distance* along the transect until the first sherd is found. Then the events "the distance to the first sherd is more than x" and "no sherds are found in the first x metres" are exactly the same. That is, the event "X is greater than x" is the same as the event "M is equal to 0". Therefore we have

$$P(X > x) = P(M = 0) = \frac{(\lambda x)^0 e^{-\lambda x}}{0!} = e^{-\lambda x}$$

and hence

$$P(X < x) = 1 - P(X > x) = 1 - e^{-\lambda x}.$$

Firstly we note that $P(X < x)$ is the cumulative distribution function of X. Secondly by looking in Table 5.4 we see that the cumulative distribution function of an Exponential random variable is of the form $1 - e^{-\lambda x}$. Therefore we can deduce that the distance, X, to the first sherd has an Exponential distribution with parameter λ.

Here we have derived the probability density function of the distance until the first sherd is found. In fact it can be shown that the distance between successive sherds also has an Exponential distribution. The parameter λ is a measure of the rate at which the artefacts are found. The mean distance between successive finds will be given by $1/\lambda$.

5.10.5 Gamma distribution

X is said to have a *Gamma* distribution with parameters α and β, denoted by

$$X|\alpha, \beta \sim G(\alpha, \beta),$$

if it has the probability density function

$$p(x|\alpha, \beta) = \frac{\beta^\alpha}{\Gamma(\alpha)} x^{\alpha-1} e^{-\beta x} \quad \text{for } 0 < x < \infty.$$

The mode, mean and variance are given by

$$\frac{\alpha - 1}{\beta}, \frac{\alpha}{\beta} \quad \text{and} \quad \frac{\alpha}{\beta^2}$$

respectively.

Table 5.4 Table of continuous probability distributions. $\Phi(x)$ denotes the cumulative distribution function of a standard Normal random variable and is equal to $\int_{-\infty}^{x} 1/\sqrt{2\pi}\exp\{-z^2/2\}dz$.

Name	Abbreviation	Probability density function	Cumulative distribution function	Range	Mean	Variance
Uniform	$U(0,1)$	1	x	$0 \leq x \leq 1$	$\frac{1}{2}$	$\frac{1}{12}$
Uniform	$U(a,b)$	$\frac{1}{b-a}$	$\frac{x-a}{b-a}$	$a \leq x \leq b$	$\frac{a+b}{2}$	$\frac{(b-a)^2}{12}$
Exponential	$M(\lambda)$	$\lambda e^{-\lambda x}$	$1 - e^{-\lambda x}$	$0 < x < \infty$	$\frac{1}{\lambda}$	$\frac{1}{\lambda^2}$
Gamma	$G(\alpha,\beta)$	$\frac{\beta^\alpha}{\Gamma(\alpha)}x^{\alpha-1}e^{-\beta x}$	—	$0 < x < \infty$	$\frac{\alpha}{\beta}$	$\frac{\alpha}{\beta^2}$
Beta	$Be(\alpha,\beta)$	$\frac{\Gamma(\alpha+\beta)}{\Gamma(\alpha)\Gamma(\beta)}x^{\alpha-1}(1-x)^{\beta-1}$	—	$0 \leq x \leq 1$	$\frac{\alpha}{\alpha+\beta}$	$\frac{\alpha\beta}{(\alpha+\beta)^2(\alpha+\beta+1)}$
Standard Normal	$N(0,1)$	$\frac{1}{\sqrt{2\pi}}\exp\left\{-\frac{x^2}{2}\right\}$	$\Phi(x)$	$-\infty < x < \infty$	0	1
Normal	$N(\mu,\sigma^2)$	$\frac{1}{\sigma\sqrt{2\pi}}\exp\left\{-\frac{(x-\mu)^2}{2\sigma^2}\right\}$	$\Phi\left(\frac{x-\mu}{\sigma}\right)$	$-\infty < x < \infty$	μ	σ^2

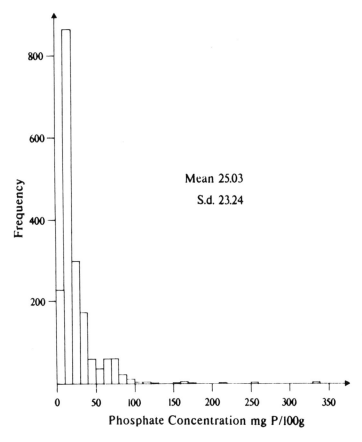

Figure 5.15 Histogram of raw soil phosphate concentrations, in mg phosphate per 100 g soil, from Laconia Survey site LS G165.

Example

The example outlined in Section 5.10.4 to illustrate the Exponential distribution can also serve here. Imagine the field walkers operating in the same fashion, but this time we need to model, say, the distance covered between every tenth artefact found (or it could be every fifth, or every twentieth or whatever). It can be shown (see Ross, 1988, p. 180–181) that the distance between every, say, tenth sherd can be modelled by a Gamma distribution — with, in this particular case, parameters α and β set equal to 10 and λ respectively. Therefore the mean distance between every tenth sherd would be $10/\lambda$.

5.10.6   Log-normal distribution

X is said to have a *log-normal* distribution with parameters λ and δ, denoted by

$$\ln X|\lambda, \delta \sim N(\lambda, \delta^2),$$

if it has the probability density function

$$p(x|\lambda, \delta) = \frac{1}{x\sigma\sqrt{2\pi}} \exp\left\{-\frac{(\ln x - \lambda)^2}{2\delta^2}\right\} \quad \text{for } 0 < x < \infty.$$

The mean and variance of X are

$$\exp\left\{\lambda + \frac{1}{2}\delta^2\right\}$$

and

$$\omega(\omega - 1)\exp\{2\lambda\}$$

respectively where $\omega = \exp\{\sigma^2\}$.

Example

Phosphate in soils is of value to the archaeologist as a chemical marker of human activity in the past; the full background to this will be explained in Chapter 10. In essence the resulting measurements are rather similar to those of the clay pipes, a series of measurements over a range, but when we plot them they do not show the classic symmetry of the Normal distribution. In fact the histogram is skewed to the right (see Figure 5.15). It is suggested that if we take the natural logarithm of the data it then appears more symmetric and could be modelled by a Normal distribution — see Figure 5.16. The rationale behind this is that where conditions are good for the accumulation of phosphate, the concentrations will increase by more than a merely linear rate (as it were by addition); in fact they multiply up.

The log-normal distribution can be treated as a Normal distribution by carrying out a *transformation* of the data, taking the logarithm of their values, rather than just the raw numbers.

5.10.7   Log-skew Laplace distribution

X is said to have a *log-skew Laplace* distribution with parameters α, β and μ if its probability density function, denoted by $p(x)$, can be expressed as

$$p(x) = \begin{cases} \alpha(1 - \beta)\exp\left\{-\alpha\frac{(1-\beta)}{\beta}(\mu - x)\right\} & \text{for } -\infty \le x \le \mu \\ \alpha(1 - \beta)\exp\left\{-\alpha(x - \mu)\right\} & \text{for } \mu < x \le \infty. \end{cases}$$

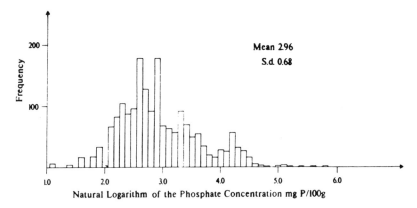

Figure 5.16   Histogram of the natural logarithm of soil phosphate concentrations in mg phosphate per 100 g soil.

Example

This model, which is not commonly encountered in other applications, has been applied to the analysis of particle size data used to characterize archaeological and geological sediments (Fieller and Flenley, 1988). Particle size can be measured in a number of different ways, but to avoid unnecessary complication here we simply mention the "rack of sieves" method, whereby the soil sample is passed through sieves of successively diminishing mesh size, and the portions trapped in each sieve are weighed. In the earlier history of research simple descriptive statistics were used to distinguish different sediments, but such techniques proved inadequate when dealing with complex situations, for example where there are mixed grain-size populations. As a result various model-based approaches are to be preferred, and that described here is illustrated graphically in Figures 5.17 and 5.18.

In Figure 5.17 the logarithms of the particle sizes are plotted against the logarithms of their proportions in the soil sample. The data have been fitted against the log-skew Laplace distribution, and it will be seen that they increase linearly on this scale up to a point, and that they then decrease linearly. These characteristics of the distribution are the basis for its use in this context. Figure 5.18, on a natural-log scale, shows that the curve fits reasonably well

In describing the use of the log-skew Laplace distribution to aid in the interpretation of particle size data, Fieller and Flenley (1988) make some interesting observations. (i) The slope of the left-hand line in the log–log plot reflects the relative proportion of finer material in the sample. (ii) The slope of the right-hand line on the log–log plot reflects the relative proportion of coarser material in the sample. (iii) The abscissa of the point of intersection

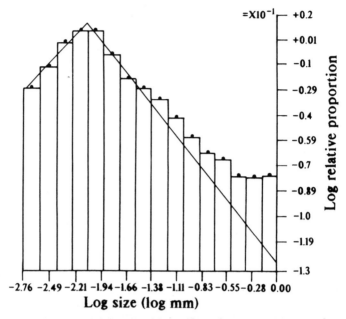

Figure 5.17 Histogram of the natural logarithm of grain size against the natural logarithm of relative proportion from Libyan sand sample 276, and fitted log-skew Laplace distribution. (After Fieller and Flenley, 1988, p. 82, fig. 10.1.)

of the two lines gives a 'typical' (or modal) logarithmic particle size. Once the form of the distribution can be interpreted this way, a clearer understanding of more complex sediment histories becomes possible — see Fieller and Flenley (1988) for more details.

5.11 Summary

We have in this chapter introduced the concepts of sample and population, random variables, probability mass functions for discrete random variables, and the corresponding probability density functions for continuous random variables. We have stated some of the most commonly occurring distributions and have illustrated them with archaeological examples. We must emphasize that this list is by no means complete. Moreover every new archaeological problem will need careful thought as to how to model the natural variation in it. It must be remembered that not all situations can be addressed using standard distributions. The choice of a suitable distribution is so important to the statistical modelling which lies at the heart of Bayesian analysis that

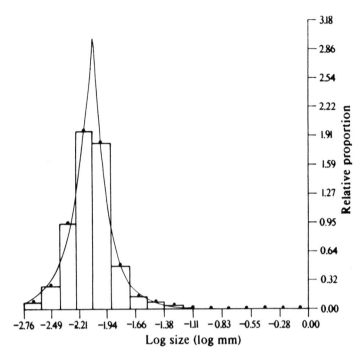

Figure 5.18   Histogram of the natural logarithm of grain size against relative proportion from Libyan sand sample 276, and fitted log-skew Laplace distribution. (After Fieller and Flenley, 1988, p. 83, fig. 10.2.)

we have found it necessary to devote two chapters to the theme. In the next we consider the more complicated bivariate and multivariate distributions.

6

Bivariate and multivariate distributions

6.1 Introduction

We have, so far, considered probability distributions for one variable at a time — the number of radioactive decay-events in a time period, soil particle sizes, the volume of clay-pipe bowls and the like. In this chapter we wish first of all to consider stochastic models for situations involving *two* random variables that may vary together. Later in the chapter we shall extend the principles already explored for the bivariate case to the multivariate case, where more than two random variables are involved.

As an example we might consider the relationship between the diameter of the base of a Hohokan bowl and its height — how well could an expert estimate the height of the pots if one had to go by base sherds alone? Or, in a study of the prehistoric agricultural economy of a region, one might be interested in how the number of sheep and the number of pigs, identified through bone samples, varied from site to site. In examples like these it is not sufficient to know the distributions of each of the variables separately. Thus in the latter case we may conjecture that the number of sheep associated with a particular site can be modelled as having a Poisson distribution with mean of 25, and the number of pigs as also Poisson but with mean of 30. However we are also interested in how these two variables interact with each other. It may be that more sheep go together with more pigs (and this could be interpreted to mean that the more prosperous the farmer the larger the stock of animals). On the other hand, it may be the more the sheep the fewer the pigs (which could be interpreted in another way, to mean that farmers diversified according to the environmental niche in which they operated).

Both random variables need to be considered and modelled simultaneously. What is needed is the *joint distribution*, and the earlier parts of this chapter are concerned with the modelling of joint distributions through bivariate probability mass functions and bivariate probability density functions. As in the previous chapter we shall first consider the case of discrete random variables and then move onto the case of continuous random variables. Later we will turn our attention to the more general multivariate case.

6.2   Discrete bivariate random variables

In this section we will show how two discrete random variables may be modelled simultaneously. To do so we will introduce the concept of a joint probability mass function. From this we will show how to calculate the marginal probability mass function of an individual variable. We will extend the ideas (introduced in Section 4.4) of conditional probability and independence to random variables. All this will emphasize that when two variables are dependent it is insufficient to consider the variables separately. Later we will introduce the expectation of jointly distributed random variables, including the terms of covariance and correlation which measure to some extent their inter-relationship. This pattern will be repeated in Section 6.3 when we consider continuous random variables.

6.2.1   Joint probability mass function

We define the *joint probability mass function* of two discrete random variables X and Y as

$$p(x, y) = P([X = x] \text{ and } [Y = y])$$

for all pairs (x, y). That is, $p(x, y)$ is the probability that X takes the value x *and* Y takes the value y. Thus $p(x, y)$ is a probability and therefore must be a number between 0 and 1 inclusive. Also the sum over all pairs (x, y) must equal 1.

Example

By way of illustration let us take the case of a surface survey where pot-sherds and roof-tile fragments have been counted in $1\,\mathrm{m}^2$ squares. To slightly simplify matters we suppose that the number of tile fragments can never exceed five and the number of pot-sherds can never be more than three (this artificial constraint is just for ease of presentation). If we let X be the number of tile fragments in a square, then the range of X is 0, 1, 2, 3, 4, 5, and if Y is the number of pot-sherds, its range is 0, 1, 2, 3. The event space of the two variables can be represented by the rows and columns of Table 6.1. For example, $X = 3$ and $Y = 2$ is such an event. Suppose that probabilities can be ascribed to the various events. Perhaps the probabilities were assessed to

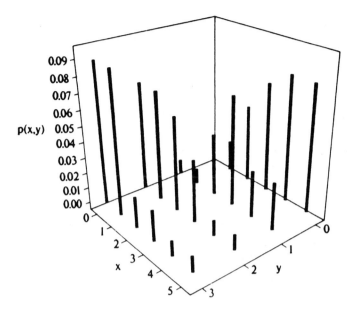

Figure 6.1 Joint probability mass function of the tile counts, X, and pot-sherd counts, Y, in $1\,\mathrm{m}^2$ sample squares from a field survey (see Table 6.1).

be the following.

$$P([X = 0] \text{ and } [Y = 0]) = 0.01, \quad P([X = 1] \text{ and } [Y = 0]) = 0.02,$$
$$P([X = 2] \text{ and } [Y = 0]) = 0.05, \quad P([X = 0] \text{ and } [Y = 1]) = 0.01,$$
$$P([X = 0] \text{ and } [Y = 2]) = 0.07, \quad P([X = 0] \text{ and } [Y = 3]) = 0.09,$$

and so forth. Each of these probabilities has been entered into the appropriate cell of Table 6.1. Note that the total of the probabilities in all the cells is 1. The information in the table can also be represented in visual terms in a three-dimensional plot, see Figure 6.1, where the heights of the bars equal the values of the probabilities.

6.2.2 *Marginal probability mass function*

The *marginal probability mass function* of X, denoted by $p_X(x)$, is defined as

$$p_X(x) = \sum_y p(x, y). \tag{6.1}$$

where the summation is over all values of y, with x as the first coordinate. This arises because

$$p_X(x) = P(X = x)$$

Table 6.1   The probabilities of tile counts, X, and pot-sherd counts, Y, in $1\,\mathrm{m}^2$ sample squares from a field survey.

		0	1	2	3	4	5	$P(Y = y)$
Pot-	0	0.01	0.02	0.05	0.07	0.08	0.08	0.31
sherd	1	0.01	0.01	0.04	0.07	0.03	0.03	0.19
counts	2	0.07	0.07	0.06	0.04	0.01	0.01	0.26
Y	3	0.09	0.09	0.02	0.02	0.01	0.01	0.24
$P(X = x)$		0.18	0.19	0.17	0.20	0.13	0.13	1.00

Tile counts X spans columns 0–5.

and by the theorem of total probability (see Section 4.4.4):

$$P(X = x) = \sum_y P(X = x \text{ and } Y = y) = \sum_y p(x, y).$$

In fact $p_X(x)$ is merely the probability mass function of X by itself. Likewise the marginal probability mass function of Y is denoted by $p_Y(y)$ and is given by

$$p_Y(y) = \sum_x p(x, y)$$

where now the summation is over all the values of x with y as the second co-ordinate.

Example

Consider the tile fragments/pot-sherds example introduced above. Let us now calculate the marginal probability mass function of X which represents the number of tile fragments found per $1\,\mathrm{m}^2$. The probability of no tile fragments is given by

$$
\begin{aligned}
p_X(0) &= \sum_{y=0}^{3} p(0, y) \\
&= p(0, 0) + p(0, 1) + p(0, 2) + p(0, 3) \\
&= 0.01 + 0.01 + 0.07 + 0.09 \\
&= 0.18.
\end{aligned}
$$

Thus there is a probability of 0.18 that a sample square contains no tile fragments. The probability of exactly one tile fragment is given by

$$
\begin{aligned}
p_X(1) &= \sum_{y=0}^{3} p(1, y) \\
&= p(1, 0) + p(1, 1) + p(1, 2) + p(1, 3) \\
&= 0.02 + 0.01 + 0.07 + 0.09 \\
&= 0.19.
\end{aligned}
$$

Using a similar method we find that the probabilities of 2, 3, 4 or 5 tile fragments are given by 0.17, 0.20, 0.13 and 0.13 respectively. In fact the marginal probability mass function is given in the bottom row of Table 6.1. Each entry in this row has been obtained by adding up the probabilities in the column immediately above. Finally notice that the marginal probabilities add up to 1, that is the values in the bottom row of the table sum to 1.

    If we wish to calculate the marginal probability mass function of Y we proceed in a similar fashion. For example the probability of no pot-sherds is given by

$$
\begin{aligned}
p_Y(0) \quad &= \quad P(Y = 0) = P([X = 0] \text{ and } [Y = 0]) + P([X = 1] \text{ and } [Y = 0]) \\
&\quad + P([X = 2] \text{ and } [Y = 0]) + P([X = 3] \text{ and } [Y = 0]) \\
&\quad + P([X = 4] \text{ and } [Y = 0]) + P([X = 5] \text{ and } [Y = 0]) \\
&= \quad 0.31.
\end{aligned}
$$

The marginal probability mass function of Y is given as the final column in Table 6.1. Each entry in this column is the sum of all the probabilities in the row alongside it.

6.2.3   Conditional distributions

Recall from Section 4.4 that, provided $P(E_2) > 0$,

$$
P(E_1 | E_2) = \frac{P(E_1 \text{ and } E_2)}{P(E_2)}.
$$

This can be applied to two discrete random variables by letting E_1 and E_2 be the events $X = x$ and $Y = y$ respectively. This leads to the following definitions.

    If X and Y have joint probability mass function $p(x, y)$ and marginal mass functions $p_X(x)$ and $p_Y(y)$, then the conditional probability mass function of X given $Y = y$ is

$$
p_{X|Y}(x|y) = P(X = x | Y = y) = \frac{P([X = x] \text{ and } [Y = y])}{P(Y = y)} \quad \text{if } P(Y = y) > 0
$$

and so may be expressed as

$$
p_{X|Y}(x|y) = \frac{p(x, y)}{p_Y(y)} \quad \text{if } p_Y(y) > 0. \tag{6.2}
$$

Likewise the conditional probability mass function of Y given $X = x$ is

$$
p_{Y|X}(y|x) = P(Y = y | X = x) = \frac{p(x, y)}{p_X(x)} \quad \text{if } p_X(x) > 0.
$$

Example

We might be interested in the conditional distribution of the number of tile fragments found given that two pot-sherds have been found in a square. That is we want the probability of $X = x$ given that $Y = 2$. For instance

$$p_{X|Y}(0|2) = P(X = 0|Y = 2) = \frac{P([X = 0] \text{ and } [Y = 2])}{P([Y = 2])} = \frac{0.07}{0.26} = 0.269.$$

Similarly

$$p_{X|Y}(2|2) = P(X = 2|Y = 2) = \frac{P([X = 2] \text{ and } [Y = 2])}{P([Y = 2])} = \frac{0.06}{0.26} = 0.231.$$

Similar calculations yield

$$p_{X|Y}(1|2) = P(X = 1|Y = 2) = 0.07/0.26 = 0.269,$$

$$p_{X|Y}(3|2) = P(X = 3|Y = 2) = 0.04/0.26 = 0.154,$$

$$p_{X|Y}(4|2) = P(X = 4|Y = 2) = 0.01/0.26 = 0.038,$$

and

$$p_{X|Y}(5|2) = P(X = 5|Y = 2) = 0.01/0.26 = 0.038.$$

The conditional probability mass function of X given $Y = 2$ is shown in Figure 6.2.

6.2.4   Interpretation of the marginal mass function

Recall from (6.1) that

$$p_X(x) = \sum_y p(x, y)$$

and from (6.2) that

$$p_{X|Y}(x|y) = \frac{p(x, y)}{p_Y(y)} \quad \text{if } p_Y(y) > 0.$$

Therefore we have

$$p_X(x) = \sum_y p_{X|Y}(x|y)p_Y(y).$$

In words this means that the marginal mass function of X is the weighted average of the conditional mass function of $X|Y = y$ with the weight attached to $Y = y$ being $p_Y(y)$, the marginal probability that Y equals y.

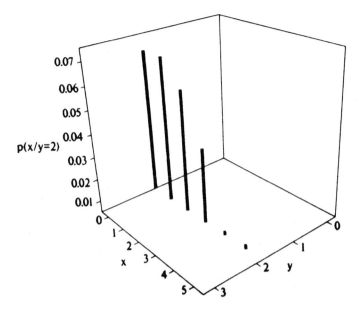

Figure 6.2   Conditional probability mass function of tile counts, X, given that the pot-sherd count is 2, that is, conditional on $Y = 2$.

6.2.5   Independence

Recall that in Section 4.4.3 we stated that events E_1 and E_2 were independent if and only if

$$P(E_1 \text{ and } E_2) = P(E_1)P(E_2)$$

or equivalently if and only if

$$P(E_1|E_2) = P(E_1).$$

The same principle applies to random variables. Let E_1 and E_2 represent the events $X = x$ and $Y = y$ respectively. Two random variables are said to be independent if and only if

$$p(x,y) = p_X(x)p_Y(y) \text{ for all values of } x \text{ and } y$$

or equivalently if and only if

$$p_{Y|X}(y|x) = p_Y(y) \text{ for all values of } x \text{ and } y.$$

Example

In the sherd-counting example $p_X(0) = 0.18$ and $p_Y(0) = 0.31$. Therefore if the variables were independent $p(0,0) = p_X(0)p_Y(0) = 0.18 \times 0.31 = 0.0558$

but $p(0,0) = 0.01$ so that the numbers of tile fragments and pot-sherds found in a square are not independent, according to the joint probability given in Table 6.1. So the number of tile fragments found in a square is related to the number of pot-sherds and *vice versa*.

This dependence can also be seen from our earlier calculations of the marginal and conditional distributions of X and Y. For example, the marginal probability of finding no tile fragments is $p_X(0) = 0.18$. The conditional probability of no tile fragments given two pot-sherds have been found is $p_{X|Y}(0|2) = 0.269$. Since these two values are not equal we can deduce that X and Y are dependent.

6.2.6 Bayes' theorem using bivariate discrete distributions

Recall from Section 4.5 that if E_1 and E_2 are events with $P(E_2) > 0$ then Bayes' theorem states that

$$P(E_1|E_2) = \frac{P(E_2|E_1)}{P(E_2)} P(E_1).$$

Let E_1 and E_2 be the events $X = x$ and $Y = y$ respectively. In this case Bayes' theorem becomes

$$P(X = x|Y = y) = \frac{P(Y = y|X = x)}{P(Y = y)} P(X = x)$$

or alternatively

$$P(X = x|Y = y) \propto P(Y = y|X = x)P(X = x).$$

In the latter case the normalizing constant is $1/P(Y = y)$.

In terms of probability mass functions we have

$$p_{X|Y}(x|y) = \frac{p_{Y|X}(y|x)}{p_Y(y)} p_X(x) \quad \text{if } p_Y(y) > 0$$

or

$$p_{X|Y}(x|y) \propto p_{Y|X}(y|x)p_X(x)$$

where the normalizing constant is $1/p_Y(y)$.

6.2.7 Expectation of discrete bivariate random variables

We have seen in Section 5.9.1 that the expectation of a single discrete random variable can be expressed as

$$E(X) = \sum_x x p(x).$$

Extending this to bivariate random variables is quite straightforward. For example, the expected values of X, Y and XY are

$$\mu_X = E(X) = \sum_x \sum_y xp(x,y) = \sum_x x \sum_y p(x,y) = \sum_x xp_X(x),$$

$$\mu_Y = E(Y) = \sum_x \sum_y yp(x,y) = \sum_y y \sum_x p(x,y) = \sum_y yp_Y(y)$$

and

$$E(XY) = \sum_x \sum_y xyp(x,y)$$

respectively, where the sums extend over the total ranges of X and Y. Note also that if we let $Z = X + Y$ then

$$
\begin{aligned}
E(Z) = E(X+Y) &= \sum_x \sum_y (x+y)p(x,y) \\
&= \sum_x \sum_y xp(x,y) + \sum_x \sum_y yp(x,y) \\
&= E(X) + E(Y).
\end{aligned}
$$

Example

We can use the bivariate distribution in Table 6.1 to illustrate how to calculate the expectations involving jointly distributed discrete random variables. In the first place we can, from the margins of Table 6.1, calculate the expected value of X as

$$
\begin{aligned}
E(X) &= \sum_{x=0}^{5} xp_X(x) \\
&= 0 \times 0.18 + 1 \times 0.19 + 2 \times 0.17 + 3 \times 0.20 + 4 \times 0.13 + 5 \times 0.13 \\
&= 0 + 0.19 + 0.34 + 0.60 + 0.52 + 0.65 \\
&= 2.30
\end{aligned}
$$

and Y as

$$
\begin{aligned}
E(Y) &= \sum_{y=0}^{3} yp_Y(y) \\
&= 0 \times 0.31 + 1 \times 0.19 + 2 \times 0.26 + 3 \times 0.24 \\
&= 0 + 0.19 + 0.52 + 0.72 \\
&= 1.43.
\end{aligned}
$$

Suppose next we wish to calculate the expected value of the total, denoted by Z, of the number of tile fragments, X, and of the number of pot-sherds, Y, in a sample square. Obviously $Z = X + Y$. Now the expected value of Z is given by

$$
\begin{aligned}
E(Z) = E(X+Y) &= E(X) + E(Y) \\
&= 2.30 + 1.43 \\
&= 3.73.
\end{aligned}
$$

Finally the expected value of XY is

$$
\begin{aligned}
E(XY) &= \sum_{x=0}^{5} \sum_{y=0}^{3} xyp(x,y) \\
&= (0 \times 0 \times 0.01) + (0 \times 1 \times 0.01) + (0 \times 2 \times 0.07) + \cdots \\
&\quad + (5 \times 1 \times 0.03) + (5 \times 2 \times 0.01) + (5 \times 3 \times 0.01) \\
&= 2.21.
\end{aligned}
$$

6.2.8   Covariance and correlation

Recall that the variance of a single random variable was defined in Section 5.5.2 as

$$
\sigma_X^2 = Var(X) = E[(X - \mu_X)^2] = \sum_x (x - \mu_X)^2 p(x)
$$

where μ_X is the expected value of X.

For two variables X and Y, the variance of $(X + Y)$ is given by

$$
\begin{aligned}
Var(X+Y) &= E[(X + Y - E(X+Y))^2] \\
&= E[(X + Y - \mu_X - \mu_Y)^2] \\
&= E[((X - \mu_X) + (Y - \mu_Y))^2] \\
&= E[(X - \mu_X)^2 + (Y - \mu_Y)^2 + 2(X - \mu_X)(Y - \mu_Y)] \\
&= Var(X) + Var(Y) + 2E[(X - \mu_X)(Y - \mu_Y)].
\end{aligned} \tag{6.3}
$$

Note that $Var(X+Y)$ is not the same as $Var(X) + Var(Y)$; there is an extra term, namely $2E[(X - \mu_X)(Y - \mu_Y)]$, at the end of (6.3).

The *covariance* of X and Y is a measure of association between the two random variables and is defined to be

$$
Cov(X,Y) = E[(X - \mu_X)(Y - \mu_Y)].
$$

This can also be written as

$$
Cov(X,Y) = E(XY) - \mu_X \mu_Y. \tag{6.4}
$$

Finally there is the *correlation* of two random variables X and Y which is defined as

$$
\rho = \rho(X,Y) = \frac{Cov(X,Y)}{\sqrt{Var(X)Var(Y)}}. \tag{6.5}
$$

The value of ρ lies between -1 and $+1$ inclusive $(-1 \le \rho \le 1)$. When the value of ρ equals either -1 or $+1$ there is a linear relationship between X and Y.

This section has been very brief. A fuller explanation together with archaeological examples of covariance and correlation can be found for instance in Doran and Hodson (1975, p. 58–61), Shennan (1988, p. 126–131) or Baxter (1994, p. 27–28).

Example

Again returning to the probabilities given in Table 6.1, it is possible to calculate the covariance of X and Y using (6.4). We have already calculated that $E(X) = 2.30$, $E(Y) = 1.43$ and $E(XY) = 2.21$ so that the covariance between the number of tile fragments and the number of pot-sherds is given by

$$\begin{aligned} Cov(X,Y) &= E(XY) - \mu_X \mu_Y \\ &= 2.21 - (2.30 \times 1.43) \\ &= -1.08. \end{aligned}$$

Next we calculate the correlation, ρ, of X and Y. From (6.5) it is clear that we need, in addition to the $Cov(X,Y)$ just calculated, $Var(X)$ and $Var(Y)$. Recall from (5.3) that $Var(X) = E(X^2) - \mu_X^2$ and to find this, first we calculate $E(X^2)$ which we do as follows.

$$\begin{aligned} E(X^2) &= \sum_{x=0}^{5} x^2 p(X = x) \\ &= 0 \times 0.18 + 1 \times 0.19 + 4 \times 0.17 \\ &\quad + 9 \times 0.20 + 16 \times 0.13 + 25 \times 0.13 \\ &= 8.00. \end{aligned}$$

Therefore the variance of X is given by

$$Var(X) = 8.00 - (2.30)^2 = 2.71.$$

Similarly we find that $Var(Y) = 1.3451$ and hence

$$\rho = \frac{Cov(X,Y)}{\sqrt{Var(X)Var(Y)}} = \frac{-1.08}{\sqrt{2.71 \times 1.3451}} = -0.566.$$

The interpretation of this is as follows. The two variables, the tile counts (X) and the sherd counts (Y), are negatively correlated: the high concentrations of tile tend to occur with the low concentrations of sherd, and the high concentrations of sherd tend to occur with the low concentrations of tile. Such a situation could arise, for example, where the high tile concentrations indicated the presence of building, which had been kept clear of broken pottery and other domestic refuse: hence high tile and low sherd counts. The high sherd counts could indicate middens consisting largely of domestic refuse, but relatively little tile.

6.3   Continuous bivariate random variables

We now move on to consider continuous random variables such as the measurements of the diameter and height of the Hohokan bowls mentioned in Section 6.1, rather than discrete variables such as counts of different types of sherds.

6.3.1   Joint probability density function

We have already encountered in Section 5.8.1 the probability density function, denoted by $p(x)$, of a single continuous random variable, X, and that it satisfies the equation

$$P(x_1 \leq X \leq x_2) = \int_{x_1}^{x_2} p(x)\,dx.$$

The total area between the curve of the probability density function and the x-axis over the whole distribution is equal to 1, and the probability over the range of X between x_1 and x_2 is found by calculating the area under the probability density function curve between those limits.

For the joint density of two continuous random variables X and Y we need to specify the *joint probability density function* which is denoted by

$$p(x, y).$$

We can picture this as something like the three-dimensional bar diagram of Figure 6.1, but instead of the separate columns we need to think in terms of a surface as in Figure 6.3, where, so to speak, a blanket has been spread over all the bars of the discrete example. Thinking in these three-dimensional terms, it is possible to conceive of a zone (say a circle) between the x and y axes as representing a statement that X and Y take the values (x, y) within the circle. In this case the volume of the cylinder which rises and meets the surface $p(x, y)$ is equal to the *probability* that X and Y take the values x and y respectively (that is to say the values within the circle). The total volume below the surface $u = p(x, y)$ down to the (x, y) plane is equal to 1.

6.3.2   Marginal probability density function

Just as for the discrete random variables where the marginal mass function of X was found by summing the joint probability mass function of X and Y over all y (see Section 6.2.2) so in the case of a continuous random variable we *integrate*. Thus the marginal probability density function of X is given by

$$p_X(x) = \int_{-\infty}^{\infty} p(x, y)\,dy \qquad \text{for} \quad -\infty < x < \infty$$

and the marginal probability density function of Y is given by

$$p_Y(y) = \int_{-\infty}^{\infty} p(x, y)\,dx \qquad \text{for} \quad -\infty < y < \infty.$$

When integrating $p(x, y)$ with respect to y, x is treated as if it were a constant.

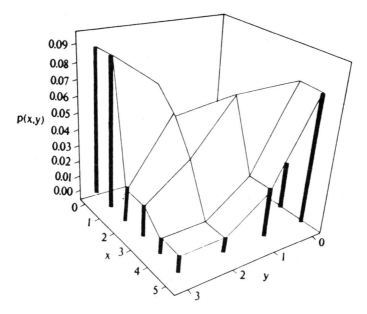

Figure 6.3 Wire diagram placed over the joint probability mass function of Figure 6.1.

6.3.3 Conditional densities and independence

We saw in Section 6.2.3 how to calculate the conditional probability mass function for discrete bivariate random variables; the form for continuous variables follows exactly the same pattern. The conditional probability density function of X given $Y = y$ is

$$p_{X|Y}(x|y) = \frac{p(x,y)}{p_Y(y)} \quad \text{for any } y \text{ where } p_Y(y) > 0$$

and similarly, for Y given $X = x$,

$$p_{Y|X}(y|x) = \frac{p(x,y)}{p_X(x)} \quad \text{for any } x \text{ where } p_X(x) > 0.$$

By analogy with discrete random variables, two continuous random variables are independent if and only if

$$p(x,y) = p_X(x)p_Y(y) \quad \text{for all } x \text{ and } y,$$

or equivalently if and only if

$$p_{X|Y}(x|y) = p_X(x) \quad \text{for all } x \text{ and } y.$$

6.3.4 Bayes' theorem for bivariate continuous distributions

It is plain to see from the definition of the conditional probability density functions of X and Y that, provided $p_X(x) > 0$ and $p_Y(y) > 0$,

$$p_{X|Y}(x|y)p_Y(y) = p(x,y) = p_{Y|X}(y,x)p_X(x).$$

Hence we have another form of Bayes' theorem:

$$p_{X|Y}(x|y) = \frac{p_{Y|X}(y|x)}{p_Y(y)}p_X(x)$$

or alternatively

$$p_{X|Y}(x|y) \propto p_{Y|X}(y|x)p_X(x)$$

where the normalizing constant is $1/p_Y(y)$.

6.3.5 Expectation of continuous bivariate random variables

Expectations are treated in exactly the same way as in the discrete case but with the summations replaced by integrals. For example, the expected values of X, Y and XY are given by the two-dimensional integrals

$$E(X) = \int_{-\infty}^{\infty}\int_{-\infty}^{\infty} xp(x,y)\,dx\,dy,$$

$$E(Y) = \int_{-\infty}^{\infty}\int_{-\infty}^{\infty} yp(x,y)\,dx\,dy$$

and

$$E(XY) = \int_{-\infty}^{\infty}\int_{-\infty}^{\infty} xyp(x,y)\,dx\,dy.$$

6.3.6 Bivariate Normal distribution

Two variables X and Y are said to have a Bivariate Normal distribution, denoted by

$$(X,Y)|\mu_X,\mu_Y,\sigma_X,\sigma_Y,\rho \sim BVN(\mu_X,\mu_Y,\sigma_X,\sigma_Y,\rho),$$

if they have a joint probability density function

$$p(x,y) = \frac{1}{2\pi\sigma_x\sigma_y\sqrt{1-\rho^2}}\exp\left\{-\frac{1}{2(1-\rho^2)}\left[\left(\frac{x-\mu_X}{\sigma_X}\right)^2\right.\right.$$
$$\left.\left.-2\rho\left(\frac{x-\mu_X}{\sigma_X}\right)\left(\frac{y-\mu_Y}{\sigma_Y}\right)+\left(\frac{y-\mu_Y}{\sigma_Y}\right)^2\right]\right\}$$
$$\text{for } -\infty < x < \infty; -\infty < y < \infty$$

where μ_X and μ_Y are the means of X and Y, σ_X^2 and σ_Y^2 are their variances, and ρ stands for the correlation between X and Y.

Useful properties of the bivariate normal distribution include the following.

(i) The marginal distribution of X is normal with mean μ_X and variance σ_X^2. A comparable result applies for Y.

(ii) The conditional distribution of X given $Y = y$ is normal with mean and variance given by

$$\mu_X + \rho\sigma_X \left(\frac{y - \mu_Y}{\sigma_Y}\right) \text{ and } (1 - \rho^2)\sigma_Y^2$$

respectively, with a similar expression for the conditional distribution of Y given $X = x$.

(iii) X and Y are independent if and only if the correlation between them is 0. That is if and only if $\rho = 0$.

6.4   Multivariate distributions

In the previous sections we have developed the modelling of two random variables X and Y. We shall now extend this to cover situations where we have n random variables which we denote by $X_1, \ldots, X_n$. Our explanation will by necessity be brief; we refer the interested reader to Mardia *et al.* (1982).

In general, the joint distribution of more than two variables is called a *multivariate distribution*. To simplify notation we let the vector $\boldsymbol{X} = (X_1, \ldots, X_n)$ and then talk in terms of the random vector $\boldsymbol{X}$. If all the X_i are discrete, then $\boldsymbol{X}$ has a discrete distribution and has a joint probability mass function given by

$$p(x_1, \ldots, x_n) = p([X_1 = x_1] \text{ and } [X_2 = x_2] \text{ and } \ldots \text{ and } [X_n = x_n])$$

or using vector notation $\boldsymbol{x} = (x_1, \ldots, x_n)$ also,

$$p(\boldsymbol{x}) = p(\boldsymbol{X} = \boldsymbol{x}).$$

If all the X_i are continuous then $\boldsymbol{X}$ is said to be continuous and has a joint probability density function denoted by $p(x_1, \ldots, x_n)$ or using vector notation $p(\boldsymbol{x})$. If $\boldsymbol{X}$ is continuous then the marginal density of X_1 is

$$f_1(x_1) = \int_{-\infty}^{\infty} \cdots \int_{-\infty}^{\infty} p(x_1, \ldots, x_n) \, dx_2 \cdots dx_n.$$

More generally the marginal joint probability density function of any m of the n variables can be found by integrating the joint probability density of all the n variables over all the possible values of the other remaining $n - m$ variables. For example, if we have five random variables X_1, X_2, X_3, X_4 and X_5 then the marginal density of X_3 and X_5 is given by

$$p_{3,5}(x_3, x_5) = \int_{-\infty}^{\infty} \int_{-\infty}^{\infty} \int_{-\infty}^{\infty} p(x_1, x_2, x_3, x_4, x_5) \, dx_1 \, dx_2 \, dx_4,$$

where the integration is over the full ranges of X_1, X_2 and X_4 respectively. If the random vector is discrete then similar results apply with the integral signs replaces by summation signs.

6.4.1   Conditional distributions

The conditional probability density of

$$X_1, \ldots, X_k \text{ given } X_{k+1} = x_{k+1}, \ldots, X_n = x_n$$

is expressed as

$$g(x_1, \ldots, x_k | x_{k+1}, \ldots, x_n) = \frac{p(x_1, \ldots, x_n)}{f_0(x_{k+1}, \ldots, x_n)}$$

where $f_0(x_{k+1}, \ldots, x_n)$ is the marginal density of $X_{k+1}, \ldots, X_n$. In particular the conditional probability density function of X_1 given that $X_2 = x_2, \ldots, X_n = x_n$ is defined as

$$h_1(x_1 | x_2, \ldots, x_n) = \frac{p(x_1, x_2, \ldots, x_n)}{p_{-1}(x_2, \ldots, x_n)}$$

where $p_{-1}(x_2, \ldots, x_n)$ is the marginal density of $X_2, \ldots, X_n$. This conditional density, $h_1(x_1 | x_2, \ldots, x_n)$, is often referred to as the *full conditional distribution* of X_1 and will be extensively used later in the book.

The concept of independence carries over to n random variables as follows. $X_1, \ldots, X_n$ are independent if and only if

$$p(x_1, \ldots, x_n) = f_1(x_1) f_2(x_2) \cdots f_n(x_n).$$

In other words, $X_1, \ldots, X_n$ are independent if and only if their joint probability density function can be expressed as the product of the marginal densities.

Finally there is the important concept of *conditional independence*. Suppose we have three random variables, X_1, X_2 and X_3. X_1 and X_2 are said to be *conditionally independent*, conditional on X_3 if

$$p(x_1, x_2 | x_3) = p(x_1 | x_3) p(x_2 | x_3) \quad \text{for all } x_1, x_2 \text{ and } x_3.$$

That is, if X_3 is fixed, then X_1 and X_2 are independent. This concept is easily generalized to more than three variables.

6.4.2   Bayes' theorem for multivariate distributions

It is quite easy to see that by using the definition of a conditional probability density of $X_1, \ldots, X_k$ given $X_{k+1} = x_{k+1}, \ldots, X_n = x_n$ then Bayes' theorem applies to groups of random variables. Let $X^{(1)} = (X_1, \ldots, X_k)$ and $X^{(2)} = (X_{k+1}, \ldots, X_n)$, then

$$p(x^{(1)}|x^{(2)}) = \frac{p(x^{(2)}|x^{(1)})}{p(x^{(2)})}p(x^{(1)}) \qquad (6.6)$$

where $p(x^{(1)}|x^{(2)})$ is the conditional density of $X^{(1)}$ given $X^{(2)}$ and $p(x^{(2)}|x^{(1)})$ is the conditional density of $X^{(2)}$ given $X^{(1)}$; $p(x^{(1)})$ and $p(x^{(2)})$ are the marginal densities of $X^{(1)}$ and $X^{(2)}$ respectively. Alternatively (6.6) can be written as

$$p(x^{(1)}|x^{(2)}) \propto p(x^{(2)}|x^{(1)})p(x^{(1)})$$

where the normalizing constant is $1/p(x^{(2)})$.

6.4.3 Expectation

The expected value of a random vector, X, is given by the vector μ where $\mu = (\mu_1, \ldots, \mu_n)$ and $\mu_i = E(X_i)$. The variance of an individual X_i will be denoted by $\sigma_{i,i}$ where $\sigma_{i,i} = E[(X_i - \mu_i)^2]$. The covariance between X_i and X_j, $E[(X_i - \mu_i)(X_j - \mu_j)]$, will be denoted by $\sigma_{i,j}$.

The variance/covariance structure of all the n random variables can be represented by the $n \times n$-dimensional matrix Σ where

$$\Sigma = \begin{bmatrix} \sigma_{1,1} & \sigma_{1,2} & \cdots & \sigma_{1,n} \\ \sigma_{2,1} & \sigma_{2,2} & \cdots & \sigma_{2,n} \\ \vdots & \vdots & & \vdots \\ \sigma_{n,1} & \sigma_{n,2} & \cdots & \sigma_{n,n} \end{bmatrix}$$

Note also that Σ is symmetric about the diagonal as $Cov(X_i, X_j) = Cov(X_j, X_i)$. Finally the correlation between X_i and X_j is denoted by $\rho_{i,j}$ where

$$\rho_{i,j} = \frac{\sigma_{i,j}}{\sigma_i \sigma_j}.$$

6.5 Examples of multivariate distributions

6.5.1 Multinomial distribution

The random vector X is said to have a Multinomial distribution if its joint probability mass function is given by

$$p(x) = p(x_1, \ldots, x_n) = \frac{m!}{x_1! x_2! \cdots x_n!} \theta_1^{x_1} \theta_2^{x_2} \cdots \theta_n^{x_n},$$

where $\theta_i > 0$ for $i = 1, \ldots, n$, $m = \sum_{i=1}^n x_i$ and $\sum_{i=1}^n \theta_i = 1$. The Multinomial distribution is a generalization of the Binomial distribution (see Section 5.6.3) to when we have k possible types of outcome of a trial (instead of two for the Binomial). The probability that the ith type occurs at an individual trial is θ_i. If we have a total of m independent trials then the probability that we observe x_1 of type 1, x_2 of type 2, ..., x_n of type k is given by $p(x_1, \ldots, x_n)$ above.

Example

In Section 5.6.3 we used the example of flat or pitched roofs to illustrate the Binomial distribution. Let us generalize the problem by assuming that there are n different categories of roof type in a particular region, with type i occurring with probability θ_i. Assuming that the type of roof on a building is independent of that of other buildings, then in a sample of size m, the probability that we have x_1 of type 1, x_2 of type 2, ..., x_n of type n is $p(x_1, \ldots, x_n)$ as stated above.

6.5.2   Multivariate Normal distribution

An n-dimensional random vector, $\boldsymbol{X}$, has a Multivariate Normal distribution (which is a generalization of the Bivariate Normal to n dimensions) with mean vector $\boldsymbol{\mu}$ and covariance matrix Ω, denoted by

$$\boldsymbol{X}|\boldsymbol{\mu}, \Omega \sim MVN(\boldsymbol{\mu}, \Omega),$$

if its joint probability density function is given by

$$p(\boldsymbol{x}|\boldsymbol{\mu}, \Omega) = c \exp\left\{ -\frac{1}{2}(\boldsymbol{x} - \boldsymbol{\mu})'\Omega^{-1}(\boldsymbol{x} - \boldsymbol{\mu}) \right\}$$

where c is the normalizing constant that ensures that the density integrates to 1 over the full ranges for all the X_i. In fact the constant c is given by

$$c = \frac{1}{(2\pi)^{n/2}|\Omega|}.$$

Here Ω^{-1} denotes the inverse of the matrix Ω, $|\Omega|$ its determinant and Ω' the transpose — readers unfamiliar with these terms should consult a book on matrix or linear algebra, such as Namboodiri (1984).

Example

As mentioned earlier in Section 3.6.2, chemical compositional analysis is a well-established scientific technique used by archaeologists to identify the source or provenance from which artefacts originated. This approach is feasible because, in general, different provenances have different chemical compositions.

For the moment consider just one source. Suppose n elements are measured and let μ_i be the underlying proportion of the ith element present at the source. Suppose we take a specimen of material from the source and carry out a chemical analysis. Let the random variable X_i represent the proportion of the total mass of the specimen that is the ith element. Due to the inherent variability of natural materials at a source and to inaccuracies in the techniques used to obtain the chemical information, there is an error component. We model this as

$$X_i = \mu_i + \text{ error term.}$$

Moreover it is quite usual to find that if the proportion of element i is above average, then so is element j but that generally element k would be lower. That is the various proportions of the elements are correlated with each other. We model this by allowing the error terms for different elements to be correlated. If we let $\boldsymbol{X} = (X_1, \ldots, X_n)$ then a possible model is that $\boldsymbol{X}$ has a Multivariate Normal distribution with mean vector $\boldsymbol{\mu}$ and covariance matrix Ω. The covariance matrix allows both for the inherent variability present and also its correlation structure.

If we have K different sources then each will have its own mean vector and covariance matrix. That is if the specimen comes from source k, then

$$\boldsymbol{X}|\boldsymbol{\mu}_k, \Omega_k \sim N(\boldsymbol{\mu}_k, \Omega_k)$$

where $\boldsymbol{\mu}_k$ and Ω_k are the mean vector and covariance matrix for the kth source. Given a specimen of *unknown* origin, the problem is from which source did it originate?

6.5.3 *Dirichlet distribution*

The random vector $\boldsymbol{X}$ has a Dirichlet distribution with parameters $\boldsymbol{\alpha} = (\alpha_1, \ldots, \alpha_{n+1})$ (where $\alpha_i > 0$), denoted by

$$\boldsymbol{X}|\boldsymbol{\alpha} \sim D(\boldsymbol{\alpha}),$$

if its joint probability density function is given by

$$p(x_1, \ldots, x_n | \boldsymbol{\alpha}) = c(1 - x_1 - \cdots - x_n)^{\alpha_{n+1} - 1} \prod_{i=1}^{n} x_i^{\alpha_i - 1}$$

over the n-dimensional region defined by the inequalities $x_i > 0$ ($i = 1, 2, \ldots, n$) and $\sum_{i=1}^{n} x_i < 1$. Here c is the normalizing constant.

Example

This distribution will be used to model the prior information in the sourcing and provenancing of archaeological artefacts — see Chapter 11.

6.5.4   Inverse-Wishart distribution

The $p \times p$ matrix Σ is said to have an *inverse-Wishart* distribution with v degrees of freedom and scale matrix V where $E(\Sigma) = V/(v-2)$, denoted by

$$\Sigma | v, V \sim W^{-1}(v, V),$$

if its joint probability density is given by

$$p(\Sigma | v, V) = c|V|^{(p+v-1)/2}|\Sigma|^{-(p+v/2)} \exp -\frac{1}{2}tr(\Sigma^{-1}V)$$

where c is the normalizing constant. Here, $|V|$ is the determinant of V and tr denotes the trace of a matrix (see any standard book on matrix algebra such as Namboodiri, 1984).

Example

This complicated distribution is used to model the complex statistical relationships found in archaeological sourcing and provenancing problems. We have already discussed the modelling of within-source and within-provenance variability using a Multivariate Normal with mean $\boldsymbol{\mu}$ and covariance Ω, but how do we model the prior beliefs about the parameters of the Multivariate Normal themselves? In the case of the mean vector it is possible for this uncertainty to be represented using another Multivariate Normal with, for example, mean vector $\boldsymbol{m}$ and covariance matrix M.

The modelling of the prior beliefs about the covariance matrix Ω is rather less straightforward since we need a distribution which allows us to represent the uncertainty about a matrix, not just a vector. This is just what the inverse-Wishart distribution does. A convenient model is that $\Omega | v, V \sim W^{-1}(v, V)$. This model will be used in our Bayesian approach to sourcing and provenancing to be described in Chapter 11.

6.6   Concluding remarks

In this chapter and the previous one, we have introduced many theoretical ideas and a multitude of technical terms. We also have given a series of examples illustrating the different types of probability distributions: for discrete and continuous variables, univariate, bivariate and multivariate. The examples are but a limited selection of all those available. The intention was to demonstrate the great variety of ways which can be used to model the uncertainty associated with our subject matter. Variety, however, is not superfluity: the choice of an appropriate distribution lies at the heart of statistical modelling. The characteristics of a distribution, its form, the number of parameters and their inter-relationships, must relate to the basic understanding and analysis of the problem at hand. The skill required is that of conveying the essence of any problem in mathematical form and relating that to the relevant data. As we have indicated at the beginning of Chapter 5, this is a demanding and technical subject. In this book we are only skimming the surface so that the reader may get an appreciation of what is involved. To acquire the necessary skills for themselves readers will need to consult other, more mathematical texts which treat the subject in more depth, such as DeGroot (1986), Ross (1988) or Mood *et al.* (1974).

Summarizing our progress so far, in the last few chapters we sketched out the requirements leading up to a Bayesian analysis: the development of a mathematical model and its place within the modelling cycle, the modelling of uncertainty by the use of probability, the assessment of probabilities and the rules governing their calculation. Then we looked at the use of statistical distributions to model complex stochastic problems. In the next chapter we take this programme further by considering how these are brought together in the Bayesian approach to data analysis.

7

Bayesian inference

7.1 Introduction

In Chapter 3 we described how mathematical models could be developed and expressed in terms of parameters. Then, having discussed the quantification of uncertainty in Chapter 4, we went on in Chapters 5 and 6 to demonstrate how these models could be extended to include any uncertainty about the values of the parameters by the use of probability and probability distributions. We come now to the core of our argument. We shall see how Bayes' theorem brings together data and prior information, and the model. These are incorporated into a single consistent analysis in order to make inferences about the parameters and hence about the archæological situation under investigation.

For example consider the use of radiocarbon dating techniques to determine the date and authenticity of the Turin Shroud (see Figure 7.1) which many people believe was used to wrap Christ's body. In an ideal world, and therefore with the ability to count the radioactive emissions with complete accuracy, it would be possible to obtain an exact radiocarbon age for a sample taken from the Shroud. However, in reality, the radiocarbon age is reported as an estimate together with a standard error which reflects how confident the laboratory is in that estimate.

In general archaeologists wish to have *point estimates*, for example that the radiocarbon age of a sample is 1000 radiocarbon years bp ("bp" denotes before present on the *radiocarbon* time-scale), and *interval estimates*, for example that the radiocarbon age is between 1200 and 800 radiocarbon years bp with probability 0.9. In addition, archaeologists will need to test a *hypothesis*, such as that the Turin Shroud is authentic. As we will see the Bayesian approach to data analysis provides a coherent and logical framework within which such questions may be addressed.

7.1.1 Outline of the Bayesian approach

As we have discussed in earlier chapters, statisticians, both Bayesian and classical, formalize problems by developing statistical models that represent,

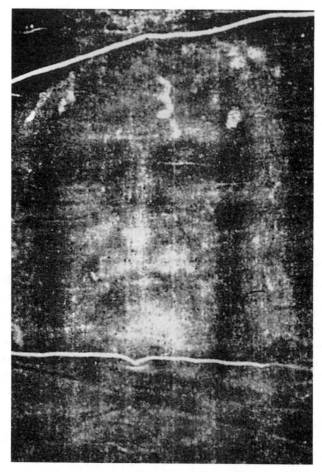

Figure 7.1   The Turin Shroud (Hulton Deutsch Collection).

in a mathematical manner, the problem in hand. Within these models, a relationship between the data and the unknown parameters is posited. In any particular problem we usually have several parameters and several pieces of data. Suppose we have k parameters denoted by θ_1, θ_2 up to θ_k. Then $\boldsymbol{\theta}$ is used as a shorthand notation for this collection (or vector) of parameters. Likewise if we have n pieces of data, denoted by x_1, x_2 up to x_n, we use $\boldsymbol{x}$ to represent all the data.

Bayesian statisticians use probabilities to express the uncertainty about the parameters, $\boldsymbol{\theta}$, before (that is, *a priori*) and after (that is, *a posteriori*) seeing the data, $\boldsymbol{x}$. The Bayesian framework offers a simple way of computing the posterior uncertainty from the prior uncertainty and the data via the model. Prior information about the problem, obtained *before* seeing the data, has to be converted into probabilistic statements about the parameters — usually in terms of a probability mass (or density) function. This is a formal way of acknowledging that no analysis is carried out in a vacuum. There is a history of previous research which precedes any particular analysis. Bayes' theorem is then used to combine the prior information and the data in order to make posterior probabilistic statements about the parameters — again in terms of a probability mass (or density) function which then needs to be interpreted in terms of the archaeological problem under study.

In broad terms, the Bayesian method requires the user to be able to supply data relevant to the problem at hand and to specify the following two components.

(i) *The prior probability mass (or density) function* — denoted by $p(\boldsymbol{\theta})$. This may be read as "how much belief do I attach to possible values of the unknown parameters before (prior to) observing the data?" A relatively large value of $p(\boldsymbol{\theta})$ for a particular value of $\boldsymbol{\theta}$ indicates that this value is thought to be relatively likely; a relatively small value that it is relatively unlikely.

(ii) *The likelihood* — a mathematical function denoted by $l(\boldsymbol{\theta}; \boldsymbol{x})$. This may be read as "how much are particular values of the parameters supported by the data?" To be more specific, consider a fixed value of the data, $\boldsymbol{x}$. On the one hand, if the value of the likelihood at a particular value of the parameters, say $\boldsymbol{\theta}_1$, is relatively large then this indicates that the data strongly support $\boldsymbol{\theta}_1$. On the other hand, if at another value of $\boldsymbol{\theta}$, say $\boldsymbol{\theta}_2$, the likelihood is relatively small, then this suggests that the support from data for $\boldsymbol{\theta}_2$ is relatively small. That is, if $l(\boldsymbol{\theta}_1; \boldsymbol{x})$ is greater than $l(\boldsymbol{\theta}_2; \boldsymbol{x})$, then $\boldsymbol{\theta}_1$ is better supported by the data than $\boldsymbol{\theta}_2$.

In practice, the likelihood function is determined from the probability distribution of $\boldsymbol{x}$ conditional on $\boldsymbol{\theta}$, which we denote by $p(\boldsymbol{x}|\boldsymbol{\theta})$. However, it is important to realize that the data, $\boldsymbol{x}$, are known and the parameters, $\boldsymbol{\theta}$, are unknown. Therefore the likelihood should be regarded as a function of the unknown parameters, $\boldsymbol{\theta}$, for observed and therefore fixed values of the data, $\boldsymbol{x}$. To emphasize this point, the likelihood is denoted by by $l(\boldsymbol{\theta}; \boldsymbol{x})$.

The primary result of a Bayesian analysis is the *posterior probability mass (or density) function* — represented by $p(\boldsymbol{\theta}|\boldsymbol{x})$. This may be read as "how much belief do I attach to possible values of the unknown parameters after (posterior to) observing the data?" This posterior probability mass (or density) function then needs to be interpreted in terms of the real problem under study. In general, a relatively large value of $p(\boldsymbol{\theta}|\boldsymbol{x})$ indicates that the corresponding value of θ is, *a posteriori*, relatively likely and a relatively small value that it is relatively unlikely.

As we have seen in earlier chapters, the uncertainty about discrete parameters is modelled using probability mass functions whereas probability density functions are used for continuous parameters. In order to simplify what follows we use the term *prior* to denote prior probability mass function if θ is discrete and prior probability density function if θ is continuous. Likewise the term *posterior* will denote posterior probability mass or density function according to whether the parameters are discrete or continuous. We hope that it will be clear to the reader in any particular context whether the parameters are discrete or continuous.

7.1.2   Dating the Turin Shroud

We clarify these ideas by the use of a simple archaeological example with one parameter, denoted by θ, and one piece of data, denoted by x. Suppose we are interested in assessing the authenticity or otherwise of the Turin Shroud using radiocarbon dating techniques. Let the unknown calendar date of the material used in its manufacture be represented by θ (measured in years BP). The four basic components of the Bayesian analysis are as follows.

(i) *The prior* — before sending the sample of linen threads from the Shroud for radiocarbon analysis, the archaeologist has information about the unknown calendar date θ. This information can be expressed in terms of a prior probability mass function, $p(\theta)$.

In the case of the Turin Shroud what is known *a priori* about θ? Very little in fact — according to Damon *et al.* (1989) written records say that it was displayed in France in the AD 1350s, brought to Turin in AD 1578 and since AD 1694 it has been kept there in a specially designed shrine. Even so there is some doubt whether the Shroud is the same as that displayed in the 1350s. In the light of all this uncertainty, a reasonable way to proceed is to say nothing is known, *a priori*, about its date. In practical terms, we work in 20-year time intervals back to say 5000 BP. (The 5000 BP is an arbitrary large date in the past since which we believe that the material from the Shroud must have died.) The date of the Shroud occurring in any of the 20-year intervals is assumed to be, *a priori*, equally likely. That is, it has a discrete uniform distribution.

(ii) *The data* — consider one radiocarbon determination made by the Oxford Radiocarbon Laboratory with identifier OX-2575. Its estimated radiocarbon age is 795 BP (with a quoted standard deviation of 65). That is we have x set equal to 795.

(iii) *The likelihood* — the likelihood, $l(\theta; x)$, captures how x, the estimated radiocarbon age, is related to the true, but unknown, calendar age θ.

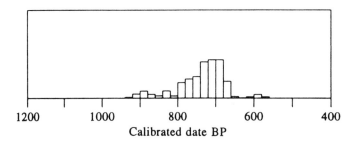

Figure 7.2   Posterior probability histogram for radiocarbon determination
OX-2575.

Given a particular radiocarbon age, the likelihood reflects how likely
each value of the calendar age, θ, is. If $l(\theta_1; x)$ is relatively large
compared with $l(\theta_2; x)$, then the date represented by θ_1 is better
supported by the radiocarbon estimate than the date represented by
θ_2. In other words, based on the data, the model tells us that θ_1 is
more likely than θ_2. In this example the likelihood brings together the
radiocarbon determination and the unknown calendar date via the model
(the radiocarbon calibration curve and the statistical model for the
uncertainty involved).

(iv) *The posterior* — after receiving the laboratory results, the posterior
information about θ is expressed in terms of the posterior probability
mass function, $p(\theta|x)$.

In Figure 7.2 we give the posterior probability histogram for the
calendar date associated with sample OX-2575. This histogram should be
compared with the corresponding histogram of the prior which gives equal
probability to all 20-year periods, that is, the prior is flat. In contrast
the posterior has a distinct mode at around 700 BP (AD 1250) and
most of the probability is between about 900 BP (AD 1050) and 600 BP
(AD 1350). Based on the evidence of this radiocarbon determination we
can say that, *a posteriori*, the organic material used in the manufacture
of the Turin Shroud is highly likely to have died between AD 1050
and AD 1350. The posterior probability that it died before AD 1050
is virtually zero thereby casting considerable doubt on the Shroud's
authenticity.

7.1.3   The Bayesian framework

The Bayesian approach to statistics allows for the updating of information or
beliefs and provides a *formal* framework for it. Consequently, by adopting
the Bayesian approach, we explicitly express our prior beliefs, collect our

data and then combine these in a logical and coherent way to obtain our posterior beliefs. In other words, the Bayesian framework provides a simple, but formal way of dealing with life's ubiquitous problem of learning from experience. In operating in this way, Bayesians accept that opinions held by experts before the data are collected can, and should, have a bearing on the data investigation.

An advantage of adopting the Bayesian approach is that within the framework it is possible to combine two (or more) sources of information. For example consider the case of ordering archaeological deposits. In the chronological ordering of a number of archaeological contexts we may have good stratigraphic, typological or stylistic information, or radiocarbon dates which suggest that one or more of them cannot be older (or younger) than one or more of the others. Rowe (1961) discusses the relationship between archaeological stratigraphic evidence and the information obtained by the seriation of contexts based upon typological evidence. He concludes that the most useful inferences will be made where both types of information are combined. The Bayesian approach does indeed permit the combination of information from different sources, although before we can do so the information from both sources needs to be quantified in terms of probability mass (or density) functions.

To summarize, a main tenet of the Bayesian philosophy is that when prior information, knowledge or beliefs exist they should be explicitly stated, modelled and then used in a coherent fashion in the inference procedure. It would, after all, be impossible to learn from experience if we could never use previous knowledge to shed light on current investigations. Of course this raises the question of how this can be achieved. More specifically, how do we assess the prior information and how do we model it in terms of a prior probability mass (or density) function? How do we update the prior in the light of the data to get the posterior probability mass (or density) function? How do we turn our posterior into posterior information or beliefs about the situation at hand? Much of the remainder of the book is devoted to answering these questions. But we start by considering how to combine the prior information with that provided by the data using Bayes' theorem.

7.2   Mechanics of the Bayesian approach

7.2.1   Use of Bayes' theorem for events

In the previous example we have described the prior, the data, the likelihood and the posterior but how do we get from the prior, data and likelihood to the posterior? The key to the Bayesian methodology is Bayes' theorem which we introduced in Section 4.5.

Consider two events E_1 and E_2 and suppose that event E_2 has occurred: how should we alter our beliefs about event E_1 occurring? Recall that if E_1 and E_2 are events with $p(E_2) > 0$ then Bayes' theorem states that

$$P(E_1|E_2) = \frac{P(E_2|E_1)}{P(E_2)} P(E_1) \qquad (7.1)$$

or alternatively

$$P(E_1|E_2) \propto P(E_2|E_1)P(E_1). \qquad (7.2)$$

In (7.1) and (7.2) $P(E_1)$ can be thought of as the prior probability of event E_1 occurring; $P(E_1|E_2)$ can be interpreted as the posterior probability of event E_1 occurring given (or in the light of) the information that event E_2 has occurred. In this context $P(E_2|E_1)$ is known as the likelihood of E_1 occurring based upon the information that event E_2 has occurred.

It is important to note that $P(E_2|E_1)$ is the conditional probability of event E_2 occurring given that event E_1 has occurred — but E_2 has occurred and we are using $P(E_2|E_1)$ to say something about the likely occurrence of E_1 based upon this fact. To emphasize the distinction between the two interpretations, the likelihood of E_1 based upon the information that E_2 has occurred will be denoted by $l(E_1; E_2)$.

Recall that in Section 4.5.3 Bayes' theorem was extended as follows. Suppose we have n events denoted by $A_1, A_2, \ldots, A_n$ where the A_i are exclusive and exhaustive. Let E be another event such that $P(E) > 0$, then we have

$$P(A_i|E) \propto P(E|A_i)P(A_i).$$

In this context the A_i can be thought of as forming a set of hypotheses, of which one and only one is true. If event E is observed, then the prior probability of A_i, $P(A_i)$, changes to the posterior probability of A_i, $P(A_i|E)$. The term $P(E|A_i)$ is known as the likelihood of hypothesis A_i based upon the information that event E has occurred. We denote now this likelihood by $l(A_i; E)$. Therefore Bayes' theorem can be written as

$$P(A_i|E) \propto l(A_i; E)P(A_i).$$

7.2.2 Making inferences about parameters using Bayes' theorem

In Section 6.2.6 we demonstrated how Bayes' theorem can be applied to bivariate discrete random variables. Now we will show how the theorem is used to make inferences about unknown parameters. Consider a problem with one discrete parameter, θ, and one piece of data, x. Suppose that x can be thought of as a realization of a discrete random variable X. Let E_1 be the event that the parameter takes the value θ and let E_2 be the event that the random variable X takes the value x. So we replace E_1 by θ and E_2 by x in

Bayes' theorem as expressed in (7.1) to get

$$p(\theta|x) = \frac{p(x|\theta)}{p(x)}p(\theta). \tag{7.3}$$

where

$$p(x) = \sum_{\theta} p(x|\theta)p(\theta)$$

is the marginal probability mass function of x. We interpret $p(\theta)$ as the prior probability mass function of θ, $p(\theta|x)$ as the posterior probability mass function of θ and $p(x|\theta)$ as the conditional probability mass function of x given a specific value of the parameter, θ. However, in any particular situation, the data x is known and the parameter θ is unknown. Therefore, as we have done above, we think of $p(x|\theta)$ as the likelihood of θ, denoted by $l(\theta; x)$, based upon the information provided by observed data x. The term $p(x)$ can be thought of as the "average value of the likelihood $l(\theta; x)$" with the weight attached to $l(\theta; x)$ being given by the prior probability of θ, namely $p(\theta)$.

We note (7.3) may be written as

$$p(\theta|x) = \frac{l(\theta; x)}{p(x)}p(\theta) \tag{7.4}$$

or alternatively as

$$p(\theta|x) \propto l(\theta; x)p(\theta). \tag{7.5}$$

As we have seen in Section 6.3.4, Bayes' theorem can be applied to bivariate continuous random variables. Thus (7.4) and (7.5) hold for continuous X and θ; we interpret $p(\theta)$ as the prior probability density function of θ and $p(\theta|x)$ as the posterior probability density function of θ. In this context, the conditional probability density function of x given θ, $p(x|\theta)$, will be regarded as the likelihood of θ based on the data x. As X is continuous, the marginal probability density function of x can be expressed

$$p(x) = \int_{-\infty}^{\infty} p(x|\theta)p(\theta)\, d\theta.$$

Later in Section 6.4.2 Bayes' theorem was further extended to the multivariate case. In other words in (7.1) we can replace E_1 by $\boldsymbol{\theta}$, the vector of parameters $\theta_1, \ldots, \theta_k$, and E_2 by $\boldsymbol{x}$, the vector of data $x_1, \ldots, x_n$, and Bayes' theorem still is true. That is we have

$$p(\boldsymbol{\theta}|\boldsymbol{x}) = \frac{p(\boldsymbol{x}|\boldsymbol{\theta})}{p(\boldsymbol{x})}p(\boldsymbol{\theta}).$$

Let $l(\boldsymbol{\theta}; \boldsymbol{x})$ denote the likelihood of $\boldsymbol{\theta}$ based upon the data $\boldsymbol{x}$. By following a similar argument to that given above we obtain

$$p(\boldsymbol{\theta}|\boldsymbol{x}) = \frac{l(\boldsymbol{\theta}; \boldsymbol{x})}{p(\boldsymbol{x})}p(\boldsymbol{\theta}) \tag{7.6}$$

where

$$p(x) = \text{Average value of } \{l(\theta; x)\}$$

with the weight given to $l(\theta; x)$ being $p(\theta)$ (see Section 6.2.4). This relationship holds for discrete or continuous parameters or data, or even when some are discrete and the remainder are continuous. Of course care must be taken when applying such a general, but simply expressed, result to a particular problem.

To summarize: the likelihood, $l(\theta; x)$, should be thought of as a function of the unknown parameters, θ, and reflects the information about θ provided by the data, x. A relatively large value of the likelihood indicates that the corresponding value of θ is supported by the data, a relatively small value suggests relatively less support. There remains the question of how to calculate the likelihood in a particular problem. The answer is simple — the likelihood is taken to be proportional to the sampling distribution of the data, x, given the parameter values θ. Thus we have

$$l(\theta; x) \propto p(x|\theta)$$

where $p(x|\theta)$ is the joint probability mass (or density) function of x given θ.

7.2.3   Interpretation of Bayes' theorem

Sometimes, the first term of the right-hand side of (7.6), the likelihood, $l(\theta; x)$, divided by $p(x)$, is called the *standardized likelihood*. We can then express Bayes' theorem in words as

Posterior belief = Standardized likelihood × Prior belief.

How does Bayes' theorem work? Firstly, $p(\theta)$ represents the prior information about θ. We then observe the data x and Bayes' theorem tells us that the posterior information, represented by $p(\theta|x)$, is equal to the prior times the standardized likelihood, $l(\theta; x)/p(x)$. By the result of this operation, we have combined two sources of information, that provided by the prior and that provided by the data.

How do we interpret these various statements of Bayes' theorem? The posterior probability of a particular value of θ will be low if either the prior probability or the standardized likelihood (or both) are low for that value of θ. Really this is saying the particular value of θ is not supported by either the prior information or the data. If a value of θ is well supported by both the prior information and the data, i.e. both the prior probability and the standardized likelihood are relatively high, then the posterior probability of that value will be relatively high.

Finally we note that (7.6) can be expressed as

$$p(\theta|x) \propto l(\theta; x)p(\theta) \tag{7.7}$$

or in words:

The posterior is proportional to the likelihood times the prior.

Thus the posterior can be obtained by multiplying the likelihood by the prior and then the result is normalized so that, in the discrete case, the posterior probabilities add up to 1, or, in the continuous case, the posterior density integrates to 1. In many of the examples that follow in this and subsequent chapters, we will use (7.2), (7.5) or (7.7) to find an expression for the posterior, up to a constant of proportionality. That is, by using one of those equations we can obtain an expression that the posterior is *proportional to*, but there still remains the problem of determining what we shall refer to as the *normalizing constant* which we denote by c. This normalizing constant satisfies the equation

$$p(\boldsymbol{\theta}|\boldsymbol{x}) = c\,l(\boldsymbol{\theta};\boldsymbol{x})p(\boldsymbol{\theta})$$

and so, by using (7.6), is equal to $1/p(\boldsymbol{x})$.

7.3   From prior to posterior

In the Turin Shroud example given above we have deliberately avoided giving the mathematical details of the analysis mainly because, despite being a rather simple archaeological example, the mathematics is quite complicated. Instead we have provided a histogram of the posterior and have tried to give an intuitive explanation of what is going on. However we cannot avoid the mathematics any longer. Consequently we are now going to present a series of examples of increasing complexity to give the reader a better idea of the mathematics involved.

7.3.1   Example 1 — Case I

Suppose we wish to estimate the prevalence of a non-contagious disease (it might be sickle cell anaemia) in an ancient society by using data arising from a large cemetery. The identification of the disease from skeletons is difficult and needs expensive scientific analysis, hence it is inconceivable that all the skeletons are analysed. Suppose that five skeletons are selected at random for scientific analysis and of these three are found to have the disease. How can we use this to estimate the incidence rate in the whole population? A crude estimate is that about 3/5 or 60% of the population suffered from the disease. This is based solely upon the data and so does not take into account any other evidence.

Let θ represent the unknown incidence rate in the whole population at the time under study. Although the disease is not contagious, there may be some genetic factor involved which means that individuals who are close blood relations may be more susceptible than those who are not. However let us agree that determining blood relationships is currently impossible (although the use of DNA fingerprinting techniques in the future may mean that this might not always be the case).

Suppose that the sample can be considered to be representative so we are able to ensure as far as possible that the skeletons can be considered exchangeable (see Section 4.6). Therefore the probability that a particular individual has the disease is θ. Also the probability that they did not suffer from the disease is $1 - \theta$. The likelihood, $l(\theta; x)$, that 3 out of the 5 have the disease and consequently 2 do not is $\theta \times \theta \times \theta \times (1 - \theta) \times (1 - \theta)$ and is given by

$$l(\theta; x) = \theta^3 (1 - \theta)^2.$$

How do we express our prior information about θ? That depends upon what information we have available at the time. Strictly we ought to decide upon this *before* collecting the samples and certainly *before* seeing the laboratory report. We need to ask ourselves or perhaps relevant subject-matter experts: before seeing the data, what do we (you) know about the incidence rate of this disease at that period? One reply could be that there is no information available in which case θ is thought *a priori* equally likely to take any value between 0 and 1. In other words the prior probability density function for θ is a Uniform distribution between 0 and 1. If this is the case then

$$p(\theta) = 1 \quad \text{for } 0 \leq \theta \leq 1.$$

Hence our posterior beliefs about the incidence rate (after seeing the data) are expressed in terms of the posterior distribution of θ which is given by

$$p(\theta|x) \propto l(\theta; x) p(\theta).$$

Thus, based upon no prior information regarding θ, we have the posterior of θ expressed as

$$p(\theta|x) \propto \theta^3 (1 - \theta)^2 \quad \text{for } 0 \leq \theta \leq 1. \tag{7.8}$$

There just remains the task of evaluating the normalizing constant, c, which satisfies the equation

$$p(\theta|x) = c\,\theta^3 (1 - \theta)^2.$$

This we can do by recognizing that the posterior probability density function $p(\theta|x)$ has the form of the probability density function of a Beta random variable. Recall from Section 5.10.2 that the random variable Y is said to have a Beta distribution with parameters α and β if its probability density function is

$$\frac{\Gamma(\alpha + \beta)}{\Gamma(\alpha)\Gamma(\beta)} y^{\alpha-1} (1 - y)^{\beta-1} \quad \text{for } 0 \leq y \leq 1. \tag{7.9}$$

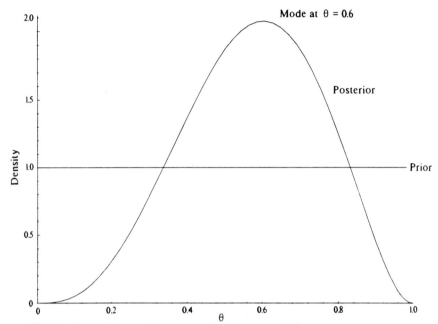

Figure 7.3   Prior and posterior densities of θ in Example 1 — Case I.

By comparing (7.8) and (7.9) we see that $\alpha = 4$ and $\beta = 3$ and that the normalizing constant, c, is equal to

$$\frac{\Gamma(\alpha + \beta)}{\Gamma(\alpha)\Gamma(\beta)} = \frac{\Gamma(4 + 3)}{\Gamma(4)\Gamma(3)}.$$

Note that if α is an integer then $\Gamma(\alpha) = (\alpha - 1)!$ and therefore c is equal to

$$\frac{6!}{3!2!} = \frac{6 \times 5 \times 4 \times 3 \times 2 \times 1}{(3 \times 2 \times 1) \times (2 \times 1)} = 60.$$

Hence the posterior distribution of θ is given by

$$p(\theta|x) = 60\,\theta^3(1 - \theta)^2 \quad \text{for } 0 \le \theta \le 1.$$

In Figure 7.3 we give graphs of the prior and posterior distributions of θ. Notice that the prior is a horizontal straight line indicating that all values of θ were thought, a priori, to be equally likely. In contrast the posterior has a distinct mode (peak) at $\theta = 0.6$ indicating, a posteriori, that the most likely value of θ is 0.6, that is we believe that the incidence rate is mostly likely to be 60%.

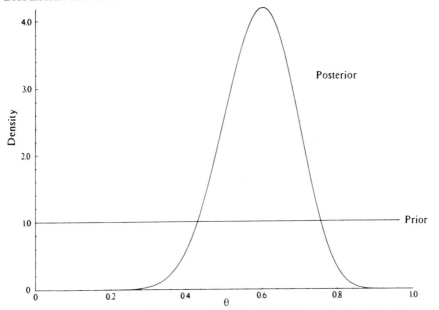

Figure 7.4 Prior and posterior densities of θ in Example 1 — Case II.

7.3.2 Example 1 — Case II

Suppose that more funds had been available so that more skeletons had been analysed. Assume that instead of 5 skeletons, 25 had been analysed and of these 15 were found to have the disease. Thus the proportion identified as having the disease remained the same as before. Using the same Uniform prior for θ, the posterior is easily shown to be

$$p(\theta|x) = 84\,987\,760\,\theta^{15}(1-\theta)^{10} \quad \text{for } 0 \le \theta \le 1.$$

In this case the normalizing constant was calculated by evaluating

$$\frac{\Gamma(27)}{\Gamma(16)\Gamma(11)} = \frac{26!}{15!10!} = 84\,987\,760.$$

The mode of the posterior is still at $\theta = 0.6$ corresponding to a 60% incidence rate for the disease but the distribution is much more tightly concentrated about this value than previously (see Figure 7.4).

7.3.3 Example 1 — Case III

Consider now the same situation as Case I but with a different, more informative prior. Suppose that the analysis of skeletons from other sites in the same broad region and of a similar period suggested that the incidence rate probably lies in the range 20% to 80%. If we interpret this as meaning that our prior belief is that the incidence rate lies between 20% and 80% with probability 0.8 then a possible prior for θ is given by

$$p(\theta) = 30\,\theta^2(1-\theta)^2 \quad \text{for } 0 \le \theta \le 1$$

since for this distribution about 80% of the probability lies between 0.2 and 0.8. Therefore, with this prior information the posterior for θ is proportional to

$$\theta^3(1-\theta)^2\theta^2(1-\theta)^2 = \theta^5(1-\theta)^4 \quad \text{for } 0 \le \theta \le 1.$$

Again we notice that the right-hand side of this equation has the form of a Beta distribution and hence deduce that the normalizing constant is

$$\frac{\Gamma(11)}{\Gamma(6)\Gamma(5)} = \frac{10!}{5!4!} = 1260.$$

Graphs of the prior, standardized likelihood and posterior are shown in Figure 7.5. Notice the posterior lies "between" the prior and the likelihood. It is a sort of compromise of the two, as one would expect. The mode of the posterior is at $\theta = 0.56$, so that the most likely value of the incidence rate is about 56%.

We see that the posterior is a compromise of the likelihood and the prior. If our sample had been larger, more weight would have been given to the data than to the prior and so the posterior would be closer to the likelihood than the prior. As our sample size increases, the data dominate more and more — a perfectly natural way of operating.

7.4   Summarizing the posterior

Bayes' theorem gives us a method for calculating a mathematical expression for the posterior probability mass (or density) function. How should we convey its meaning to the non-statistical reader? In a way we need to be able to invert the process we have already followed of converting prior information into a mathematical form expressed as a prior probability mass (or density) function. Instead we need to summarize the posterior probability mass (or density) function in an everyday form, readily understandable by the non-statistician.

7.4.1   Measures of posterior information

The best way to proceed is to treat the posterior distribution as any other distribution. In Chapter 5 we described common ways of summarizing distributions both discrete and continuous. Therefore all we need to do is to calculate and report the appropriate summary measures of the posterior such

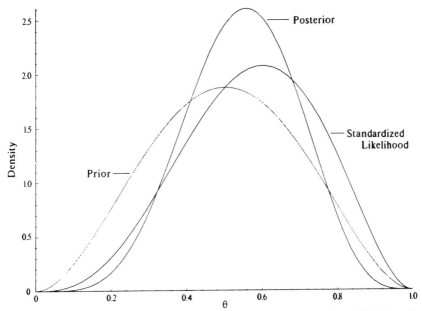

Figure 7.5   Prior and posterior densities of θ and the standardized likelihood in Example 1 — Case III.

as the *posterior mean, posterior mode, posterior median, posterior standard deviation* and the like. Of course some readers of the reports will be unfamiliar with the terminology and this must be borne in mind when writing the reports.

If we return to Example 1 — Case I, the posterior distribution of θ has a Beta distribution with parameters $\alpha = 4$ and $\beta = 3$. Therefore referring to Table 5.4 (page 108) we have

$$\text{posterior mean} = \frac{\alpha}{\alpha+\beta} = \frac{4}{4+3} = 0.57$$

$$\text{posterior mode} = \frac{\alpha-1}{\alpha+\beta-2} = \frac{3}{5} = 0.60$$

$$\text{posterior standard deviation} = \sqrt{\frac{\alpha\beta}{(\alpha+\beta)^2(\alpha+\beta+1)}} = \sqrt{\frac{12}{49\times8}} = 0.17.$$

If, as in Example 1 — Case II, 25 skeletons had been examined of which 15 had the disease, corresponding to $\alpha = 16$ and $\beta = 11$, we have

$$\text{posterior mean} = \frac{16}{16+11} = 0.59$$

$$\text{posterior mode} = \frac{15}{25} = 0.60$$

$$\text{posterior standard deviation} = \sqrt{\frac{176}{729\times28}} = 0.09.$$

As we can see, examining 5 or 25 skeletons but having the same proportion with the disease results in the mode being the same. The posterior mean

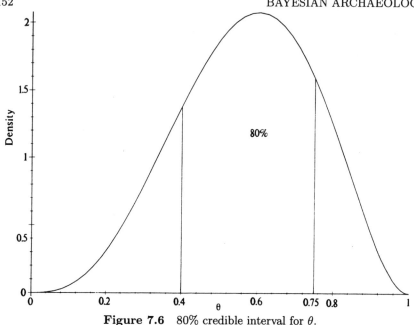

Figure 7.6   80% credible interval for θ.

is affected slightly whereas the standard deviation using the large sample is about half that of the former. This reflects the greater knowledge that has accrued from a larger sample, that is to say more precise inferences can be made. Even so we still need to consider how to report a likely range of values for the incident rate and we turn to this in the next section.

7.4.2   Credible intervals and highest posterior density regions

Summarizing a posterior distribution by its posterior mean, mode or other similar measures provides only part of the picture. It is far better to report where most of the posterior probability lies, that is where we have the greatest belief in the values of θ. To do so, Bayesians use the concepts of *credible intervals* and *highest posterior density regions*.

Credible intervals are obtained directly from the posterior distribution and so are based upon the current knowledge. With a credible interval we refer directly to the posterior probability of θ being in a particular interval conditional upon the data and the prior information. For example, we may wish to know what is the posterior probability that more than 40% but less than 75% of the population suffered from the disease. In other words we need to evaluate the posterior probability that θ is greater than 0.4 and less than 0.75; this is represented by the area under the posterior probability density function between $\theta = 0.4$ and $\theta = 0.75$ (see Figure 7.6).

For Example 1 — Case I, this can be evaluated (using calculus) to be 0.8. Hence we can say that, *a posteriori*, we believe that θ lies between 0.4 and 0.75 with probability 0.8. The interval 0.4 to 0.75 is called an *80% credible interval*. A 49% credible interval would be from 0.5 to 0.75; a 93% interval would be from 0.3 to 0.95.

However, notice from Figure 7.6 that the density at $\theta = 0.41$ is equal to $60 \times (0.41)^3 \times (0.59)^2 = 1.4395$ whereas at $\theta = 0.76$ the value of the density is $60 \times (0.76)^3 \times (0.24)^2 = 1.51710$. The former value is in our 80% credible interval and the latter is not. In simple terms this means that some values of θ that are *outside* our credible interval are *more likely* to occur than other values that are in the interval. A strange state of affairs!

Moreover, if we calculate the probability that θ lies between 0.2 and 0.74 we find that it is equal to 0.80 also. Both intervals, 0.46 to 0.75 and 0.26 to 0.74, express the same confidence in θ but they are different. Note that the first is shorter than the second. Another 80% credible interval is from 0.0 to 0.73. Which is the best one to report? By and large it is preferable to have as short an interval as possible as then we can make the sharpest possible inferences.

To summarize: we would like to report intervals that

(i) are, for a given probability level, as short as possible and
(ii) contain values of the parameter that are more likely than those outside the interval.

Is it possible to calculate intervals that have both these properties? Fortunately the answer is yes and, after some reflection, it should be clear to the reader that the two conditions are equivalent.

Bayesians call such intervals *highest posterior density* (HPD) regions. Consider a uni-modal (one peak) posterior density. The interval from a to b is a 90% HPD interval if the posterior probability that θ lies between a and b is equal to 0.90 and, for any θ between a and b, the density is higher than for any θ outside the interval. In some special cases, the HPD region may be easily calculated using pencil and paper but, more often than not, we have to resort to using a computer.

Returning to Example 1 — Case I, the 90% HPD interval is 0.29 to 0.86 (see Figure 7.7). This may appear quite a wide range but this is not surprising as only five skeletons were examined. For Case II, the corresponding interval is from 0.39 to 0.78 (see Figure 7.8). For Case III, a 90% interval estimate would be from about 31% to 78% (see Figure 7.9).

Here we have calculated credible intervals and HPD regions for a uni-modal distribution. For multi-modal distributions, like those arising in radiocarbon dating, we sometimes obtain the HPD region being composed of a collection of disjoint intervals, that is to say the HPD region consists of two or more intervals that do not overlap.

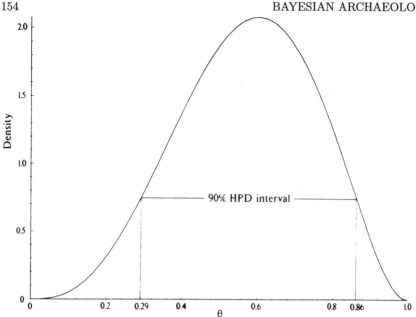

Figure 7.7   90% highest posterior density interval for Example 1 — Case I.

7.4.3   Example 2

In the previous analyses of the data from Consumerland (see Sections 4.6.3 and 5.7) we have taken a very simplistic view. For example, it is unlikely that any expert would be willing to state figures as precise as 80% for the proportion of mugs in a bar and 40% for a restaurant. More realistically they may believe that these are the most likely (modal) values but would also be prepared to accept that the proportions would vary from one establishment to another depending upon local customs and preferences. The question is "how should this variability be assessed and modelled?"

In our case, the expert not only was regarded as one of the world's authorities on the decoding of *The Book of the Golden Pages* but had been the director of several similar earlier excavations. Even so the expert found it difficult, despite all their knowledge about the customs and ritualistic practices of the society, to assess how the proportions varied from bar to bar and from restaurant to restaurant. Eventually after much thought the expert came up with the following.

For a bar, the expert believed that the proportion would hardly ever rise above 95% and was unlikely to fall much below 65%. For a restaurant the proportion was felt to vary between 20% and 60%. How confident was our expert about these "informed guesses"? When pressed by the statistician, the expert admitted that they were not totally confident that the proportions

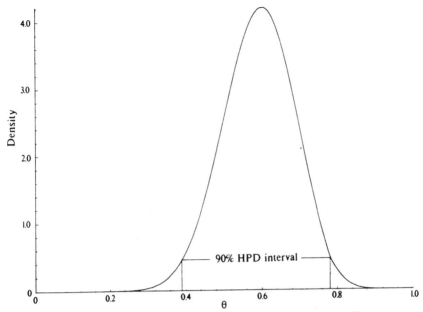

Figure 7.8   90% HPD interval for Example 1 — Case II.

must be inside these ranges. They were prepared to admit that whereas they were 90% sure of the range for a bar they were less confident for a restaurant as there appeared to be several different subtypes in the region. Consequently the expert was only 80% confident for the range for restaurants. Moreover within the ranges he felt that the proportions were not distributed uniformly but fell away to either side of the modal values given above.

Faced with this sort of quantitative information, the statistician needs to convert it into a more mathematical form. A natural way to proceed for a Bayesian is to model the variability in the proportion of mugs in a bar by one Beta distribution and use another (different) Beta distribution for a restaurant.

Let $p_B(\theta)$ be the probability density function representing the expert's beliefs about the proportion of mugs, θ, in a bar. Using a Beta distribution to model these beliefs, we then have

$$p_B(\theta) = \frac{\Gamma(\alpha_1 + \beta_1)}{\Gamma(\alpha_1)\Gamma(\beta_1)}\theta^{\alpha_1 - 1}(1 - \theta)^{\beta_1 - 1} \quad \text{for } 0 \leq \theta \leq 1 \qquad (7.10)$$

where α_1 and β_1 are parameters that need to be determined. Similarly, to model the expert's belief's about restaurants we have

$$p_R(\theta) = \frac{\Gamma(\alpha_2 + \beta_2)}{\Gamma(\alpha_2)\Gamma(\beta_2)}\theta^{\alpha_2 - 1}(1 - \theta)^{\beta_2 - 1} \quad \text{for } 0 \leq \theta \leq 1$$

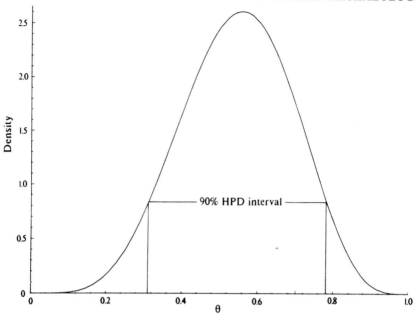

Figure 7.9   90% HPD interval for Example 1 — Case III.

where α_2 and β_2 are parameters to be determined. We will use the expert's opinions as expressed above to determine which values should be given to α_1, β_1, α_2 and β_2.

For a bar, the expert's prior beliefs can be interpreted as

$$P(0.65 \leq \theta \leq 0.95) = \int_{0.65}^{0.95} p_B(\theta)\, d\theta = 0.9$$

and for a restaurant

$$P(0.2 \leq \theta \leq 0.6) = \int_{0.2}^{0.6} p_R(\theta)\, d\theta = 0.8.$$

In addition, we have the information that the modes are at 0.8 and 0.4 for bar and restaurant respectively. All this information can be used to determine (possibly using a computer) appropriate values of α_1, β_1, α_2 and β_2.

For a bar, if we set $\alpha_1 = 13$ and $\beta_1 = 4$ then we have the mode of the distribution at $\theta = 0.8$ (as required) and a highest density interval given by

$$P(0.61 \leq \theta \leq 0.93) = 0.9.$$

The interval 0.61 to 0.93 is not exactly the same as the expert's 0.65 to 0.95 but is fairly close.

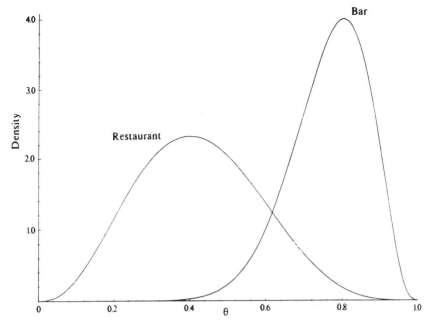

Figure 7.10 Prior densities representing the expert's beliefs about the proportion of mugs in a bar and a restaurant.

For the restaurant, setting $\alpha_2 = 11/3$ and $\beta_2 = 5$ gives the mode at 0.4 and

$$P(0.20 \leq \theta \leq 0.62) = 0.8$$

which we believe is an adequate representation of the prior judgement of the expert (0.20 to 0.60). Plots of the densities are given in Figure 7.10. As can be seen, although the two modes are quite distinct, the two distributions do overlap. The expert believes that it is possible for a bar to have a proportion of mugs as low as 0.6 or even lower; also it is possible for a restaurant to have a proportion of 0.6 or higher. In other words classifying a site on the basis of the numbers of mugs and goblets may lead to erroneous conclusions. However if that is all the excavator finds, then we must do the best we can.

How do we use this extra information? Recall that we use the notation B and R to represent the event that the site was a bar or restaurant respectively. Also X is the random variable used to represent the number of mugs found. The posterior probability that the site is a bar given we find n glasses of which x are mugs is denoted by $P(B|X = x)$. As before, by using Bayes' theorem we have

$$P(B|X = x) \propto P(X = x|B)P(B)$$

where $P(B)$ is the prior probability that it is a bar.

We turn our attention to calculating $P(X = x|B)$. Given that the site is a bar and the proportion of mugs is fixed at θ, X has a Binomial distribution and so the probability that x of the n glasses are mugs is

$$P(X = x|B \text{ and } \theta) = \binom{n}{x}\theta^x(1 - \theta)^{n-x}. \tag{7.11}$$

We need to "average" this probability over all values of θ, with the "weight" given to $P(X = x|B \text{ and } \theta)$ being in accord with the prior beliefs about that value of θ, that is by $p_B(\theta)$. This is expressed mathematically by the following equation:

$$P(X = x|B) = \int_0^1 P(X = x|B \text{ and } \theta)p_B(\theta)\,d\theta. \tag{7.12}$$

By substituting in (7.12) for $p_B(\theta)$ and $P(X = x|B \text{ and } \theta)$ from (7.10) and (7.11) respectively, we get

$$P(X = x|B) = \int_0^1 \binom{n}{x}\theta^x(1 - \theta)^{n-x}\frac{\Gamma(\alpha_1 + \beta_1)}{\Gamma(\alpha_1)\Gamma(\beta_1)}\theta^{\alpha_1-1}(1 - \theta)^{\beta_1-1}\,d\theta$$

and then, using the result of Section 5.10.2, we obtain

$$P(X = x|B) = \binom{n}{x}\frac{\Gamma(x + \alpha_1)\Gamma(n - x + \beta_1)}{\Gamma(n + \alpha_1 + \beta_1)}\frac{\Gamma(\alpha_1 + \beta_1)}{\Gamma(\alpha_1)\Gamma(\beta_1)}.$$

In this particular example we have $n = 5$, $x = 2$, $\alpha_1 = 13$ and $\beta_1 = 4$ so that

$$P(X = 2|B) = \frac{5!}{3!2!}\frac{\Gamma(15)\Gamma(7)}{\Gamma(22)}\frac{\Gamma(17)}{\Gamma(13)\Gamma(4)} = \frac{7280}{81\,396} = 0.0894.$$

Similarly, given that the site is a restaurant, we get

$$P(X = x|R) = \binom{n}{x}\frac{\Gamma(x + \alpha_2)\Gamma(n - x + \beta_2)}{\Gamma(n + \alpha_2 + \beta_2)}\frac{\Gamma(\alpha_2 + \beta_2)}{\Gamma(\alpha_2)\Gamma(\beta_2)}.$$

With $\alpha_2 = 11/3$ and $\beta_2 = 5$ we obtain

$$P(X = 2|R) = \frac{5!}{3!2!}\frac{\Gamma(17/3)\Gamma(8)}{\Gamma(41/3)}\frac{\Gamma(26/3)}{\Gamma(11/3)\Gamma(5)} = \frac{59\,049}{44\,811\,728}\times\frac{150\,535}{729} = 0.2721.$$

Recall that, a priori, our expert thought that the site was twice as likely to be a bar than a restaurant, that is

$$P(B) = \frac{2}{3} \text{ and } P(R) = \frac{1}{3}.$$

Hence the posterior probability that the building is a bar is given by

$$P(B|X = 2) \propto P(X = 2|B)P(B) = 0.0894 \times \frac{2}{3} = 0.0596$$

and the posterior probability that it is a restaurant is

$$P(R|X = 2) \propto P(X = 2|R)P(R) = 0.2721 \times \frac{1}{3} = 0.0907.$$

Since it is either a bar or a restaurant we have

$$P(B|X = 2) = \frac{0.0596}{0.0596 + 0.0907} = 0.40$$

and

$$P(R|X = 2) = 1 - 0.40 = 0.60.$$

That is to say, based on the evidence of two mugs and three goblets and the expert's prior knowledge, we believe, a *posteriori*, that there is about a 60% chance that the site is a restaurant and a 40% chance that it is a bar. Notice that using this more sophisticated model the chance that the building is a restaurant is 60% which is below the 77% chance we calculated using the simpler model of Section 4.6.3. This has arisen despite incorporating more knowledge into the analysis. Naively one would expect a "better", more precise answer using a more complex model. In fact this may happen in some instances. But in this case, the more sophisticated model of the prior has actually captured more of the expert's uncertainty about the problem and this is reflected in the final results and conclusions. The new model blurs the distinction between a bar and a restaurant made in the simplistic analysis — as a result we should not expect to get such a clear-cut answer.

7.5 Bayes' theorem applied to multi-parameter problems

Up to now we have considered the Bayesian analysis of problems with a single parameter which we have denoted by θ. However most models of real-life situations have several parameters. Suppose we have a model with k parameters, which we denote by θ_1, θ_2 and so on up to θ_k. We represent the collection of all these parameters by $\boldsymbol{\theta}$. If we return to Bayes' theorem as expressed in (7.7) we see that it applies when we have more than one parameter. That is

$$p(\boldsymbol{\theta}|\boldsymbol{x}) \propto l(\boldsymbol{\theta}; \boldsymbol{x})p(\boldsymbol{\theta}).$$

Consequently when we have more than one parameter we proceed in exactly the same manner as before. We have to specify a likelihood function which specifies how the data, $\boldsymbol{x}$, are related to the parameters, $\boldsymbol{\theta}$, of the model. We have to represent our prior information or prior beliefs about these parameters by a joint prior, denoted by $p(\boldsymbol{\theta})$. Bayes' theorem gives us a method for calculating the joint posterior of $\boldsymbol{\theta}$. Finally this joint posterior has to be interpreted.

7.5.1   Formulating the likelihood and prior

There are no hard and fast rules that can followed in specifying the likelihood and eliciting the prior information. Both operations are very much dependent upon the problem in hand but some points to look out for include:

(i) Are the data, $x_1, x_2, \ldots, x_n$, conditionally independent of each other given $\boldsymbol{\theta}$? (See the concept of conditional independence introduced in Section 6.4.1.) If they are, the likelihood based on $\boldsymbol{x}$ is the product of the likelihoods based on the individual data values. That is

$$l(\boldsymbol{\theta}; \boldsymbol{x}) = l(\boldsymbol{\theta}; x_1) l(\boldsymbol{\theta}; x_2) \cdots l(\boldsymbol{\theta}; x_n).$$

(ii) Are the x_i independent and does x_i only depend upon θ_i? In this case the likelihood may be expressed as

$$l(\boldsymbol{\theta}; \boldsymbol{x}) = l(\theta_1; x_1) l(\theta_2; x_2) \cdots l(\theta_n; x_n).$$

(iii) Are the parameters $a\ priori$ independent of each other? If so the joint prior will be the product of the priors of the individual parameters so that we have

$$p(\boldsymbol{\theta}) = p(\theta_1) p(\theta_2) \ldots p(\theta_k).$$

(iv) If there are just two parameters, θ_1 and θ_2, then their joint prior can be expressed in the form

$$p(\theta_1, \theta_2) = p(\theta_1 | \theta_2) p(\theta_2)$$

which may prove useful in the elicitation process.

The possibilities are endless. In later chapters we will analyse case studies in which we hope to give the reader a feel for how we have arrived at particular models, likelihoods and priors.

7.5.2   The posterior and its interpretation

By using Bayes' theorem the joint posterior of $\theta_1, \ldots, \theta_k$, $p(\boldsymbol{\theta}|\boldsymbol{x})$, is simply calculated as the likelihood, $l(\boldsymbol{\theta}; \boldsymbol{x})$, multiplied by the joint prior, $p(\boldsymbol{\theta})$. It is how to interpret the joint posterior that is our particular concern here. We have seen in Chapter 6 how to summarize and interpret joint probability distributions, and we need to apply the same techniques to the joint posterior. However the interpretation of a multi-dimensional probability distribution can be difficult and is a skill that comes from long experience.

To start with it is reasonable to look at the parameters one by one. That is we calculate the marginal posterior distribution of θ_1, then of θ_2 and so on. These marginals tell us how each individual parameter behaves by "averaging" out the influence of the remaining parameters. However the marginals give no information as to how the parameters are related to or interact with each other. To examine any interaction, we can look at the joint posterior distribution of each pair of parameters. With k parameters we must examine $k(k-1)/2$ joint distributions. If we want to study how three parameters interact then we must look at and interpret three joint posterior distributions — not an easy task for even an experienced statistician.

7.5.3   Analysis of a simplified version of St Veit-Klinglberg

To illustrate the Bayesian analysis of an archaeological problem involving more than one parameter we return to the St Veit-Klinglberg example introduced in Section 1.1.1. In Section 2.2.2 we developed a mathematical model for the problem and in Section 2.4.2 we modelled the prior information. As set up in those sections, the mathematical model of St Veit-Klinglberg has ten parameters of interest but that is far too many for illustrative purposes. To reduce the model to a more manageable size we discuss the interpretation of just three parameters.

The archaeological problem

We consider the dating of only three contexts 358, 813 and 1210 from St Veit-Klinglberg. For illustrative purposes we are deliberately ignoring any information regarding the other contexts and how they relate to the chosen contexts. We also ignore radiocarbon determinations from the remaining seven contexts.

The model

Using the notation introduced in Section 2.2.2, θ_i represents the (unknown) calendar date of the organic material found in context i. Thus θ_{358}, θ_{813} and θ_{1210} denote the calendar dates of the organic material sampled from contexts 358, 813 and 1210 respectively.

Let x_{358} denote the radiocarbon determination for context 358 and σ_{358} the standard deviation reported by the radiocarbon laboratory. Now x_{358} is a realization of a random variable, denoted by X_{358}. As we will see in Section 9.2.4, a suitable statistical model for relating X_{358} and the unknown calendar date, θ_{358}, is that X_{358} has a Normal distribution with a mean which is dependent upon the calendar date, θ_{358}, and the standard deviation set equal to the value reported by the laboratory, σ_{358}. The dependence of the mean on the calendar date is represented by $\mu(\theta_{358})$ where $\mu(\cdot)$ denotes the calibration curve. Written mathematically we have

$$X_{358}|\theta_{358} \sim N(\mu(\theta_{358}),\ \sigma_{358}^2).$$

By a similar argument we have

$$X_{813}|\theta_{813} \sim N(\mu(\theta_{813}),\ \sigma_{813}^2)$$

and

$$X_{1210}|\theta_{1210} \sim N(\mu(\theta_{1210}),\ \sigma_{1210}^2).$$

From Section 5.10.3 we see that the probability density function of X_{358} is given by

$$\frac{1}{\sigma_{358}\sqrt{2\pi}} \exp\left\{-\frac{(x_{358}-\mu_{358}(\theta_{358}))^2}{2\sigma_{358}^2}\right\} \quad \text{for } -\infty < x_{358} < \infty \qquad (7.13)$$

and similar expressions hold for the densities of X_{813} and X_{1210}.

The data

The radiocarbon determinations for St Veit-Klinglberg have already been given in Table 1.1. Only three concern us here: $3340 \pm 80, 3270 \pm 75$ and 3200 ± 70 for contexts 358, 813 and 1210 respectively. Thus we have $x_{358} = 3340$, $\sigma_{358} = 80$, $x_{813} = 3270$, $\sigma_{813} = 75$, $x_{1210} = 3200$ and $\sigma_{1210} = 70$.

The prior

As we have said in Section 2.4.2, there is stratigraphic evidence that the organic material sampled from context 813 was deposited before that in context 1210 but after that in context 358. Since all the dates are measured BP we have the following set of inequalities relating the parameters of the model to each other:

$$\theta_{358} > \theta_{813} > \theta_{1210}.$$

The above mathematical relationship summarizes the prior archaeological information. So long as the dates satisfy it the archaeologist will be contented. However, dates that do not satisfy these constraints would conflict with prior evidence and therefore are impossible on archaeological grounds. Apart from this information about the order of the dates, nothing else is known.

For the moment, let us suppose that the dates are modelled as being discrete. One way of modelling the prior information is to say that any set of dates which satisfies the stratigraphic sequence is equally likely; any set of dates which does not satisfy the stratigraphic sequence is considered to be impossible and therefore is given a prior probability of zero. For example, the set of dates $\theta_{358} = 2600$, $\theta_{813} = 2500$ and $\theta_{1210} = 2700$ will have a probability of zero since $\theta_{358} < \theta_{1210}$ is in contradiction of the archaeological information. On the other hand, the set $\theta_{358} = 3000$, $\theta_{813} = 2900$ and $\theta_{1210} = 2800$ is a possibility, as is the set $\theta_{358} = 3100$, $\theta_{813} = 2900$ and $\theta_{1210} = 2856$. Both sets of dates satisfy the constraints imposed by the archaeology but, a priori, we have no preference between the two. As a consequence we take them to be equally likely. For practical purposes, θ_{1210} must be greater than zero, and it is convenient, although not essential, to place some arbitrary large upper limit on the possible value of θ_{358}, say $\theta_{358} < 4000$. Therefore, the joint prior probability mass function of $\boldsymbol{\theta} = (\theta_{358}, \theta_{813}, \theta_{1210})$ may be expressed as

$$p(\boldsymbol{\theta}) = p(\theta_{358}, \theta_{813}, \theta_{1210}) = \begin{cases} c & \text{for } 0 < \theta_{1210} < \theta_{813} < \theta_{358} < 4000 \\ 0 & \text{otherwise} \end{cases}$$

where c is the normalizing constant.

We can express $p(\boldsymbol{\theta})$ more succinctly by using the following notation which will also be of use later in the book. Let A represent the set or collection of dates satisfying

$$4000 > \theta_{358} > \theta_{813} > \theta_{1210} > 0.$$

Then we say that

$$p(\boldsymbol{\theta}) \propto \mathrm{I}_A(\boldsymbol{\theta})$$

with

$$\mathrm{I}_A(\boldsymbol{\theta}) = \begin{cases} 1 & \boldsymbol{\theta} \in A \\ 0 & \boldsymbol{\theta} \notin A \end{cases}$$

where "$\boldsymbol{\theta} \in A$" should be interpreted as the dates satisfying the above inequality, and "$\boldsymbol{\theta} \notin A$" as the dates not satisfying the above inequality.

If we model $\boldsymbol{\theta}$ as being continuous, a similar argument will apply and so the joint prior probability density function of $\boldsymbol{\theta}$ can be represented as

$$p(\boldsymbol{\theta}) \propto \mathrm{I}_A(\boldsymbol{\theta}).$$

The likelihood

The likelihood based on all the radiocarbon determinations is given by

$$l(\boldsymbol{\theta}; \boldsymbol{x}) = l(\theta_{358}, \ \theta_{813}, \ \theta_{1210}; x_{358}, \ x_{813}, \ x_{1210}).$$

Since, conditional on the θ_i, the X_i are independent, we have that

$$l(\theta_{358}, \ \theta_{813}, \ \theta_{1210}; x_{358}, \ x_{813}, \ x_{1210}) = l(\theta_{358}; x_{358})l(\theta_{813}; x_{813})l(\theta_{1210}; x_{1210}).$$

That is the joint likelihood is the product of the individual likelihoods. Furthermore, we have $X_i|\theta_i \sim N(\mu(\theta_i), \sigma_i^2)$. Hence, using (7.13), the likelihood of X_{358} is

$$l(\theta_{358}; x_{358}) \propto \exp\left\{ -\frac{(x_{358} - \mu_{358}(\theta_{358}))^2}{2\sigma_{358}^2} \right\}.$$

Therefore the (joint) likelihood of x_{358}, x_{813} and x_{1210} is given by

$$l(\boldsymbol{\theta}; \boldsymbol{x}) \propto \prod_i \exp\left\{ -\frac{(x_i - \mu(\theta_i))^2}{2\sigma_i^2} \right\}$$

where the product is over $i = 358, 813$ and 1210.

The posterior

By Bayes' theorem, the joint posterior probability density function of $\boldsymbol{\theta}$ is given by

$$p(\boldsymbol{\theta}|\boldsymbol{x}) \propto \left\{ \prod_i \exp\left\{ -\frac{(x_i - \mu(\theta_i))^2}{2\sigma_i{}^2} \right\} \right\} I_A(\boldsymbol{\theta}) \qquad (7.14)$$

where the product is over $i = 358$, 813 and 1210.

The calculations

The joint posterior given in (7.14) is a complicated mathematical expression involving the three parameters of interest. Ideally we want to visualize how the posterior varies as the three parameters take different values but in practice this is not feasible as in order to view the posterior we need four dimensions.

What we can do is divide the calendar time-scale into 100-year blocks from 4000 BP to 3000 BP. The choice of 100-year blocks is rather arbitrary; a smaller block size may be more difficult to interpret; a larger block size may not give sufficient detail. We then evaluate the posterior using numerical methods (the details of which need not concern us here — they will be described in Sections 8.5 and 8.6).

The results

In Table 7.1 we give the joint posterior probability mass function of θ_{358}, θ_{813} and θ_{1210}. Even when the problem is simplified as much as it has been the actual interpretation is not easy. However a detailed examination of the results reveals that there is, *a posteriori*, a 61% chance that θ_{358}, θ_{813} and θ_{1210} lie in the intervals 3700 to 3500 BP, 3600 to 3400 BP and 3500 to 3300 BP respectively.

In Table 7.2 are the marginal posterior probability mass functions of the three calendar dates. We see that the posterior probability mass function of θ_{1210} is concentrated in the period 3500 to 3300 BP (probability 0.89) with just over a 50% chance that it lies in the interval 3500 to 3400 BP. Similarly there is an 88% chance that θ_{813} is in the period 3600 to 3400 BP. The posterior distribution for θ_{358} is a little more widespread. There is about a 76% chance that it is in the interval 3700 to 3500 BP but there is some evidence that it could be earlier.

Finally in Tables 7.3, 7.4 and 7.5 we give the joint posteriors of the calendar dates taken two at a time. For example there is a 68% chance that θ_{358} and θ_{813} lie in the intervals 3700 to 3500 BP and 3600 to 3400 BP respectively.

Table 7.1 Joint posterior probability mass function of θ_{358}, θ_{813} and θ_{1210}.

θ_{358}	θ_{813}	θ_{1210}	Probability
3500 to 3400	3500 to 3400	3400 to 3300	0.02
3500 to 3400	3500 to 3400	3500 to 3400	0.02
3600 to 3500	3400 to 3300	3400 to 3300	0.01
3600 to 3500	3500 to 3400	3300 to 3200	0.01
3600 to 3500	3500 to 3400	3400 to 3300	0.06
3600 to 3500	3500 to 3400	3500 to 3400	0.08
3600 to 3500	3600 to 3500	3300 to 3200	0.01
3600 to 3500	3600 to 3500	3400 to 3300	0.05
3600 to 3500	3600 to 3500	3500 to 3400	0.08
3600 to 3500	3600 to 3500	3600 to 3500	0.01
3700 to 3600	3400 to 3300	3400 to 3300	0.01
3700 to 3600	3500 to 3400	3300 to 3200	0.01
3700 to 3600	3500 to 3400	3400 to 3300	0.06
3700 to 3600	3500 to 3400	3500 to 3400	0.07
3700 to 3600	3600 to 3500	3300 to 3200	0.01
3700 to 3600	3600 to 3500	3400 to 3300	0.07
3700 to 3600	3600 to 3500	3500 to 3400	0.13
3700 to 3600	3600 to 3500	3600 to 3500	0.03
3700 to 3600	3700 to 3600	3400 to 3300	0.02
3700 to 3600	3700 to 3600	3500 to 3400	0.03
3700 to 3600	3700 to 3600	3600 to 3500	0.01
3800 to 3700	3500 to 3400	3400 to 3300	0.02
3800 to 3700	3500 to 3400	3500 to 3400	0.02
3800 to 3700	3600 to 3500	3400 to 3300	0.02
3800 to 3700	3600 to 3500	3500 to 3400	0.04
3800 to 3700	3600 to 3500	3600 to 3500	0.01
3800 to 3700	3700 to 3600	3400 to 3300	0.01
3800 to 3700	3700 to 3600	3500 to 3400	0.01
3800 to 3700	3700 to 3600	3600 to 3500	0.01
3900 to 3800	3500 to 3400	3400 to 3300	0.01
3900 to 3800	3500 to 3400	3500 to 3400	0.01
3900 to 3800	3600 to 3500	3400 to 3300	0.01
3900 to 3800	3600 to 3500	3500 to 3400	0.02
3900 to 3800	3700 to 3600	3500 to 3400	0.01

Table 7.2   Marginal posterior probability mass functions.

Calendar date BP	θ_{358}	θ_{813}	θ_{1210}
3300 to 3200	0.00	0.00	0.04
3400 to 3300	0.00	0.02	0.37
3500 to 3400	0.04	0.39	0.52
3600 to 3500	0.31	0.49	0.07
3700 to 3600	0.45	0.10	0.00
3800 to 3700	0.14	0.00	0.00
3900 to 3800	0.06	0.00	0.00

Table 7.3   Joint posterior probability mass function of θ_{358} and θ_{813}.

		θ_{813}			
		3400 to 3300	3500 to 3400	3600 to 3500	3700 to 3600
	3500 to 3400	0.00	0.04	0.00	0.00
	3600 to 3500	0.01	0.15	0.15	0.00
θ_{358}	3700 to 3600	0.01	0.14	0.24	0.06
	3800 to 3700	0.00	0.04	0.07	0.03
	3900 to 3800	0.00	0.02	0.03	0.01

Table 7.4   Joint posterior probability mass function of θ_{813} and θ_{1210}.

		θ_{1210}			
		3300 to 3200	3400 to 3300	3500 to 3400	3600 to 3500
	3400 to 3300	0.00	0.02	0.00	0.00
θ_{813}	3500 to 3400	0.02	0.17	0.20	0.00
	3600 to 3500	0.02	0.15	0.27	0.05
	3700 to 3600	0.00	0.03	0.05	0.02

Table 7.5 Joint posterior probability mass function of θ_{358} and θ_{1210}.

		θ_{1210}			
		3300 to 3200	3400 to 3300	3500 to 3400	3600 to 3500
	3500 to 3400	0.00	0.02	0.02	0.00
	3600 to 3500	0.01	0.12	0.16	0.01
θ_{358}	3700 to 3600	0.02	0.16	0.23	0.04
	3800 to 3700	0.00	0.05	0.07	0.02
	3900 to 3800	0.00	0.02	0.04	0.00

7.6 Sequential updating of beliefs

Suppose we have carried out a Bayesian analysis. What happens next, that is, if and when further related data, denoted by y, are obtained? Again, it is very straightforward. Basically, today's posterior becomes tomorrow's prior. In other words: our new "current" beliefs about θ are $p(\theta|\text{previous data } (x))$; these will be revised via Bayes' theorem to:

$$p(\theta|\text{previous and new data } (x \text{ and } y));$$

and so it goes on. Bayes' theorem is a perfectly natural tool for successive revision of beliefs as each piece of new data comes in. We simply "learn from experience" so that, provided the old and new data are conditionally independent,

$$p(\theta|\text{previous and new data}) \propto l(\theta; \text{new data})p(\theta|\text{previous data}).$$

Mathematically we have

$$p(\theta|x \text{ and } y) \propto l(\theta; y)p(\theta|x).$$

7.6.1 The Turin Shroud continued

In Section 7.1.2 we calculated the posterior distribution of the date of the Turin Shroud based on the evidence of the radiocarbon determination OX-2575 carried out by the Oxford Radiocarbon Laboratory. In fact the same laboratory carried out several replicated measures on the same sample. Moreover, material was sent for analysis at several other laboratories worldwide. Let us now imagine that the OX-2575 result was the first to arrive for statistical analysis. After the analysis of this first piece of data our current beliefs about the date of the Shroud are as expressed in Section 7.1.2.

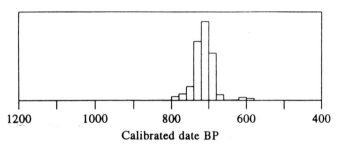

Figure 7.11   Posterior probability histogram for radiocarbon determination
ETH-3883 using the posterior for OX-2575 as prior.

Suppose that another result arrives, say ETH-3883 from the Zurich
laboratory. The estimated radiocarbon age is 733 radiocarbon years bp with
a reported standard deviation of 61. We can use the posterior produced using
OX-2575 as the prior for analysing ETH-3883. If we do so, the resulting
posterior is as given in Figure 7.11. In broad terms, based upon the evidence
provided by the two radiocarbon determinations OX-2575 and ETH-3883, the
material from the Turin Shroud most likely died between 800 BP and 660 BP.
On the conventional calendar scale that is between AD 1150 and AD 1290.

7.6.2   Today's posterior is tomorrow's prior

We mentioned above that the Bayesian approach is a formal way of learning
in a sequential manner as today's posterior becomes tomorrow's prior. Why
is this? Suppose we initially have a prior, $p(\boldsymbol{\theta})$, and we collect some data,
denoted by $\boldsymbol{x}$. Then the posterior conditional on these data is given by

$$p(\boldsymbol{\theta}|\boldsymbol{x}) \propto l(\boldsymbol{\theta};\boldsymbol{x})p(\boldsymbol{\theta}),$$

that is

{Posterior of $\boldsymbol{\theta}$ given $\boldsymbol{x}$}
is proportional to
{Likelihood of $\boldsymbol{\theta}$ based on $\boldsymbol{x}$ } times {Prior of $\boldsymbol{\theta}$}.

Suppose that we subsequently collect some more data, denoted by $\boldsymbol{y}$, which
are independent of the first. Then the posterior conditional on both $\boldsymbol{x}$ and $\boldsymbol{y}$
is given by

$$p(\boldsymbol{\theta}|\boldsymbol{x},\boldsymbol{y}) \propto l(\boldsymbol{\theta};\boldsymbol{x},\boldsymbol{y})p(\boldsymbol{\theta}).$$

In other words we have

{Posterior of $\boldsymbol{\theta}$ given $\boldsymbol{x}$ and $\boldsymbol{y}$}
is proportional to
{Likelihood of $\boldsymbol{\theta}$ based on $\boldsymbol{x}$ and $\boldsymbol{y}$} times {Prior of $\boldsymbol{\theta}$}.

However if the two sets of data, x and y, are independent of each other given θ, the likelihood of θ based on x *and* y is the likelihood of θ based on x multiplied by the likelihood of θ based on y. That is

$$l(\theta; x, y) = l(\theta; x)l(\theta; y).$$

Thus we have

$$p(\theta|x, y) \propto l(\theta; x)l(\theta; y)p(\theta).$$

Notice that the product of the first and third terms on the right-hand side of the above expression is $l(\theta; x)p(\theta)$ which is proportional to the posterior of θ given x, $p(\theta|x)$. In other words we have,

$$p(\theta|x, y) \propto l(\theta; y)p(\theta|x).$$

As we have said above

$$p(\theta|\text{previous and new data}) \propto l(\theta; \text{new data})p(\theta|\text{previous data}).$$

That is "today's posterior becomes tomorrow's prior".

7.6.3   Example 1 — Case IV

Suppose another cemetery is found in the same region as the one considered in Example 1 — Case I. Moreover it is thought to be of the same period. Twenty skeletons are examined of which 12 are found to have the disease. For our prior we use the posterior from Example 1 — Case I. Here the data from Example 1 — Case I are denoted by x. Let y denote the new data. Recall that the posterior from that example was given by

$$p(\theta) = p(\theta|x) \propto \theta^3(1 - \theta)^2 \quad \text{for } 0 \leq \theta \leq 1.$$

This now is used as our *prior*. The likelihood of the new sample is

$$l(\theta; y) \propto \theta^{12}(1 - \theta)^8 \quad \text{for } 0 \leq \theta \leq 1$$

and therefore the posterior of θ conditional on both x and y is given by

$$p(\theta|x, y) \propto \theta^{15}(1 - \theta)^{10} \quad \text{for } 0 \leq \theta \leq 1$$

which is exactly the same as we obtained in Example 1 — Case II. Notice that if the order in which the cemeteries were discovered had been reversed, we would have still arrived at the same posterior and therefore made the same inferences.

7.7   How to represent prior ignorance

In some situations, the archaeologist will claim that they have no prior information, knowledge or beliefs about a particular parameter. Often, when pressed, they admit there is some evidence that is relevant, and perhaps they believe the parameter lies in a fairly wide range. In the case of the Turin Shroud we should perhaps have insisted that the date could be any year before 0 BP but being pragmatic we took the range 0 to 6000 BP. Evidence, however slight, can be used to formulate a suitable prior for the parameter although the prior will be flat or almost flat. However very occasionally the archaeologist will persist in their claim of "no evidence". What should be done?

The concept of *prior ignorance* is very appealing. For if we can model prior ignorance by what is called a *non-informative prior* or *vague prior*, then the posterior must truly represent the information carried by the data and is not "contaminated" by any prior beliefs. Also, this concept may be useful when the prior information is perhaps open to debate — any resulting posterior would not include this controversial evidence. We are "letting the data speak for themselves". However the modelling of prior ignorance does cause some problems, mainly because the priors may be *improper*.

7.7.1   Improper priors

Suppose we wished to model our prior ignorance about the date of manufacture of the Turin Shroud. Ideally we want to say that every year before AD 1950 was equally likely. Therefore we want the prior probability mass function to be of the form

$$p(\theta) = k \ \text{ for } \theta = 1, 2, \ldots,$$

that is k is a constant. To find k, we recall that the probability mass function must add up to 1. After a little thought we realize that it is impossible to find such a k. Such a probability mass function is said to be *improper*, because it does not sum to 1.

Similarly, if we modelled θ as being continuous we would require a prior probability density function of the form

$$p(\theta) = k \ \ \text{ for } 0 < \theta.$$

Therefore we need to find a value of k so that the density integrates to 1. That is

$$\int_0^\infty p(\theta)\, d\theta = 1$$

but again no k can be found.

Despite being improper, such distributions can often be used in Bayesian analysis although care must be taken to ensure that the posterior is not improper also. (See O'Hagan, 1994, p. 78–80 for a full discussion of the problems that may arise.) One concern is that by implying prior ignorance about θ, we are not implying prior ignorance about, say, θ^2. However if our prior evidence is genuinely weak compared to that provided by the data then using an improper prior should cause no problems.

7.7.2 Vague and reference priors

There is an undeniable attraction to the idea of a non-informative or vague prior, the idea being regarded as almost synonymous with the making of "objective" inferences. However, whatever prior we use, we *are* inputting *some* information into the analysis. In a way, what we want is to input the minimum information possible. It is in this context that Bayesians use the term *vague prior*, so chosen as to represent their minimal beliefs about a parameter. In the case of the Turin Shroud, every year is equally likely, so we used a discrete Uniform distribution over some very wide range.

However one person's vague prior may be different from a second person's. This is all perfectly natural and permitted within the Bayesian framework, but is it possible to have a generally agreed "conventional" prior that is non-informative, that is a prior that is accepted by all Bayesians as a reasonable way of representing prior ignorance in a particular situation? Bayesians call such a prior a *reference prior* (see Bernardo and Smith, 1994, p. 298–339) and the subsequent data analysis a *reference analysis*. The topic can be very technical and so we will not pursue it any further in this book.

7.8 The Bayesian versus classical debate

Throughout this book we have adopted the Bayesian approach to data analysis and have not considered the classical (or frequentist) approach. Applications of Bayesian statistics to archaeological data interpretation are relatively new, and the differences between the Bayesian and classical approaches are not widely discussed in the archaeological literature. Therefore we shall briefly outline the the arguments for and against the two approaches.

7.8.1 Prior probability and subjectivity

Classical statisticians see the inclusion, in the Bayesian approach, of subjective information as a serious drawback. Classical statistical methods, it is held, allow objective conclusions based only on the current data and not on subjectively assessed prior information or prior beliefs. For the moment we put aside arguments concerning the subjective nature of probability assessments. Bayesians would reply that prior information exists regardless of the statistical framework adopted. In this case, it is surely better to be explicit about the prior information and include it coherently in the analysis than to go ahead as if there were no experience which has a bearing on the problem in hand.

Of course, it is difficult to quantify one's beliefs prior to the analysis of data; nor would we deny that personal judgement is part of the process of defining a prior. However, classical methods of data analysis cannot usually escape an element of personal judgement either, but in their case judgement enters at the stage of interpreting the results, after processing the data. The danger then arises of *ad hoc* argument and special pleading. From a Bayesian stance, we would argue that it is more rigorous to state any information about the problem from the start and to include it in a logical and coherent fashion into the analysis.

Let us explore these points through a particular example. Archaeologists investigating patterns of trade use chemical analysis to identify the provenance of artefacts. The underlying assumption is that chemical concentrations of certain elements vary from provenance to provenance. Thus the chemical composition of an artefact can be used to "fingerprint" it as coming from one particular source.

Often techniques of cluster analysis are used to identify groups of samples having similar chemical compositions, although the algorithms require no prior definition of what is meant by a cluster or group. It is left to the *subjective judgement* of the user to decide, *after* they have seen the results, what constitutes a cluster and how many clusters there are. In such circumstances, any decision can be open to question; one analyst might judge there to be three clusters but another might claim there are four or even more depending upon their own viewpoint.

We need to distinguish between "exploratory" methods of statistical analysis (Baxter, 1994) and model-based techniques which can be used to make inferences. A model-based solution requires a definition of the nature of a cluster, which in turn can arise only from the expertise (or prior knowledge) of those who have worked in the area. Therefore, we believe that it is better to define what is meant by a cluster *before* the statistical analysis is carried out. This demands that statistical algorithms should be designed to allow the incorporation of a prior definition of a cluster. Otherwise we are in danger of finding patterns in tea-leaves or similar random data sets!

7.8.2 The classical or frequentist approach

The classical approach to statistics holds that the data are the only source of relevant information. It adopts a "frequency" view of probability by conjecturing an *infinite* series of *identical, repeatable* experiments. Thus, when reporting a 95% confidence interval produced by a classical method, the correct interpretation is that 95% of intervals constructed in that manner will include the true value of the unknown parameter. However it tells us nothing about the actual situation under consideration, only what will happen in the long term, if the experiment were repeated in exactly the same way time after time.

The long-term approach of classical methods may be applicable to some situations such as the monitoring of the diameters of ball bearings made by the thousands in a factory, but does it apply to archaeology? Usually, archaeologists get one, and only one, chance to excavate a site and to take samples for scientific analysis; thus many archaeological "experiments" are unrepeatable under identical conditions, let alone an infinite series of experiments! In contrast to the classical concept of a confidence interval, Bayesians produce 95% posterior credible intervals which clearly express the posterior belief that the unknown parameter is included in that interval *with probability* 0.95 (or in other words that the unknown parameter lies in the interval with odds of 19 to 1). Moreover, the credible intervals refer to *the* unique situation under study.

7.8.3   Hypothesis testing

Hypothesis testing is another area where the contrast between the two approaches can be brought out. Classical statisticians work in terms of the probability of observing the data assuming that the null hypothesis is true. (That is, assuming that the Turin Shroud is authentic, what is the probability that the observed radiocarbon determination would have arisen by chance?) What is actually required is the probability that the null hypothesis is true *given* the observed data. (Given the radiocarbon result, what is the probability that the Turin Shroud is authentic?)

Let θ be the unknown calendar date of the Turin Shroud and let x be the radiocarbon determination. The null hypothesis is that the Shroud dates to the time of Christ's death, say between AD 20 and AD 80. That is θ should lie between 1930 BP and 1870 BP. Classical statistical hypothesis testing involves calculation of

$$P(x|1870 < \theta < 1930).$$

This clearly tells us how likely the radiocarbon determination, x, is if the Shroud was actually manufactured around the time of Christ's death. However, we really are interested in

$$P(1870 < \theta < 1930|x),$$

the probability of the manufacture of the Shroud being about the time of Christ's death given that we have observed the data. Of course these two are easily linked via Bayes' theorem since

$$P(1870 < \theta < 1930|x) = c\,P(x|1870 < \theta < 1930)P(1870 < \theta < 1930)$$

where $P(1870 < \theta < 1930)$ is the prior probability that the Shroud was made about the time of Christ's death and c is the normalizing constant. In terms of the logic of the argument, the Bayesian approach seems natural whereas the classical approach has an unnatural, almost topsy-turvy feel to it.

7.8.4   Comment

All these points add up to fundamental differences between the classical and Bayesian perspectives. The debate has been fiercely contested over the last two decades and continues, if somewhat less vehemently, to the present day. For a balanced and wide-ranging discussion of the points raised here and other more subtle ones relating to implementation, we refer the reader to Barnett (1982) and Stuart and Ord (1991, chapter 31). A review of the topic in the context of the history of the philosophy of science is found in Howson and Urbach (1993).

7.9   Discussion

To sum up, the Bayesian philosophy holds that any uncertainty about parameter values should be expressed in terms of probabilities. Moreover where prior information is available, it should be included explicitly and coherently into the inference procedure. After all, archaeologists will inevitably make heuristic use of their previous knowledge, both in making inferences and in planning field and laboratory projects. The nature of the Bayesian approach is such that experts undertaking data collection must be clear about their goals and must also make their prior information explicit before any data analysis begins. Such explicit statements have obvious advantages in situations where several different groups of workers are tackling the same problem since their results can then be directly compared. Such direct comparisons are not always possible when classical inference procedures are used since expert opinion is not included in the data analysis and is consequently not always (if ever) clearly stated. In fact, when these inferential procedures are used, prior information can only be used retrospectively as part of a discussion of the results after the statistical investigation is complete.

In later chapters we will discuss several Bayesian analyses of archaeological data. However, before we do so, in the next chapter we give some indication of how the Bayesian approach may be implemented in practice. As we will see this is not an easy process, as it requires considerable thought and mathematical expertise. In fact there are some challenging and unsolved problems associated with model formulation, eliciting and specifying the prior, computing and summarizing the posterior, and interpreting the posterior which still need to be resolved by the Bayesian community.

8

Implementation issues

8.1 Introduction

The Bayesian philosophy provides a formal framework within which to update information or beliefs. It is both conceptually simple and intuitively plausible; it may be viewed as a way of formalizing a common sense approach to learning from data. If this is the case, why are not all data analyses carried out using Bayesian methods? Smith (1991, p. 369) gives a clue with the use of "deceptively" in his statement

> In formal terms, this Bayesian inference prescription is deceptively simple ...

Dempster (1980, p. 273) provides a stronger indication of the reasons why the adoption of Bayesian techniques is (not yet anyway) commonplace:

> The application of (Bayesian) inference techniques is held back by conceptual factors and computational factors. I believe that Bayesian inference is conceptually much more straightforward than non-Bayesian inference ... Hence, I believe that the major barrier to much more widespread application of Bayesian methods is computational ...

To what does "computational" refer? Recall that we are taking an archaeological problem and developing a mathematical model expressed in terms of a likelihood and a prior. Bayes' theorem tells us that the posterior is proportional to the product of the likelihood and the prior. Mathematically this is expressed as

$$p(\boldsymbol{\theta}|\boldsymbol{x}) \propto l(\boldsymbol{\theta};\boldsymbol{x})p(\boldsymbol{\theta})$$

where $\boldsymbol{\theta}$ represents the parameters and $\boldsymbol{x}$ the data. So what is the difficulty?

To be able to make archaeological inferences about the problem in hand, we need to:

(i) determine the normalizing constant, c, such that

$$p(\boldsymbol{\theta}|\boldsymbol{x}) = c\,l(\boldsymbol{\theta};\boldsymbol{x})p(\boldsymbol{\theta}),$$

(ii) summarize the posterior in a mathematical manner, and then
(iii) interpret this in language readily understandable by the archaeological
community.

It is the first two parts of this summarization process from the posterior
(expressed as proportional to likelihood times prior) that pose many of the
difficulties encountered in the routine application of the Bayesian paradigm.
This is not to say that the third is not without its challenges.

In the next section we will examine these difficulties in more detail and
in subsequent sections briefly describe how Bayesians have attempted to
overcome them. However, the whole area presents an extremely difficult
mathematical problem, which it is the responsibility of the mathematical
community to solve. It is still very much an ongoing topic for mathematical
and statistical research and is likely to remain so for several years to come.
The reader should be aware that whatever we write here may soon become
superseded as novel mathematical techniques are developed and applied.

8.2   Evaluation of the posterior

Suppose we have constructed a suitable statistical model, formulated the prior
knowledge and collected the data. What do we actually expect in the way of
posterior information? At the most basic level the answer is archaeologically
interpretable representations of the posterior. But what does this mean in
practice? For most real-life problems, the model will be multi-parameter,
the posterior will be multi-parameter also and as a result will be (almost
invariably) difficult to interpret. Even if it were possible to interpret such a
posterior, Bayes' theorem only tells us that the posterior is *proportional* to
the likelihood times the prior; for this statement to be useful, we need to be
able to evaluate the normalizing constant, c. In other words we must be able
to determine c which satisfies the equation

$$p(\boldsymbol{\theta}|\boldsymbol{x}) = c\,l(\boldsymbol{\theta};\boldsymbol{x})p(\boldsymbol{\theta}).$$

In our simple examples of Chapter 7 with one parameter of interest, we were
able to identify the constant c by recognizing the form of $l(\boldsymbol{\theta};\boldsymbol{x})p(\boldsymbol{\theta})$. Even
with one parameter this can be done only in a small minority of special cases
(see Section 8.3). Once the number of parameters increases there is even less
chance of obtaining a posterior whose form is immediately recognizable.

Moreover, even if we can evaluate the normalizing constant, there
still remains the problem of summarizing and interpreting a complicated
mathematical expression involving several parameters. With one parameter,
at least we can look at the posterior probability mass (or density) function
and see where most of the probability lies, calculate the posterior mean,

the posterior mode and other summary statistics. With two parameters we can obtain contour plots of the joint posterior distribution and see how the parameters are related. But the extension to three or more parameters is difficult even for the most experienced professional statistician. At best we can look at the marginal distributions for each parameter and at all the joint distributions.

As we progress through this chapter it will become ever clearer that, apart from a relatively few simple special cases (see Sections 8.3 and 8.4), the evaluation and interpretation of the posterior cannot be undertaken without the use of computer software which uses either mathematical techniques for numerical integration (see Section 8.5) or Markov chain Monte Carlo methods (see Section 8.6).

8.3   Conjugate prior distributions

Recall that in Section 7.3.3 we discussed the problem of making inferences about the incidence rate, denoted by θ, of a particular disease. We let X be the random variable representing the number of skeletons with the disease out of a total of n. We modelled X as having a Binomial distribution with parameters n and θ. In the particular example we had $n = 5$ and $x = 3$, and so the *likelihood* was given by

$$l(\theta; x) \propto \theta^3 (1 - \theta)^2 \quad \text{for } 0 \leq \theta \leq 1.$$

We represented our prior knowledge about θ by a Beta distribution with parameters $\alpha = 3$ and $\beta = 3$ (see Section 5.10.2). That is the *prior* probability density was

$$p(\theta) = 30 \, \theta^2 (1 - \theta)^2 \quad \text{for } 0 \leq \theta \leq 1.$$

Hence the *posterior* probability density was given by

$$p(\theta|x) \propto \theta^5 (1 - \theta)^4 \quad \text{for } 0 \leq \theta \leq 1.$$

By recognizing the form of the posterior as a Beta distribution with parameters $\alpha = 6$ and $\beta = 5$ we then were able to make posterior inferences.

This is a particular example of what Bayesians call *conjugacy*. We started with a model, a Binomial with parameter θ, in this case. We chose a particular form of prior, a Beta distribution, to represent our prior information about θ. We calculated the posterior as being proportional to the likelihood times the prior. We recognized the posterior distribution of θ as being a Beta distribution with different parameter values. With this model and a carefully chosen form of prior, the mathematics worked out "nicely". For how many other situations will this occur?

In Bayesian statistics there are what are called *conjugate families* of prior distributions. The whole topic of conjugate families can be very mathematical and we do not intend in this book to explore the topic in any great depth. The reader is referred to Bernardo and Smith (1994, p. 265–285) or O'Hagan (1994, p. 146–147). Many books on Bayesian statistics give lists of conjugate families, see for example Bernardo and Smith (1994, p. 436–442) or Smith (1988, p. 98).

8.3.1 Estimating sherd intensity

Before we leave the topic of conjugate priors, we give another example, this time from the analysis of field survey data. Suppose an archaeologist wishes to estimate the underlying intensity of sherds per unit area in a region. Because of financial and time constraints it is impossible to survey completely the whole area and so a partial survey is carried out by counting the number of sherds found in n areas of unit size. Assume that the sherds can be considered to be randomly scattered about the countryside at unknown rate θ per unit area. Let X_i represent the number of sherds found in the ith unit, then a possible model is that X_i has a Poisson distribution (see Section 5.6.4) with unknown mean θ. Therefore we have

$$p(X_i = x_i) = \frac{\theta^{x_i} \exp\{-\theta\}}{x_i!}.$$

Assuming that, conditional on θ, the counts are independent, the likelihood is given by

$$l(\theta; x_1, \ldots, x_n) = \prod_{i=1}^{n} \frac{\theta^{x_i} \exp\{-\theta\}}{x_i!} \propto \theta^s \exp\{-n\theta\}$$

where $s = \sum_{i=1}^{n} x_i$.

The conjugate prior for a Poisson distribution is a Gamma distribution so that we take as the prior probability density function of θ,

$$p(\theta) \propto \theta^\tau \exp\{-\zeta\theta\}$$

(see Section 5.10.5) where τ and ζ are parameters chosen so that one particular Gamma distribution represents our prior information about θ. (For instance we may believe that the prior mean should be m_0 and that the prior variance should be v_0. Referring to Table 8.4 we note that the mean and variance of a Gamma random variable with parameters τ and ζ are τ/ζ and τ/ζ^2 respectively. Therefore we could set

$$m_0 = \frac{\tau}{\zeta} \text{ and } v_0 = \frac{\tau}{\zeta^2}$$

and choose τ and ζ to be

$$\frac{m_0^2}{v_0} \text{ and } \frac{m_0}{v_0}$$

respectively.)

The posterior probability density function of θ is proportional to the likelihood times the prior and so is given by

$$p(\theta|x_1, \ldots, x_n) \propto \theta^{\tau+s} \exp\{-(\zeta + n)\theta\}.$$

Because we have chosen a conjugate prior we recognize the posterior as being a Gamma distribution with parameters

$$(\tau + s) \text{ and } (\zeta + n).$$

8.3.2   Comments

The major attraction of using a conjugate prior is that it provides a mathematically tractable answer. Against this must be weighed the disadvantage of forcing prior information into a particular family of priors. However by varying the parameters of the conjugate prior, the ζ and τ above, a rich class of priors is produced that can often adequately model many possible forms of prior information. If necessary, *mixtures* of conjugate priors may prove a more suitable modelling device (see Bernardo and Smith, 1994, p. 279–285 or O'Hagan, 1994, p. 160–163).

To summarize: a modelling process based upon conjugate priors may be adequate to enable sensible and robust inferences to be made. However for many real-life situations, suitable conjugate priors do not exist.

8.4   Approximations

Much research has been devoted to finding ways of approximating either the posterior distribution itself or posterior summary statistics, such as the posterior mean. Some methods of approximation do exist but generally are suitable only for low-dimensional problems and as a result suitable approximations may not be available in all circumstances. Moreover they appear to be superseded by the computationally intensive methods to be described in Sections 8.5 and 8.6.

A simple example of an approximation is the following. If the sample size is very large, then the parameter of interest has, approximately, a Normal distribution with a mean and a variance that are functions of the data. Applying the result to our sherd-counting example of Section 8.3 we find that the posterior distribution of θ is approximately Normal with mean equal to the sample mean, $\bar{x}$, and variance equal to $\bar{x}/n$. Notice that the posterior does not depend upon the prior! Quite rightly so in this case, for when we have a large amount of data, our prior beliefs should not influence our posterior beliefs.

This type of approximation does rely on having a large amount of data, whereas most researchers are trying to make inferences from as little data as possible either because of the costly nature of data collection or because large amounts of data are unobtainable. We refer the interested reader to Smith (1991) and Press (1989, chapter III) for further details of the wide variety of approximate methods that have been developed.

8.5   Numerical integration

8.5.1   A simple example

Recall that in Section 7.3.3 we found that the posterior probability density function of the incidence, θ, of the disease was proportional to $\theta^5(1 - \theta)^4$. We recognized this as a Beta distribution with parameters 6 and 5 and hence we were able to evaluate the normalizing constant as being equal to 1260. Suppose now that we were not aware of its being a Beta distribution. Is there some other method by which we could evaluate this constant?

In Section 5.8.1 it was noted that the area under a probability density function, in this case $p(\theta|x)$, must equal 1. Hence the area under the curve $c\theta^5(1-\theta)^4$ between 0 and 1 must equal 1 (where c is the normalizing constant). Consider now the curve $f(\theta) = \theta^5(1 - \theta)^4$ and let A denote the area under this curve between 0 and 1. Then it is easy to see that c is equal to $1/A$.

Mathematicians can, in this example, evaluate A exactly by using calculus (although for the majority of real-life problems this will not be the case). Alternatively we can use numerical methods to obtain an approximate answer. Here we present just one method for illustrative purposes; we make no claims at all about it being the best, quickest or most accurate method.

Divide the area A into a number, say five, of separate areas denoted by A_1, A_2, A_3, A_4 and A_5 (see Figure 8.1). Approximate each area by the area of a trapezium as shown in Figure 8.2.

Now the area of a trapezium with parallel sides of lengths a and b, and with width w is equal to

$$\frac{a + b}{2} \times w.$$

Thus we may approximate area A_2 by

$$\frac{f(0.2) + f(0.4)}{2} \times 0.2 = \frac{(0.2^5 0.8^4 + 0.4^5 0.6^4)}{2} \times 0.2 = 0.0001458$$

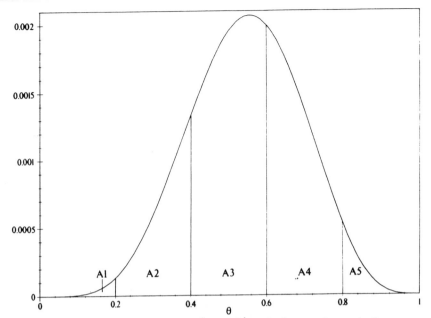

Figure 8.1 Graph of the curve $\theta^5(1-\theta)^4$ with the area beneath the curve divided into five areas denoted by A_1, A_2, A_3, A_4 and A_5.

and A_3 by

$$\frac{f(0.4) + f(0.6)}{2} \times 0.2 = \frac{(0.4^5 0.6^4 + 0.6^5 0.4^4)}{2} \times 0.2 = 0.0003317.$$

By carrying out similar calculations (noting that A_1 and A_5 are approximated by triangles rather than trapezia) we approximate A_1, A_4 and A_5 by 0.0000131, 0.0002515 and 0.0000524 respectively. Since

$$A = A_1 + A_2 + A_3 + A_4 + A_5$$

an approximate value for A is given by

$$\begin{aligned}
\hat{A} &= 0.0000131 + 0.0001458 + 0.0003317 + 0.0002515 + 0.0000524 \\
&= 0.0007945.
\end{aligned}$$

Therefore an estimate, call it $\hat{c}$, of the required constant is

$$\hat{c} = \frac{1}{\hat{A}} = \frac{1}{0.0007945} = 1258.6532,$$

not quite the correct answer of 1260 but reasonably close considering it is based on a crude method. It is easy to see that in this case $\hat{A}$ can be calculated

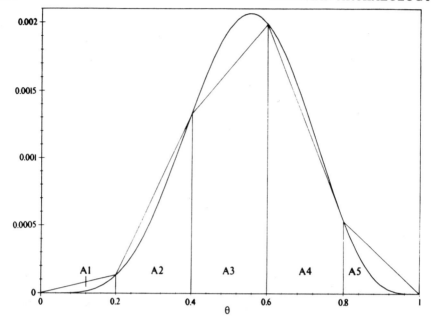

Figure 8.2   Each of the areas A_2 to A_4 is approximated by a trapezium whereas A_1 and A_5 are approximated by triangles.

using the formula

$$\hat{A} = \frac{f(0.0) + 2f(0.2) + 2f(0.4) + 2f(0.6) + 2f(0.8) + f(1.0)}{2} \times 0.2.$$

This formula easily generalizes to the case when the interval is divided into n areas of equal width $w = 1/n$, in which case we have

$$\hat{A} = \sum_{i=0}^{n} w_i f(\theta_i)$$

where

$$w_i = \begin{cases} 1/2n & i = 0 \text{ or } n \\ 1/n & i = 1, \dots, n-1. \end{cases}$$

If we increase the number of partitions of A to 10 each of width 0.1, $\hat{A}$ and $\hat{c}$ become 0.0007937 and 1259.9219 respectively. Increasing to 100 partitions results in $\hat{c} = 1260.0000$.

The posterior mean of a continuous random variable is given by

$$E(\theta|x) = \int_{-\infty}^{\infty} \theta p(\theta|x) \, d\theta.$$

If we wish to use numerical methods to estimate the posterior mean, we could proceed as follows. Using five partitions we would calculate our estimate using the expression

$$\widehat{E(\theta|x)} = (0.1 \times A_1) + (0.3 \times A_2) + (0.5 \times A_3) + (0.7 \times A_4) + (0.9 \times A_5)$$

and with the A_i as above $\widehat{E(\theta|x)}$ equals 0.546 39. This should be compared with the correct value of $6/11 = 0.545\,45$. Approximations based on 10, 20 and 100 partitions are 0.545 47, 0.545 46 and 0.545 45 respectively. It is possible to estimate the posterior variance by a similar method.

8.5.2 The general methodology

In the previous section we have used a very simple example to illustrate what mathematicians refer to as *numerical integration* or *quadrature*. We are the first to admit that our method of approximation is very crude; much more accurate (and quicker) methods do exist. However we hope that the reader will have gained some insight into the general area of numerical integration. Readers who wish to learn more about the subject of numerical integration or numerical methods in general should perhaps consult Köckler (1994).

We turn to the more general problem of approximating numerically integrals of the form

$$I = \int_a^b f(\phi)\, d\phi$$

by summations of the form

$$\hat{I} = \sum_{i=0}^{n} w_i f(\phi_i)$$

where the ϕ_i are called *abscissae* and the w_i are *weights*. Various methods have been developed to choose the ϕ_i and w_i in some optimal manner but a full treatment is well beyond the scope of this book. More details can be found in (Köckler 1994, chapter 6).

In the above example using a Beta distribution the range was from $a = 0$ to $b = 1$. Initially this interval was divided into 5 areas of equal width (equal to 0.2) by taking the abscissae as $\phi_0 = 0$, $\phi_1 = 0.2$, $\phi_2 = 0.4$, $\phi_3 = 0.6$, $\phi_4 = 0.8$, $\phi_5 = 1.0$; and the weights as $w_0 = 0.1$, $w_1 = 0.2$, $w_2 = 0.2$, $w_3 = 0.2$, $w_4 = 0.2$, $w_5 = 0.1$.

The general approach outlined above extends to two dimensions by calculating volumes rather than areas. In this case a simple way of approximating the integral

$$I = \int_a^b \int_c^d f(\phi_1, \phi_2)\, d\phi_2\, d\phi_1$$

is to take the one-dimensional values for $\phi_{1,i}$ and w_i for the interval a to b, and the values $\phi_{2,j}$ and v_j for the interval c to d, and hence use

$$\hat{I} = \sum_{i=0}^{n} \sum_{j=0}^{n} w_i v_j f(\phi_{1,i}, \phi_{2,j}).$$

The extension to three or more dimensions is harder to visualize but can be carried out numerically. The drawback is that as the number of parameters of the model increases, the number of dimensions of the numerical calculations increases also. By the time one has, say, ten parameters, the computer time needed may be so enormous that it renders crude methods unusable. Much research has been concerned with the development of efficient algorithms both to speed up the process and to allow the extension to problems with even more parameters. Stroud (1971) compares and contrasts the various methods that are available whereas Smith (1991) gives a review of those that have been proved useful for Bayesian statistics. However, despite being implemented on the most modern of computer workstations, even the latest highly efficient algorithms do have limitations in terms of the complexity of problems that can be analysed.

8.5.3   St Veit-Klinglberg by numerical integration

In Section 7.5.3 we reported the joint posterior probability of the dates, represented by θ_{358}, θ_{813} and θ_{1210}, of samples from three stratified contexts from St Veit-Klinglberg. One way of calculating these probabilities is by using numerical integration. (Another method of doing so will be described in Section 8.6.2.)

For example, suppose we are interested in calculating the probability that θ_{358} lies between 3700 and 3600 BP, θ_{813} lies between 3600 and 3500 BP, and θ_{358} lies between 3500 and 3400 BP. Mathematically we need to find

$$P(A_{358} \text{ and } A_{813} \text{ and } A_{1210}|x)$$

where A_{358}, A_{813} and A_{1210} are the events $3600 < \theta_{358} < 3700$, $3500 < \theta_{813} < 3600$ and $3400 < \theta_{1210} < 3500$ respectively. This probability is given by the triple-integral

$$I = \int_{3600}^{3700} \int_{3500}^{3600} \int_{3400}^{3500} p(\theta_{358}, \theta_{813}, \theta_{1210}|x_{358}, x_{813}, x_{1210})\, d\theta_{358}\, d\theta_{813}\, d\theta_{1210}$$

where

$$p(\theta_{358}, \theta_{813}, \theta_{1210}|x_{358}, x_{813}, x_{1210})$$

is the joint posterior probability density function of θ_{358}, θ_{813} and θ_{1210}. Returning to (7.14) we see that this joint posterior probability density function is *proportional to*

$$g(\theta_{358}, \theta_{813}, \theta_{1210}) = \begin{cases} \prod_i \exp\left\{ \frac{-(x_i - \mu(\theta_i))^2}{2\sigma_i^2} \right\} & \text{for } \theta_{1210} < \theta_{813} < \theta_{358} \\ 0 & \text{otherwise} \end{cases}$$

where the product is for $i = 358, 813, 1210$. Therefore we need to evaluate the normalizing constant, denoted by c, which too can be expressed in terms of a triple-integral as

$$c^{-1} = I_1 = \int_0^\infty \int_0^{\theta_{358}} \int_0^{\theta_{813}} g(\theta_{358}, \theta_{813}, \theta_{1210}) \, d\theta_{1210} \, d\theta_{813} \, d\theta_{358}.$$

Hence the required probability is given by

$$\frac{I_2}{I_1}$$

where

$$I_2 = \int_{3600}^{3700} \int_{3500}^{3600} \int_{3400}^{3500} g(\theta_{358}, \theta_{813}, \theta_{1210}) \, d\theta_{358} \, d\theta_{813} \, d\theta_{1210}.$$

Now I_2 can be evaluated, albeit crudely, by

$$\hat{I}_2 = \sum_{i=0}^n \sum_{j=0}^n \sum_{k=0}^n w_k v_j u_i g(\theta_{358,k}, \theta_{813,j}, \theta_{1210,i})$$

where $\theta_{1210,i} = 3400 + (100i/n)$, $\theta_{813,j} = 3500 + (100j/n)$ and $\theta_{358,k} = 3600 + (100k/n)$. Since we are dividing an interval of 100 years into n partitions of equal width, the common width being $100/n$, the weights are taken as

$$w_k = \begin{cases} 50/n & k = 0 \text{ or } n \\ 100/n & k = 1, \ldots, n-1 \end{cases}$$

and similarly for the v_j and u_i. Each 100-year interval has been divided into n parts of equal width.

I_1 can be calculated in a similar fashion but we must note that the integral is over an infinite range. In practice we take a sufficiently wide range outside of which the probability is negligible. In fact we use the 1000-year period from 4000 to 3000 BP. Thus we have

$$\hat{I}_1 = \sum_{i=0}^m \sum_{j=0}^m \sum_{k=0}^m w_k^* v_j^* u_i^* g(\theta_{358,k}, \theta_{813,j}, \theta_{1210,i})$$

where $m = 10n$, $\theta_{1210,i} = 3000 + (1000i/m)$, $\theta_{813,j} = 3000 + (1000j/m)$ and $\theta_{358,k} = 3000 + (1000k/m)$ and where weights are similar to those used in the previous approximation. That is

$$w_k^* = \begin{cases} 500/m & k = 0 \text{ or } m \\ 1000/m & k = 1, \ldots, m-1 \end{cases}$$

and similarly for the v_j^* and u_i^*.

Taking $n = 10$ corresponds to evaluating I_1 and I_2 (and therefore $g(\theta_{358}, \theta_{813}, \theta_{1210})$) at 10-year intervals. In this case the posterior probability is approximately 0.13253. Taking a finer division of 5-year intervals results in a value of 0.13308 and 2-year intervals a value of 0.13317. From an interpretative viewpoint, quoting that the posterior probability is about 0.13 will suffice. That is, there is about a 1 in 8 chance that the dates fall in the specified interval.

8.5.4   Comments

To sum up: techniques that rely on numerical methods for (multi-dimensional) integration can handle many problems of up to 10 or even 20 parameters. However they are much less popular than they were in the 1970's and 1980's probably because they depend on expert mathematical knowledge even when incorporated into computer packages and so appear to be only suitable for experienced (Bayesian) statisticians. Readers wishing to discover more about Bayesian packages using numerical integration techniques should consult Press (1989).

8.6   Markov chain Monte Carlo methods

Simulation-based techniques, called Markov chain Monte Carlo methods or MCMC in short, for estimating marginal posterior density and mass functions have become increasingly popular. Although the theory underlying the methods is sometimes difficult to follow, their implementation is usually straightforward. Moreover the methods can be broken down into a series of steps that can readily be understood by the non-statistician.

The most popular of the simulation approaches to posterior evaluation is known as the Gibbs sampler. A simple account of the Gibbs sampler, its properties and illustrative examples is given by Casella and George (1992). Its application to radiocarbon dating is discussed in Buck *et al.* (1992). Another, more general technique that readers may come across is the *Metropolis–Hastings algorithm* (Metropolis *et al.*, 1953; Hastings, 1970; O'Hagan, 1994, p. 237–238). Gilks *et al.* (1995) gives a review of these two methods, and others, together with numerous case studies. Here we will concentrate our efforts on describing the Gibbs sampler and its application to archaeological problems.

8.6.1 The Gibbs sampler

Suppose we have k continuous parameters denoted by $\theta_1, \theta_2, \theta_3$ up to θ_k or more simply by $\boldsymbol{\theta}$. Their joint posterior probability density function is denoted by

$$p(\boldsymbol{\theta}|\boldsymbol{x}) = p(\theta_1, \theta_2, \ldots, \theta_k|\boldsymbol{x})$$

where $\boldsymbol{x}$ denotes the data. Of interest are the marginal posterior probability density functions, denoted by $p(\theta_i|\boldsymbol{x})$.

The Gibbs sampler is a method of estimating these marginal posterior distributions. It involves choosing initial values for $\theta_2, \theta_3, \ldots, \theta_k$, say $\theta_2^{(0)}, \theta_3^{(0)}, \ldots, \theta_k^{(0)}$. Then first we obtain a "sample" value of θ_1 (how in practice we do so will be explained in Section 8.7). Let us denote this sampled value by $\theta_1^{(1)}$. It is obtained from the full conditional density function of θ_1, namely

$$p(\theta_1|\theta_2^{(0)}, \theta_3^{(0)}, \ldots, \theta_k^{(0)}, \boldsymbol{x}).$$

Next, $\theta_2^{(1)}$ is sampled from the full conditional probability density function of θ_2, namely

$$p(\theta_2|\theta_1^{(1)}, \theta_3^{(0)}, \ldots, \theta_k^{(0)}, \boldsymbol{x}).$$

This process continues with sampling $\theta_3^{(1)}$ and so on until $\theta_k^{(1)}$ is sampled from the full conditional probability density function of θ_k, that is

$$p(\theta_k|\theta_1^{(1)}, \theta_2^{(1)}, \ldots, \theta_{k-1}^{(1)}, \boldsymbol{x}).$$

This completes one iteration of the Gibbs sampler; we have a new realization for $\boldsymbol{\theta}$ which is given by $\theta_1^{(1)}, \theta_2^{(1)}, \ldots, \theta_k^{(1)}$.

Using this new realization of $\boldsymbol{\theta}$, the process is then repeated so that we obtain a second realization denoted by $\theta_1^{(2)}, \theta_2^{(2)}, \ldots, \theta_k^{(2)}$. This is obtained first by sampling $\theta_1^{(2)}$ from the full conditional density function of θ_1:

$$p(\theta_1|\theta_2^{(1)}, \theta_3^{(1)}, \ldots, \theta_k^{(1)}, \boldsymbol{x}).$$

Next, $\theta_2^{(2)}$ is sampled from the full conditional probability density function of θ_2:

$$p(\theta_2|\theta_1^{(2)}, \theta_3^{(1)}, \ldots, \theta_k^{(1)}, \boldsymbol{x}).$$

This process continues until $\theta_k^{(2)}$ is sampled from the full conditional probability density function:

$$p(\theta_k|\theta_1^{(2)}, \theta_2^{(2)}, \ldots, \theta_{k-1}^{(2)}, \boldsymbol{x}).$$

This completes the second iteration.

This whole process is repeated for a total of r iterations. At the final iteration we will have the realization $\theta_1^{(r)}, \theta_2^{(r)}, \ldots, \theta_k^{(r)}$. A series of r iterations is often referred to as a *run*.

The statistical theory underlying this approach (see Geman and Geman, 1984) shows that, for very large values of r, the distribution of the sample values $\theta_i^{(r)}$ becomes extremely close to the marginal posterior distribution of θ_i. In practical terms this means that we simulate the process until it reaches some equilibrium or *steady-state* — the initial period is called the *burn-in* period. Suppose we are convinced that equilibrium has been achieved after, say, s iterations. We then simulate the process for a further t iterations. The values $\theta_i^{(s+1)}$, $\theta_i^{(s+2)}$, ...,$\theta_i^{(s+t)}$ are used to form a histogram which is our estimate of the posterior marginal distribution of θ_i. Critical questions here include: how do we know when the simulation has reached the steady-state and how large should t be?

8.6.2   St Veit-Klinglberg using the Gibbs sampler

In Section 7.5.3 we analysed a simplified version of the St Veit-Klinglberg data set by considering in isolation three radiocarbon determinations corresponding to contexts 358, 813 and 1210. Later in Section 8.5.3 we demonstrated how to calculate by numerical integration various posterior probabilities of interest. In this section we will describe how Gibbs sampling can be used to do the same.

In Equation (7.14) the joint posterior of $\boldsymbol{\theta}$ was given by

$$p(\boldsymbol{\theta}|\boldsymbol{x}) \propto \left\{ \prod_i \exp\left\{ -\frac{(x_i - \mu(\theta_i))^2}{2\sigma_i^2} \right\} \right\} I_A(\boldsymbol{\theta})$$

where the product is over $i = 358,\ 813$ and 1210 and

$$I_A(\boldsymbol{\theta}) = \begin{cases} 1 & 0 < \theta_{1210} < \theta_{813} < \theta_{358} < 4000 \\ 0 & \text{otherwise.} \end{cases}$$

The full conditional density of θ_{358} can be determined by fixing the values of the other two parameters, θ_{813} and θ_{1210}. Then we have

$$p(\theta_{358}|\theta_{813}, \theta_{1210}, \boldsymbol{x}) \propto \exp\left\{ -\frac{(x_{358} - \mu(\theta_{358}))^2}{2\sigma_{358}^2} \right\} \quad \text{for } \theta_{813} < \theta_{358} < 4000.$$

Note that the range is for all values of θ_{358} larger (earlier) than θ_{813}.

Also note that given θ_{1210} and θ_{358}, θ_{813} must lie between them; whereas, given θ_{358} and θ_{813}, θ_{1210} must be later than both of them. Hence the full conditional densities of θ_{813} and θ_{1210} are

$$p(\theta_{813}|\theta_{358}, \theta_{1210}, \boldsymbol{x}) \propto \exp\left\{ -\frac{(x_{813} - \mu(\theta_{813}))^2}{2\sigma_{813}^2} \right\} \quad \text{for } \theta_{1210} < \theta_{813} < \theta_{358}$$

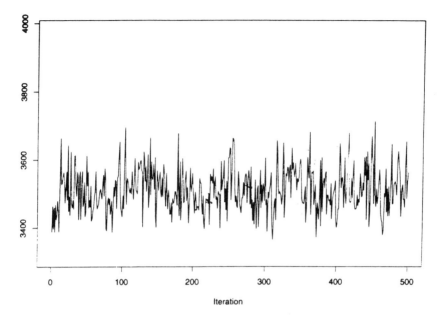

Figure 8.3   Plot of 500 simulated values of θ_{813} obtained using the Gibbs sampler with the constraint $0 < \theta_{1210} < \theta_{813} < \theta_{358} < 4000$. Notice how quickly, within about 10 iterations, the values settle down to an equilibrium level.

and

$$p(\theta_{1210}|\theta_{358}, \theta_{813}, x) \propto \exp\left\{-\frac{(x_{1210} - \mu(\theta_{1210}))^2}{2\sigma_{1210}^2}\right\} \quad \text{for } 0 < \theta_{1210} < \theta_{813}$$

respectively.

In Figure 8.3 is a plot of the sampled values for the three parameters from a run of 1000 iterations. Notice how quickly the values settle down to an equilibrium level — only a very short burn-in period is needed in this example. In contrast, Figure 8.4 shows two plots for the same problem but with some artificial constraints added. In these two cases convergence is not so quick.

8.6.3   One run or several?

Since we are evaluating each $\theta_i^{(r)}$ in a sequential fashion, they are clearly dependent upon one another; the current value of $\theta_i^{(r)}$ is usually correlated with the previous ones $\theta_i^{(r-1)}$, $\theta_i^{(r-2)}$, $\theta_i^{(r-3)}$ and so on — that is they are not independent of each other. There are two popular approaches to take account

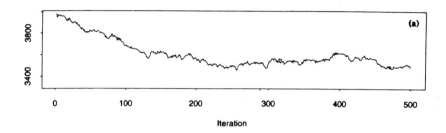

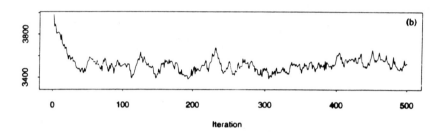

Figure 8.4 Plot of 500 simulated values of θ_{813} obtaining using the Gibbs sampler. Here we use an extra, artificial constraint that the θ_i must be within (a) 40 years of each other, that is $\theta_{358} - \theta_{1210} < 40$ and (b) 100 years of each other, that is $\theta_{358} - \theta_{1210} < 100$. Notice that in (a) equilibrium is achieved after about 200 iterations whereas in (b) about 50 iterations are required.

of this: one is to use a large number of parallel runs of the Gibbs sampler and then to form a sample by collecting the final $\boldsymbol{\theta}$ values from each. The other is to perform one long run of a single iterative chain, discarding iterations from the initial burn-in period while the results are unstable, and then forming a sample by collecting equally spaced outcomes thereafter (possibly every observation).

When considering the merits of the two approaches, Smith and Roberts (1993) note that

> In theory, the single-chain approach appears to be more efficient, in that only one transient phase is involved. However, particularly during the first tentative examination of a new problem, it can be argued that monitoring the evolutionary behaviour of several runs of the chain starting from a wide range of initial values is necessary.

In other words, the mathematical theory tells us that it is better to use *one* very long run instead of many shorter runs. However, in practice, it would be a very unusual, trusting applied statistician who relied solely on one run. Many would be inclined to carry out, at least initially, several different runs from a variety of starting positions. Perhaps in their final report only one or two extremely long runs would be used. For further discussion of the debate between one long run versus several short runs see Gelman and Rubin (1992a, 1992b).

8.6.4 How many iterations?

The theory of the Gibbs sampler says that the distribution of the sample values $\theta_i^{(1)}, \ldots, \theta_i^{(r)}$ converges to the true marginal posterior distribution of θ_i. In any practical application, we must assess when r is large enough. Naturally, this varies from problem to problem depending upon the nature of the prior distribution and the complexity of the model.

The issue of determining convergence of the Gibbs sampler (and other such Markov chain Monte Carlo simulation techniques) has received considerable attention in the Bayesian literature. The various methods that have been adopted are reviewed in Smith and Roberts (1993). Most commonly, workers using the Gibbs sampler either rely on running the simulation several times and checking for consistency between runs or carry out a very long run until it appears stable. This is, as Smith and Roberts (1993) note, because

> although there is a reassuring theoretical literature concerning the convergence of Markov chain Monte Carlo methods (for example Tierney 1994 and Besag and Green 1993), results do not easily translate into clear guide-lines for the practitioner.

Smith and Roberts (1993) also list a number of special cases for which the convergence rate of the Gibbs sampler can be predicted *a priori*, but these are not numerous and most practical convergence measures monitor the simulation results. Smith and Roberts highlight a number of such methods, including those of Gelman and Rubin (1992a), Raftery and Lewis (1992) and Roberts (1992).

One practical point worth briefly mentioning is that convergence will normally be slow if the parameters are highly correlated with each other. Therefore one means of speeding up the Gibbs sampler is to *reparametrize* the model so that the new parameters are less correlated with each other. Incidentally, reparametrization is also of great relevance in numerical quadrature. The whole area is a highly technical one which we will not pursue in this book and refer the interested reader to Bernardo and Smith (1994, p. 347–348) for a discussion of the issues involved.

8.7   Simulation techniques

In the previous section we have described the Gibbs sampler algorithm and its use in Bayesian inference. Within the algorithm, we have to "sample" from the conditional distribution of, say, θ_1 given $\theta_2, \ldots, \theta_k$ and the data. What do we mean by "sample"? Put simply it means that we generate an "observation" from the specified distribution, usually with the help of a computer. If we generated many observations and constructed a histogram, then as the sample size became very large, the histogram would become close to the underlying distribution from which we are sampling. In this section we will give a brief insight into how to "sample" from a distribution. For more detailed discussion of the techniques involved we suggest the reader consults specialized texts on the subject such as Morgan (1984) or Devroye (1986).

Most computers and hand calculators have the ability to generate a "random" number between 0 and 1. That is we can easily sample from a uniform distribution on the interval 0 to 1. In fact provided we have a supply of such random numbers, we can simulate observations from almost any random variable, discrete or continuous, by using an appropriate algorithm. Some algorithms have been developed for sampling from specific distributions but general methods suitable for a wide range of distributions are available. Here we will describe the "table look-up" method for discrete random variables, the "inversion" method and "rejection sampling" for continuous random variables.

8.7.1   Table look-up methods for discrete random variables

Suppose that we have a discrete random variable denoted by X which takes a finite number of values which we assume to be 0, 1, 2, ..., l. For ease of notation we let $p(X = x) = p_x$. Then a general algorithm for sampling or simulating an observation from X is as follows.

Let u denote a random number from a $U(0, 1)$ distribution.

$$
\begin{aligned}
\text{Set } X &= 0 \quad \text{if } 0 \le u < p_0. \\
\text{Set } X &= 1 \quad \text{if } p_0 \le u < p_0 + p_1. \\
\text{Set } X &= 2 \quad \text{if } p_0 + p_1 \le u < p_0 + p_1 + p_2. \\
\text{Set } X &= 3 \quad \text{if } p_0 + p_1 + p_2 \le u < p_0 + p_1 + p_2 + p_3. \\
&\cdots \quad \cdots \quad \cdots \\
&\cdots \quad \cdots \quad \cdots \\
\text{Set } X &= l \quad \text{if } p_0 + \cdots + p_{l-1} \le u < 1.
\end{aligned}
$$

Example. Suppose that X has a binomial distribution with parameters $n = 6$ and $p = 0.3$. Then

$$
p_x = p(X = x) = \binom{6}{x} \times 0.3^x \times 0.7^{6-x} \quad \text{for } x = 0, 1, \ldots, 6
$$

and so we have

$$
\begin{array}{llll}
p_0 & = & 0.1176 & \quad p_1 & = & 0.3025 \\
p_2 & = & 0.3241 & \quad p_3 & = & 0.1852 \\
p_4 & = & 0.0602 & \quad p_5 & = & 0.0102 \\
p_6 & = & 0.0007.
\end{array}
$$

Then the above algorithm becomes

$$
\begin{array}{lllll}
\text{Set } X & = & 0 & \text{if } 0 \le u < 0.1176. \\
\text{Set } X & = & 1 & \text{if } 0.1176 \le u < 0.4201. \\
\text{Set } X & = & 2 & \text{if } 0.4201 \le u < 0.7442. \\
\text{Set } X & = & 3 & \text{if } 0.7442 \le u < 0.9294. \\
\text{Set } X & = & 4 & \text{if } 0.9294 \le u < 0.9896. \\
\text{Set } X & = & 5 & \text{if } 0.9896 \le u < 0.9998. \\
\text{Set } X & = & 6 & \text{if } 0.9998 \le u < 1.0000.
\end{array}
$$

Suppose the random number were $u = 0.357$. We see that it satisfies the second of these inequalities and hence our simulated value of X is 1. If instead $u = 0.954$, then X would have been 4 and so on.

8.7.2   Inversion method for continuous random variables

The equivalent of the table look-up method for continuous random variables is the *inversion method*. Suppose we wish to sample from a continuous random variable X which has probability density function $p(x)$ and cumulative distribution function $F(x)$. Recall from Section 5.4.2 that the cumulative distribution function is the probability that the random variable X takes values less than or equal to x. That is $F(x) = P(X \le x)$.

The inversion method is quite simple. It says suppose u is a random number from a $U(0,1)$ distribution. To sample from X having cumulative distribution function $F(x)$ we need to solve for x the equation

$$
F(x) = u.
$$

In Figure 8.5 we give the cumulative distribution function of an Exponential random variable (see Section 5.10.4) with $\lambda = 2.0$. We operate the inversion algorithm by marking-off on the vertical axis u and drawing a horizontal line until it meets $F(x)$. We then draw a vertical line down to the horizontal axis. Where this line meets the horizontal axis gives the simulated value for x.

Example Consider sampling values from a random variable having an Exponential distribution with parameter λ. In this case the probability density function and cumulative distribution function are given by

$$
p(x) = \lambda e^{-\lambda x} \quad \text{for } 0 < x
$$

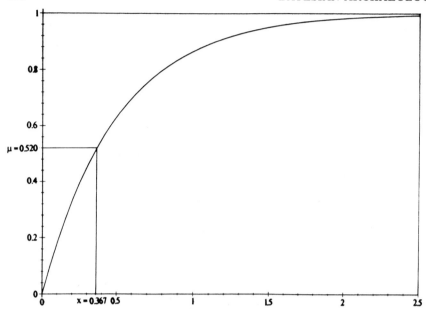

Figure 8.5   Graph of the cumulative distribution function of an Exponential random variable with parameter $\lambda = 2.0$. The random number from a $U(0,1)$ distribution is $u = 0.520$ and the corresponding simulated value of the Exponential random variable is $x = 0.367$.

and

$$F(x) = 1 - e^{-\lambda x} \quad \text{for } 0 < x$$

respectively. To sample a value for X, we set $F(x) = u$ and solve for u. That is,

$$1 - e^{-\lambda x} = u$$

and solving for x gives

$$x = -\frac{1}{\lambda} \ln(1 - u).$$

Thus if $\lambda = 2.0$ and $u = 0.520$ then the simulated value of X is $-0.5 \ln(1 - 0.520) = 0.3670$. If $u = 0.954$ then the simulated value is 1.540.

The inverse method is very simple and for every random number u it guarantees to generate a number x from the required distribution. However it does have a major practical limitation, namely that we need to be able to write down an equation relating u to the cumulative distribution function, $F(x)$, of X and be able to solve this equation by expressing x in terms of a function of u. In many cases, including one of the most important distributions — the Normal — this is impossible and other methods need to be found. Standard books on simulation, such as Morgan (1984), discuss this restriction and how it may be overcome. In the next section we will describe one technique that does so but the reader should be aware that there are many others.

8.7.3 Rejection methods

Suppose we have a continuous random variable with probability density function $p(x)$ which is non-zero over a finite range and has a finite mode. The Beta distribution with parameters 4 and 7 is an example of this. For this example let us consider how to sample from a $Be(4,7)$ random variable. Recall from Section 5.10.2 that the probability density function of a $Be(4,7)$ is given by

$$p(x) = 840\,x^3(1-x)^6 \quad \text{for } 0 \le x \le 1.$$

Suppose we "box-in" the density as in Figure 8.6. It is relatively simple to sprinkle points randomly over the rectangle by taking their horizontal and vertical coordinates as realizations of Uniform random variables. In the case of Figure 8.6, the horizontal coordinate of a random point would be simulated from a $U(0,1)$ distribution and the vertical coordinate from a $U(0,M)$ distribution. Here M represents the value of the density at the mode of the distribution. The mode is at

$$\frac{\alpha - 1}{\alpha + \beta - 2} = \frac{4 - 1}{4 + 7 - 2} = \frac{1}{3}.$$

Therefore the value of $f(x)$ at the mode is given by

$$M = 840\left(\frac{1}{3}\right)^3 \left(\frac{2}{3}\right)^6 = 2.731.$$

The rejection method says that we reject points falling above the curve $p(x)$. Points landing below $p(x)$ are accepted and their horizontal coordinate is taken as a realization of X. To implement this method first we need two random numbers from a $U(0,1)$ distribution. Suppose these are 0.847 and 0.259. To convert the latter number to be an observation from a $U(0,2.731)$ distribution we multiply it by 2.731. Thus the horizontal component of our randomly chosen point is 0.847 and its vertical component is $0.259 \times 2.731 = 0.707$. The question is "does this point lie above or below the probability density function $p(x)$?" Now

$$p(0.847) = 840(0.847)^3(1-0.847)^6 = 0.0065$$

which is less than 0.707. Hence we reject this point.

Suppose we have another pair of random numbers 0.157 and 0.354. In this case the vertical component is $0.354 \times 2.731 = 0.967$. Then as

$$p(0.157) = 840(0.157)^3(1-0.157)^6 = 1.167 > 0.967$$

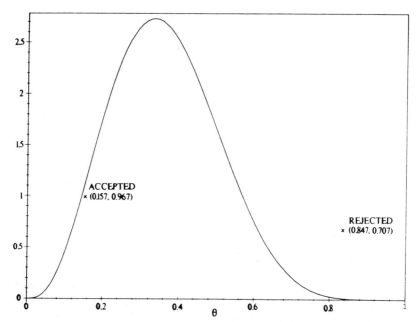

Figure 8.6   Graph of the probability density function $p(x) = 840\,x^3(1-x)^5$. The mode of the distribution is at $\theta = 1/3$ and the modal value is $M = 2.731$. The curve is boxed-in as shown. Points are randomly scattered within the box. Points below $p(x)$ are accepted but points above are rejected.

we accept this point. Thus our simulated value of X is 0.157.

In this example it is easy to see that the proportion of points accepted by the algorithm is given by

$$\frac{\text{Area under } p(x)}{\text{Area of rectangle}} = \frac{1}{1 \times M} = \frac{1}{M}$$

and therefore the proportion rejected is $(M-1)/M$.

With $M = 2.731$ this means we accept only about 37% of the points. The other 63% are rejected which is extremely wasteful, especially when the algorithm is being used several thousand times within the Gibbs sampler framework. Morgan (1984) describes methods that can overcome this and also discusses how the rejection methodology can be extended to random variables that have infinite ranges.

8.7.4 Comments

It is not the purpose of this book to give a complete description of all the many algorithms developed for computer simulation. Suffice it to say here that the literature is vast but good introductory texts include those by Morgan (1984) and Devroye (1986). The reader should appreciate that within the Gibbs sampling framework, an algorithm is not just used once but many times, on occasions up to several million. Therefore it is important that any method used should be as efficient as possible.

8.8 Discussion

We have tried in this chapter to give a summary and simplified overview of the mathematical and computer techniques used to implement the Bayesian methodology. As we said at the start of the chapter, this is a rapidly developing field, in terms of both the constant improvement in the efficiency of the algorithms and the computer power available. If past experience is a pointer to the future, a problem that currently stretches the techniques and computers to the limit will, within a decade or even less, be carried out in a few seconds.

9

Interpretation of radiocarbon results

9.1 Dating in archaeology

A colleague once said "without dating there would be no archaeology". this is rather an overstatement but almost all archaeologists are passionately interested in dates, whether their site is Lower Paleolithic or nineteenth century or any period between. Thus to be able to date a site or an event of interest as accurately as possible is an almost essential feature of archaeological research.

Dating techniques are grouped into relative and absolute chronological methods (fuller and more precise details can be found in standard books on archaeology such as Renfrew and Bahn, 1991; Aitken, 1990). Relative methods give evidence only of sequence, that event B preceded event A, but do not, of themselves, allow dating in terms of years; that is the role of absolute dating methods. Stratigraphy is one of the best-known relative methods, and it has featured a number of times in the discussion thus far: observations in the course of excavation will indicate that later archaeological sediments accumulate over earlier ones, or that later pits cut into earlier levels, thus enabling the events to be ordered. Typology is another relative method relying on the stylistic development of artefacts to place them into a chronological sequence. Seriation, about which more will be said in Chapter 12, relies on the relative similarity/dissimilarity of the contents of levels or graves, or of artefacts themselves, to place the sediments or burials or objects into a relative sequence. Occasionally other phenomena are used in relative dating, for example particular volcanic eruptions will spread ash distinctive enough to establish the contemporaneity of different archaeological contexts.

Methods of absolute dating, however, aim to provide not merely a sequence but dates in years. Thus, for the more recent past, historical events, such as the reigns of kings, emperors or pharaohs, can be used to put a date on an archaeological context; though this can often be via a rather tortuous route. Sometimes the dates, like those of the ancient Egyptian calendar, are not absolutely certain, and scholars will disagree over a low, middle or high chronology. The indirect connection between a historical event and an archaeological layer, and the disagreements over dates, lead inevitably to uncertainty.

For more remote time periods, and for areas where historical chronologies are of no use, archaeologists tend to rely on methods which exploit radioactive decay or other natural clocks. Radiocarbon dating is the most commonly used, and its principles will be explained presently. Other methods using radioisotope clocks include the potassium-argon method, used to give dates from man's remotest past; it relies on the isotope ^{40}K, which has a half-life of 1250 million years. When a volcanic flow is molten, it is assumed, virtually no argon is retained; on cooling argon will accumulate from the decay of ^{40}K to ^{40}Ar, and from this the date of an eruption can be calculated. Levels containing the archaeological traces of fossil man can be dated if lying over and under different flows. Without going into the details we shall simply mention that there are other methods based upon radioactivity which, like K-Ar, can go back much earlier than radiocarbon, but without its precision and accuracy, methods such as Electron-Spin-Resonance dating and Uranium Series dating (see Aitken, 1990).

In addition there are other natural phenomena which, whilst not as predictable as radioactive decay, are nonetheless regular enough to provide a date, so long as they can be calibrated to a known scale. This includes archaeomagnetic dating, which uses variations, through time, in the position of magnetic north. Kilns, at their last firing, will "freeze" the inclination and declination of the earth's field at that time. If the changing positions of magnetic north have been worked out for the region and for the given period, then the date of the kiln can be read off. The obsidian hydration method relies on the phenomenon that chipped tools, made from this volcanic glass, will, over the centuries, absorb water to form a thin hydrated skin. If the rate of hydration (which depends on complex factors such as the type of obsidian, the ambient temperature and humidity and so forth) can be estimated, the obsidian tools can then be dated.

Perhaps the most accurate of dating methods is tree-ring dating, or dendrochronology. This relies on matching tree-ring width variations; if one tree's pattern of thicker and thinner rings sufficiently matches another's, then they are taken to overlap in date. With a large number of samples going back through time it is possible to build up a master chronology from the present back to the earliest sample. A section of timber of unknown date can then be compared to find the most satisfactory match against the master. Provided that we have the outermost ring of the tree that the timbers came from, the precise year can be calculated in which the tree was felled. More will be said about dendrochronology in Chapter 12.

It will be evident that elements of imprecision and uncertainty affect all dating methods (whether relative or absolute) — there are archaeological uncertainties, limitations to instruments and measurement, unavoidable imprecisions in the laboratory, inherent uncertainties in the methods. To deal with these complexities it is plainly necessary to appeal to statistics, and, as we hope to show, the Bayesian method is especially good for combining the evidence of different dating techniques, for example that from absolute and relative chronological methods.

9.1.1   Structure of the chapter

In this chapter we will focus attention upon radiocarbon dating. We do so for two reasons. Firstly it is one of the most widely used scientific dating techniques. Secondly, at the time of writing, in the mid-1990's, it is an area that has seen a number of highly successful applications of Bayesian methods. In the next section we will give a brief overview of radiocarbon dating and then go through (in some detail) the Bayesian approach to the analysis of a single radiocarbon determination. In subsequent sections we will use five different case studies to illustrate the wide variety of problems tackled within the Bayesian framework. These range from dating the stratigraphy of an early Bronze Age site, St Veit-Klinglberg in Austria, to providing dates for the development of several successive cultures spanning the last 4000 years in the Jama River Valley of Central Ecuador.

9.2   Introduction to radiocarbon dating

9.2.1   The carbon cycle

We have already described radiocarbon dating and how it can be modelled mathematically in Chapter 3. To refresh the reader's memory, recall that carbon occurs naturally in the form of three isotopes, carbon-12, carbon-13 and carbon-14 (denoted by ^{12}C, ^{13}C and ^{14}C respectively). The isotope ^{14}C, which forms about one part per million million of modern carbon, is radioactive and so it decays with time. Moreover ^{14}C is continually being formed in the upper atmosphere by the interaction of cosmic rays with nitrogen atoms. After formation, ^{14}C (along with ^{12}C and ^{13}C) is, as a result of photosynthesis, eventually taken into the food chain (see Figure 9.1).

Whilst alive, plants and animals constantly exchange carbon with the carbon dioxide in the atmosphere. Upon their death, however, no more new carbon is taken up and consequently the amount of radioactive carbon (^{14}C) held in their surviving tissue begins to decrease. In other words, during the lifetime of an organism, the proportion of ^{14}C to ^{12}C and ^{13}C remains constant within its cells, but after its death this proportion decreases at a rate which

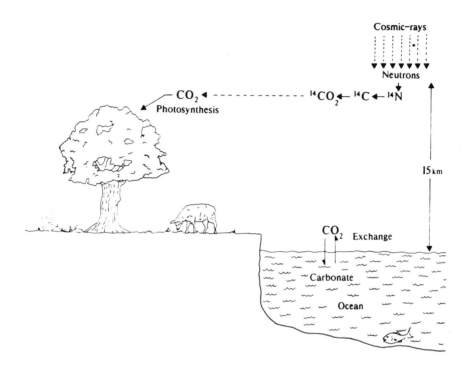

Figure 9.1   The carbon cycle. (After Aitken, 1990, p. 57, fig. 3.1.)

is dependent upon the decay rate and the proportion of ^{14}C remaining.

9.2.2   The law of radioactive decay

As we have seen in Chapter 3 the decomposition of the ^{14}C (like other radioisotopes) obeys the law of radioactive decay which may be expressed as

$$N(t) = N(0)e^{-\lambda t} \qquad (9.1)$$

where M is the amount of ^{14}C remaining after time t, $N(0)$ is the initial unknown amount of ^{14}C present upon death and λ is the decay rate. Even this law is an "average result". In theory two specimens which started off with exactly the same number of ^{14}C atoms will not have precisely the same number when they come to be measured, because of the inherent randomness of the radioactive decay process. However the numbers will be so close together that any differences are undetectable using even the most modern equipment.

If we can assume that the ^{14}C concentration in the atmosphere and living organic material is constant through time, then, provided we can measure $N(t)$ relative to a standard $N(0)$, the time elapsed since the material died can be estimated by

$$t = -\frac{1}{\lambda} \ln \left(\frac{N(t)}{N(0)} \right). \tag{9.2}$$

Now suppose we take a series of organic samples of known date (for example some wood that has been dated accurately by dendrochronology). If the model as conveyed by (9.2) were a true representation of what actually happens in nature, then a plot of the radiocarbon age against the known calendar age of the samples should be roughly a straight line. We say "roughly" because measuring the radiocarbon age is subject to some experimental and laboratory error for a variety of reasons. If the slope of the line were 1, then one year measured on the radiocarbon time-scale would be equivalent to one calendar year. If the slope were 0.5, then one radiocarbon year would be equivalent to two calendar years.

In Figure 9.2 we give a plot of the results from several dendrochronologically dated samples. The samples were taken at 20-year intervals and then submitted to several different laboratories worldwide for radiocarbon dating. In fact each laboratory made several replicated measurements using wood of the same date. As one can see, even allowing for some experimental measurement error and for some variation between laboratories, the results do not lie on a straight line. *The model is not good enough!* Hence we need to re-enter the modelling cycle.

Why is the model inadequate? In broad terms for the above model, which arises from the law of radioactive decay, to be applied in practice we need to be able to measure the amount of ^{14}C present in the organic material *when* it died. An impossibility! To get round this obstacle, radiocarbon laboratories measure the amount of ^{14}C relative to that of the isotopes ^{12}C and ^{13}C. In other words the $N(t)$ and $N(0)$ in (9.1) and (9.2) should be thought of as *proportions*, say $^{14}C/^{13}C$ or $^{14}C/^{12}C$. Then the major assumption made in order to use (9.1) is that $N(0)$, the $^{14}C/^{12}C$ ratio in organisms at death, has remained *constant* throughout time. That is a seed should contain exactly the same $^{14}C/^{12}C$ proportion at death regardless of when it died, be it AD 1959 or 3000 BC. All the evidence (see Figure 9.2) points to this assumption not being true.

There are many reasons for this, most being too complex to discuss here and we refer the interested reader to Aitken (1990) or Bowman (1990). In broad terms due to changes in atmospheric conditions as a result of sunspot activity, to changes in the geomagnetic field and the like, the ^{14}C "equilibrium" level fluctuates slightly from year to year. Thus $N(0)$ in the above equations varies depending upon when the organism died (in other words, $N(0)$ depends upon x which we are trying to estimate). The major consequence of the inadequacies

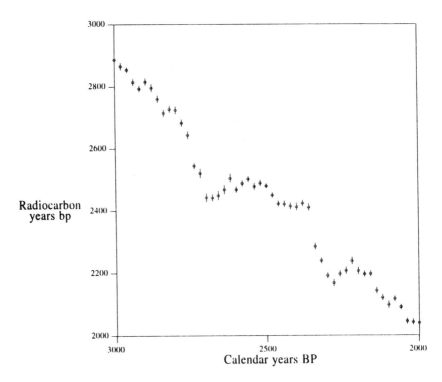

Figure 9.2 Plot of radiocarbon age against known calendar age. The vertical bars indicate the 1σ ranges where σ is the standard deviation reported by the laboratory.

of the model as given in (9.2) is that the radiocarbon laboratories are unable to estimate the calendar age of a specimen directly. Instead the laboratories report an estimate of the age of the sample on the radiocarbon time-scale. To emphasize the difference between dates on the two time-scales we use t to represent a date "Before Present", that is on the calendar time-scale, and x for a date "before present", that is on the radiocarbon time-scale (see Section 1.1.1).

9.2.3   Modelling the relationship between radiocarbon and calendar years

This need to calibrate provided the impetus for the radiocarbon community to undertake worldwide collaboration during the 1970's and 1980's. The result was the internationally agreed high-precision calibration data (Stuiver and Pearson, 1986; Pearson and Stuiver, 1986; Pearson *et al.*, 1986), which were established by determining the ^{14}C content of timber samples already accurately dated on the calendar time-scale using dendrochronology (for the science of dating wood samples using their annual growth rings see Baillie, 1982). These data have been updated by Pearson and Stuiver (1993) and Stuiver and Pearson (1993). These latest high-precision data consist of estimates of the radiocarbon age corresponding to samples whose calendar ages extend from the present day back through time to about 6000 BC, the samples being taken at 10-year or 20-year intervals. A lower-precision curve has been constructed for the period 6000 BC back to around 7900 BC (Pearson *et al.*, 1993) and work proceeds to push the curve back even earlier.

Because of the fluctuations in the ^{14}C equilibrium level, the calibration curve relating the radiocarbon age to the calendar age is not monotonic and exhibits a significant number of "wiggles" (see Figure 9.2). Therefore the question is "how to model the radiocarbon calibration curve". One could draw a "smooth" curve through the points, perhaps using some statistical technique, although the question remains as to how "smooth" the curve should be, that is, how closely the curve should follow the data points. An alternative method is simply to join adjacent data points by straight lines. This is the method that we will use in this chapter. However we are well aware that in the future it could be found to be inadequate as the radiocarbon laboratories' precision increases and we would have to carry out another iteration of the modelling cycle (see Section 9.6.4).

Let $\mu(\theta)$ represent the calibration curve. That is the underlying radiocarbon age of organic material dying θ years ago is denoted by $\mu(\theta)$. A suitable working approximation is to express $\mu(\theta)$ in a piece-wise linear form:

$$\mu(\theta) = \begin{cases} a_0 + b_0\theta & (\theta \leq t_0) \\ a_l + b_l\theta & (t_{l-1} < \theta \leq t_l, l = 1, 2, \ldots, L) \\ a_L + b_L\theta & (\theta > t_L) \end{cases} \qquad (9.3)$$

where t_l are called the "knots" of the calibration curve, $L + 1$ is the number of knots, and a_l and b_l are assumed to be known constants which ensure continuity at the knots (see Figure 9.3). In non-mathematical terms, we mark on a graph, by a series of dots, the estimated radiocarbon age and the corresponding known calendar age of the samples. We then join the adjacent dots by straight lines. Mathematically such a curve is known as *piece-wise linear* as it consists of a series of straight lines that are joined together. The "dots" are known as the "knots" of the curve. The positions of the knots on the calendar time-scale are denoted by $t_0, t_1, t_2, \ldots, t_L$.

For example consider the most recent part of the calibration curve from AD 1950 to 1840 or, in terms of years "Before Present", 0 BP to 110 BP. (For radiocarbon dating purposes, the "present" is taken as AD 1950. Note that since AD 0 does not exist, 35 BC is equivalent to 1984 BP.) The data from which the radiocarbon calibration curve is calculated consist of pairs of numbers: one is the calendar year measured BP and the other is the

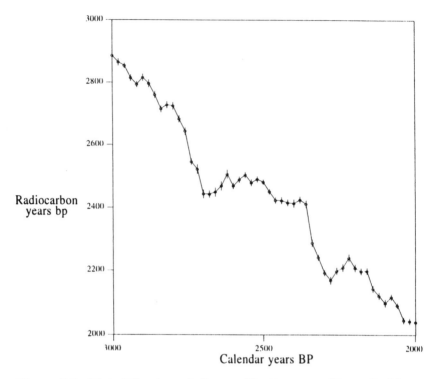

Figure 9.3 Plot of the piece-wiselinear calibration curve. The vertical bars indicate the 1σ ranges where σ is the standard deviation reported by the laboratory.

radiocarbon age. Corresponding to the calendar ages 10, 30, 50, 70, 90 and 110 BP are the estimated radiocarbon ages 178, 126, 85, 113, 120 and 118 respectively. That is, the estimated true radiocarbon age of an organic object that died in 1940 AD (10 BP) is 178, if it had died in 1920 (30 BP) then it would be 126, and so on. The first six knots are at the above six calendar ages and so are $t_0 = 10$, $t_1 = 30$, $t_2 = 50$, $t_3 = 70$, $t_4 = 90$ and $t_5 = 110$.

To make use of (9.3) we must obtain the a_i and the b_i for each interval between the knots. In other words we must calculate the slope (b_i) and intercept (a_i) terms of the straight line joining the $(i - 1)$th knot to the ith knot.

The slope of the straight line between 10 BP and 30 BP is calculated as

$$b_1 = \frac{178 - 126}{20} = 2.6$$

and the intercept term, a_1, as

$$a_1 = 178 - (2.6 \times 10) = 152.$$

Thus the equation of the line between the first two knots is

$$\mu(\theta) = 152 + 2.6\,\theta.$$

Likewise we can calculate the equation of the line between 30 BP and 50 BP. Its slope is

$$b_2 = \frac{126 - 85}{20} = 2.05,$$

its intercept term is

$$a_2 = 126 - (2.05 \times 30) = 64.5$$

and hence the equation of the line is

$$\mu(\theta) = 64.5 + 2.05\,\theta.$$

In a similar way we can calculate $b_3 = -1.4$, $a_3 = 155$, $b_4 = -0.35$, $a_4 = 137.5$, $b_5 = 0.1$ and $a_5 = 111$. This method can be extended to the whole curve back to 6000 BC. As AD 1950 is taken as 0 BP it is conventional that the curve passes through this point. Thus we take $a_0 = 0$ and $b_0 = -17.8$.

Figure 9.4 shows a plot of the piece-wise linear curve for the period 820 BP to 980 BP. Notice that some parts of the curve are steep, so that one year on the calendar scale corresponds to less than a year on the radiocarbon scale. Here radiocarbon dating can be quite accurate. On the flatter parts, the opposite is true: one calendar year corresponds to more than one radiocarbon year so that dating is less precise. Note also the famous "wiggles"; in some places one radiocarbon year corresponds to three or more calendar years. What has started out as a very simple idea leading to a simple mathematical model expressed in (9.1) and (9.2) is suddenly much more complicated.

9.2.4 Statistical model of radiocarbon dating

Consider a sample of organic material that died in calendar year θ BP. Associated with this unknown calendar date is an "age", related to the amount of ^{14}C actually present; we represent this "radiocarbon age" by $\mu(\theta)$. All organic material that died in that year, θ, will, theoretically at least (and with certain exceptions such as marine samples which need not delay us here), have the same radiocarbon age (apart from some minor variations caused by the natural variations in the decay process). In practice, due to our inability to measure the radiocarbon age accurately enough, this does not happen. Thus,

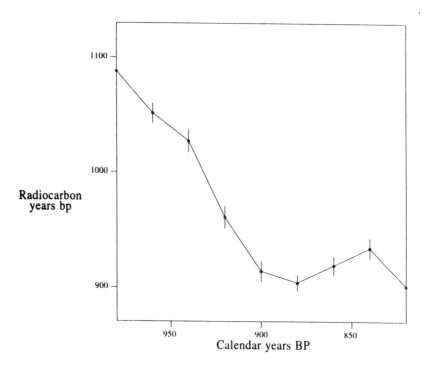

Figure 9.4 Plot of the piece-wise linear curve for the period 980–820 BP, showing the relationship between calendar and radiocarbon years.

due to the nature of the samples and the chemical and physical techniques used, the *experimentally derived* values available for $\mu(\theta)$ are not totally accurate or precise. They provide an observation, x, which is a realization of a random variable, X, which is an estimate of $\mu(\theta)$. Conventionally, X is assumed to have a Normal distribution (see Section 5.10.3) with mean $\mu(\theta)$ and variance σ^2. Therefore we have

$$X|\theta \sim N(\mu(\theta), \sigma^2),$$

where σ is the laboratory's quoted standard deviation (see Bowman, 1990 for details on how the laboratories assess σ). Since we are assuming that σ is known, we condition X only on the unknown parameter θ.

For example, suppose a plant died in year 3173 BP (1224 BC). This is between the knots of the curve occurring at 3159 BP (1210 BC) and 3179 BP (1230 BC). The equation of the piece-wise linear calibration curve between these knots is

$$\mu(\theta) = -1436.6 + 1.4\,\theta$$

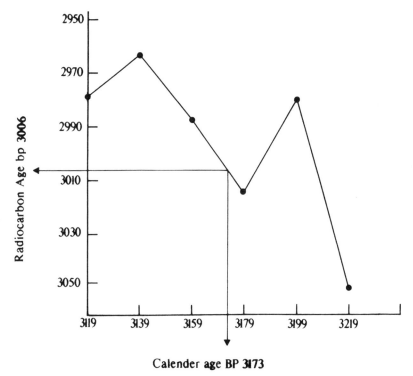

Figure 9.5 Plot of piece-wise linear curve for the period 3219–3119 BP, illustrating how a radiocarbon estimate of 3006 bp can be read off against the calendar age of 3173 BP.

and so the plant's true radiocarbon age should be

$$\mu(\theta) = -1436.6 + (1.4 \times 3173) = 3006$$

to the nearest integer. Alternatively this value can be found by reading off the appropriate part of the curve seen in Figure 9.5.

However, as we have tried to emphasize, the measurement of the radiocarbon age is subject to some random or stochastic error which we can never determine exactly. As indicated above, the generally agreed modelling assumption is that the error has a Normal distribution having a mean of 0 and a standard deviation of σ. Suppose σ is assessed to be 60 which is a typical value seen in the literature. The error will be a value drawn from a Normal distribution with mean 0 and a standard deviation of 60. A method of simulating such an error is to sample a value, z, from a Normal distribution with a mean of 0 and a standard deviation of 1, and then to multiply z by 60.

A value for z may be obtained by using an appropriate computer algorithm or using random number tables. Suppose the value of z is -1.3164. Then the "unobservable" error is $-1.3164 \times 60 = -79$ (to the nearest integer). Thus the observed radiocarbon determination is given by

$$x = 3006 - 79 = 2927.$$

If the sample were reanalysed, another, unknown error term would arise, say this time $+27$, and so another radiocarbon determination would result. In this case we would have

$$x = 3006 + 27 = 3033.$$

If a further analysis were carried out on the same sample, another, different result would be obtained and so on.

(If the precision of the laboratory were greater, a smaller standard deviation would result. For example, if the standard deviation were 40 instead of 60, then the "unobserved" error would be $-1.3164 \times 40 = -53$ and so the observed radiocarbon determination would be $3006 - 53 = 2953$. In this case the observation is closer to the true radiocarbon age. If, on the other hand, the standard deviation were 100, the error would be -132 and the observation would be 2874, much further away from the true value.)

9.2.5   A simple example

Suppose that a radiocarbon determination is reported as being 2927 bp with a standard deviation of 60. What inferences can be made about the calendar date of the sample and any associated archaeological event?

The data

In this case we have $x = 2927$. We assume that the standard deviation associated with this result is known and is taken to be that reported by the radiocarbon laboratory. Thus we take $\sigma = 60$.

The prior

For most archaeological problems little will be known *a priori* about the absolute date of a single event (though if pressed archaeologists could give wide limits within which they think the true date will lie). This means that we consider every calendar year to be as likely as any other and so we take the prior, $p(\theta)$, as

$$p(\theta) \propto 1 \quad \text{for } 0 < \theta. \tag{9.4}$$

That is, we are assuming a vague (and in this case, improper) prior for θ (see Section 7.7).

The likelihood

As X has a Normal distribution with mean $\mu(\theta)$ and variance σ^2 its probability density function is given by (see Section 5.10.3)

$$p(x|\theta) = \frac{1}{\sigma\sqrt{2\pi}} \exp\left\{-\frac{(x-\mu(\theta))^2}{2\sigma^2}\right\}$$

where $\mu(\theta)$ is given by (9.3). Therefore the likelihood is given by

$$l(\theta; x) \propto \exp\left\{-\frac{(x-\mu(\theta))^2}{2\sigma^2}.\right\} \tag{9.5}$$

Then, substituting for $\mu(\theta)$ into (9.5), the likelihood can be expressed as

$$l(\theta; x) \propto \begin{cases} \exp\left\{-\frac{(x-a_0-b_0\theta)^2}{2\sigma^2}\right\} & (\theta \le t_0) \\ \exp\left\{-\frac{(x-a_l-b_l\theta)^2}{2\sigma^2}\right\} & (t_{l-1} < \theta \le t_l, l = 1, 2, \ldots, L) \\ \exp\left\{-\frac{(x-a_L-b_L\theta)^2}{2\sigma^2}\right\} & (\theta > t_L). \end{cases} \tag{9.6}$$

The posterior

With the prior for θ and the likelihood given in (9.4) and (9.6) respectively, and by using Bayes' theorem, the posterior of θ, $p(\theta|x)$, is given by

$$p(\theta|x) \propto \begin{cases} \exp\left\{-\frac{(x-a_0-b_0\theta)^2}{2\sigma^2}\right\} & (\theta \le t_0) \\ \exp\left\{-\frac{(x-a_l-b_l\theta)^2}{2\sigma^2}\right\} & (t_{l-1} < \theta \le t_l, l = 1, 2, \ldots, L) \\ \exp\left\{-\frac{(x-a_L-b_L\theta)^2}{2\sigma^2}\right\} & (\theta > t_L). \end{cases} \tag{9.7}$$

The posterior probability density function of the calendar date, θ (given in (9.7) in a mathematical form), needs to be evaluated numerically to be of any practical use (see Section 8.5). In Figure 9.6(a) we plot the distribution in 20-year intervals. As one can see most of the probability lies between 3220 BP and 2900 BP. Moreover the distribution is not symmetric and does not have one mode but several. This makes reporting one "typical" date, such as the posterior mode, the posterior median or the posterior mean, meaningless. It is more appropriate to express our posterior beliefs about the date, θ, by means of an HPD region (see Section 7.4.2).

In Figure 9.6(b) and (c) the distributions are plotted at 10-year and 1-year intervals. The tendency, as the interval is reduced, is to make the distribution look more spiky and less smooth. This is a direct consequence of the abrupt changes in the slope of the piece-wise linear model of the calibration curve. Since the calibration curve is only estimated at 20-year intervals (on the

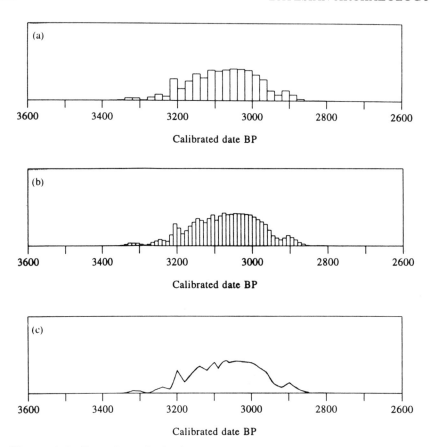

Figure 9.6 Posteriors of calendar age showing multi-modal distributions (a) plotted at 20-year intervals, (b) plotted at 10-year intervals and (c) plotted at 1-year intervals.

calendar scale) it seems unrealistic to report results calculated on a finer scale than this. Even so 20-year intervals may be regarded as giving a false impression of accuracy.

Interpretation

In interpreting the results it is probably best to report an HPD region (see Section 7.4.2). The 90% HPD region actually consists of three disjoint intervals, namely 3260 to 3240, 3220 to 2940 and 2920 to 2900 BP. Because the gap between successive intervals is relatively small, in practice we would suggest reporting these as one interval, from 3260 to 2900 BP. The 80% HPD region consists of the two intervals 3220 to 3200 and 3180 to 2960 BP. The 50% HPD region also consists of two intervals, 3160 to 3140 and 3120 to 3000 BP.

Comments

(i) The most well-known and commonly used computer program, called CALIB, for the calibration of radiocarbon determinations (see Stuiver and Reimer, 1993), calculates the posterior expressed in (9.7) although it is not called a posterior (and nowhere is Bayes mentioned!).

(ii) Typically, because of the wiggles in the calibration curve, the posterior density is not symmetric and is often multi-modal which makes interpretation difficult. For example consider the plots given in Figure 9.7 of the posteriors corresponding to radiocarbon results 1500, 2500, 3500 and 4500, all having the same standard deviation of 60. Despite the standard deviations being the same, the resulting distributions on the calendar time-scale show significant differences. These differences are a direct result of the different behaviour of the calibration curve at different points in time. At times it is very wiggly, in others it is steep, and sometimes it is flat.

9.3   Archaeological problems and questions

Whenever possible archaeologists use material from short-lived organisms, such as seeds, to provide dates for precise moments in the past. Charcoal from large timbers will at best provide an "average" date for the time at which the tree was growing (if the sample contains growth rings from the whole lifetime of the tree), but might provide a date for some arbitrary moment within the life of the tree. If the sample is of reused timber it might even date to a period considerably before the one in which we are actually interested.

There are several other important factors that need to be considered. Perhaps the most fundamental of these is that archaeologists are rarely interested in the calendar date of the organic sample they send for analysis. Most commonly, radiocarbon dating is used to date events or periods with which the samples can be directly or indirectly associated: the date of a feature such as a post-hole, a hearth or a human burial. (An obvious exception to this is the dating of specific items such as the Turin Shroud.)

Commonly, then, archaeologists obtain radiocarbon determinations from samples associated with specific excavated features. On the basis of such determinations, the types of questions posed include "when did event A occur?", "what is the time period between event A and event B?", "what is the likely order of events C, D and E?" and so on.

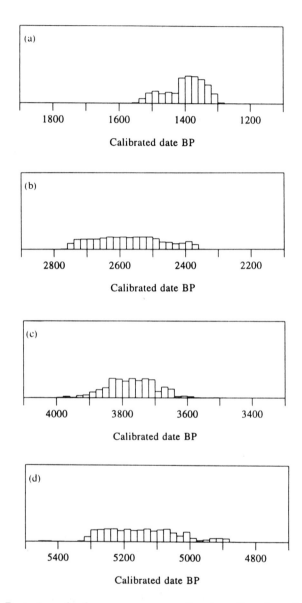

Figure 9.7   Posterior calendar years corresponding to radiocarbon dates of (a)
1500, (b) 2500, (c) 3500 and (d) 4500.

During the 1980's it soon became clear that questions of this type could only occasionally be answered using the calibration of single radiocarbon determinations. Increasingly, archaeologists were encouraged, by funding bodies and radiocarbon laboratories alike, to submit collections of related samples, for example several samples for a single event, or groups of samples whose relative chronological relationships were at least partially understood. In the case studies which follow, we will see examples of exactly this kind of inter-relationship. Frequently relative dating methods, such as stratigraphy or typology, can provide *a priori* information for the sequence of radiocarbon dates.

From the nineteenth century onwards scholars have built up and developed complex systems of periods and *phases* based on artefact typologies and stratigraphy. Indeed in some cases, like that of Egypt, the archaeology has been tied in with ancient dating systems such as dynasties and kingdoms. On the other hand the classic sequence of neolithic phases in Bulgaria is based on the massive stratified site of Karanovo; the earliest neolithic culture in Bulgaria is called Karanovo I, the much later Copper Age phase is often referred to as Karanovo VI. If radiocarbon determinations are available from such phases, archaeologists begin to ask questions such as "when did phase A stop and phase B begin?", "do phases A and B overlap?", "do they abut?" or "is there a hiatus between them?" and so on (see Figure 9.8).

So, what types of prior knowledge can we expect to have available to us when we approach a new radiocarbon calibration problem? In a few cases there may be information about the time period between successive events; for examples of this see Case Studies III and IV that follow below. But more commonly, the prior information takes the form of chronological orderings (or partial orderings) of events (see Case Study I) or phases (see Case Study II) which result in constraints on the dates and therefore on the parameters of the models. See Litton and Leese (1991) for further discussion of the types of problem which present themselves.

Moreover, although radiocarbon dating is a well-used archaeological dating tool, it does have its difficulties. For various reasons, such as sample contamination, even with the greatest of care being taken at all stages, the method may give an erroneous result. Earlier carbon (for example from other archaeological organic matter or from petrochemicals) can be accidentally incorporated into a sample in ways for which no allowance can be made, thus making the sample appear older than it really is. Likewise, later carbon (such as more recent organic matter) can infiltrate earlier contexts, making them appear younger than they really are. That is to say the reported radiocarbon age might be wrong and therefore the associated estimated calendar age will be in error also. Case Study V demonstrates how the Bayesian methodology can deal with sets of radiocarbon determinations containing the occasional "rogue" measurement.

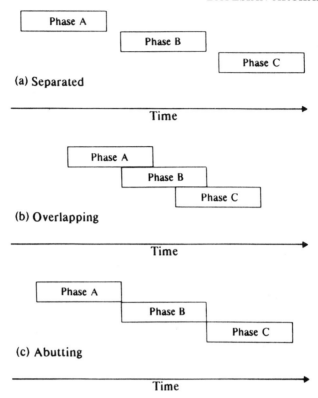

Figure 9.8   Schematic representation of three different models of archaeological phases. (a) Phases separated by time intervals. (b) Phases overlapping in time. (c) Phases abutting.

9.4   Case Study I — St Veit-Klinglberg, Austria — stratification

9.4.1   The archaeological problem

Consider the statistical analysis and subsequent archaeological interpretation of radiocarbon determinations from the early Bronze Age settlement of St Veit-Klinglberg, Austria reported by Buck *et al.* (1994a). We have mentioned this site in Chapters 1 and 2 and have used a subset of the data to illustrate the application and implementation of Bayesian paradigm in Chapters 7 and 8. Now we are in a position to discuss the analysis of the data in more detail.

The primary objective of excavating this site was to investigate the organization of a community involved in the production and distribution of copper. The dating of these settlements is particularly problematical because they lack the metalwork, found in hoards and graves, which often are used to establish typological chronologies; moreover the domestic pottery (which is also often used to provide such information) is not distinctive enough to allow fine chronological divisions. Therefore a further objective was to obtain a series of radiocarbon determinations which would date the site absolutely.

9.4.2 The model

As explained in Chapter 1, ten contexts could be partially ordered on stratigraphic grounds. A further five contexts were from deposits which had suffered erosion and could not be placed reliably in the sequence. (See Figure 1 of Chapter 1 for a schematic representation of the archaeological interpretation of the stratigraphy.) Contexts 493, 758, 814, 923, 925, 1168 and 1235 are associated with Phase I of the occupation and contexts 358, 813 and 1210 with Phase II. Contexts 45, 119, 460, 1308 and 1319 could not be assigned on the basis of the stratigraphic evidence alone to either of the phases.

Let θ_i denote the calendar date BP of event i. Then the archaeological information may be expressed in the form of the following inequalities.

$$\theta_{758} > \theta_{814} > \theta_{1235} > \theta_{358} > \theta_{813} > \theta_{1210},$$

$$\theta_{493} > \theta_{358},$$

$$\theta_{925} > \theta_{923} > \theta_{358} \text{ and}$$

$$\theta_{1168} > \theta_{358},$$

where the subscripts are the context's numbers allocated during excavation. At St Veit-Klinglberg, archaeological interest focused upon the dates of the ten events with secondary interest being placed upon adding the five unordered events to the same sequence.

9.4.3 The data

The data consist of sixteen radiocarbon determinations corresponding to the fifteen contexts (there are two determinations from context 925). These determinations are given in Table 9.1 together with information regarding the phases. Note here that each radiocarbon determination, x_i, has its own standard deviation, σ_i, reported by the radiocarbon laboratory.

9.4.4 The prior information

We observe that nothing is known a priori about the time periods between the dates of the ten contexts. All we can say is that dates which do not satisfy the above constraints, and so do not agree with the stratigraphy, are impossible on

Table 9.1 The radiocarbon determinations from the fifteen contexts at St Veit-Klinglberg.

Context number	Radiocarbon determination	Laboratory identifier	Phase
758	3275±75	OxA-3899	I
814	3270±80	OxA-3897	I
1235	3400±75	OxA-3900	I
493	3190±75	OxA-3898	I
925	3420±65	OxA-3882	I
925	3370±75	OxA-3883	I
923	3435±60	OxA-3881	I
1168	3160±70	OxA-3901	I
358	3340±80	OxA-3903	II
813	3270±75	OxA-3902	II
1210	3200±70	OxA-3904	II
460	3390±80	OxA-3905	Unknown
45	3480±75	OxA-3906	Unknown
119	3250±75	OxA-3907	Unknown
1308	3115±70	OxA-3908	Unknown
1319	3460±70	OxA-3909	Unknown

archaeological grounds. For example any set of dates containing $\theta_{758} = 2600$ and $\theta_{814} = 2700$ will be impossible since $\theta_{758} < \theta_{814}$ is in contradiction of the archaeological information. In contrast the set of dates

$$\theta_{758} = 3000, \theta_{814} = 2900, \theta_{1235} = 2800, \theta_{358} = 2700, \theta_{813} = 2600,$$

$$\theta_{1210} = 2500, \theta_{493} = 2778, \theta_{925} = 2900, \theta_{923} = 2800, \theta_{1168} = 2756$$

is a possibility as is the set

$$\theta_{758} = 3100, \theta_{814} = 3000, \theta_{1235} = 2900, \theta_{358} = 2800, \theta_{813} = 2700,$$

$$\theta_{1210} = 2600, \theta_{493} = 2878, \theta_{925} = 3000, \theta_{923} = 2900, \theta_{1168} = 2856.$$

Both sets of dates satisfy the constraints imposed by the archaeology but, *a priori*, we have no preference between the two. As a consequence we take them to be equally likely. Therefore, the θ_i are assumed to have a jointly uniform prior subject to the above inequalities. The (joint) prior for $\boldsymbol{\theta}$ may be expressed as

$$p(\theta) \propto I_A(\boldsymbol{\theta})$$

where

$$I_A(\boldsymbol{\theta}) = \begin{cases} 1 & \boldsymbol{\theta} \in A \\ 0 & \boldsymbol{\theta} \notin A \end{cases}$$

and A is the set of values of $\boldsymbol{\theta}$ satisfying the above inequalities.

9.4.5 The likelihood

We have that given θ_i, X_i has a Normal distribution with mean $\mu(\theta_i)$ and variance σ_i^2. Conditional on the θ_i, the X_i are independent, therefore the likelihood is proportional to the product of the probability densities of the X_i. Therefore the likelihood is given by

$$l(\boldsymbol{\theta}; \boldsymbol{x}) \propto \prod_i \exp\left\{ -\frac{(x_i - \mu(\theta_i))^2}{2\sigma_i^2} \right\}. \tag{9.8}$$

9.4.6 The posterior

By Bayes' theorem, the (joint) posterior of $\boldsymbol{\theta}$ is given by

$$p(\boldsymbol{\theta}|\boldsymbol{x}) \propto \left\{ \prod_i \exp\left\{ -\frac{(x_i - \mu(\theta_i))^2}{2\sigma_i^2} \right\} \right\} I_A(\boldsymbol{\theta}).$$

9.4.7 Computational details

Were it not for the constraints provided by the archaeological prior information, each of the θ_i would be, a posteriori, independent and the marginal information could be easily evaluated numerically as in Section 8.5. Indeed, even with these constraints, evaluation of the marginal posteriors of interest is not impossible using conventional numerical integration techniques. However, as has been demonstrated in Chapter 8, the implementation of the Gibbs sampler when order constraints are present is straightforward. This was the approach adopted by Buck et al. (1994a). See also Buck et al. (1992) for details of the application of the Gibbs sampler to problems in radiocarbon dating.

At each iteration of the Gibbs sampler, a value for θ_i is sampled from its full conditional density which is proportional to

$$\exp\left\{ -\frac{(x_i - \mu(\theta_i))^2}{2\sigma_i^2} \right\} I_{(c_i, d_i)}(\theta_i)$$

where c_i and d_i possibly depend upon the values of the other θ_i. Here

Table 9.2 95% highest posterior density regions for the calendar dates of the samples from St Veit-Klinglberg. All dates are measured BP. Column (i) was calculated without using the archaeological information. Column (ii) was calculated using the archaeological information.

Context number	95% Highest posterior density region	
	(i)	(ii)
758	3685 to 3375	3825 to 3755
		3725 to 3515
814	3695 to 3365	3685 to 3675
		3655 to 3485
1235	3835 to 3475	3625 to 3465
493	3625 to 3605	3625 to 3415
	3595 to 3265	
925	3825 to 3785	3835 to 3625
	3775 to 3745	
	3735 to 3545	
	3505 to 3485	
923	3865 to 3565	3805 to 3785
		3775 to 3545
		3505 to 3485
1168	3555 to 3245	3625 to 3605
	3235 to 3215	3595 to 3415
358	3825 to 3785	3535 to 3385
	3765 to 3745	
	3735 to 3395	
813	3685 to 3665	3515 to 3365
	3655 to 3365	
1210	3625 to 3605	3475 to 3265
	3595 to 3325	
	3295 to 3265	

$I_{(c_i,d_i)}(\theta_i)$ is defined to be

$$I_{(c_i,d_i)}(\theta_i) = \begin{cases} 1 & c_i < \theta_i < d_i \\ 0 & \text{otherwise.} \end{cases}$$

For example, consider just the sample from context 358. The calendar date for this event is restricted to be earlier than the date of event 813, but later than events 1235, 493, 923 and 1168. Thus, c_{358} is set equal to the current value of θ_{813} and d_{358} to the smallest of the current values of $\theta_{1235}, \theta_{493}, \theta_{923}$ and

θ_{1168}. If we consider context 493, it must be before context 358 but nothing else is known; therefore in this case c_{493} is set to the current value of θ_{358} and d_{493} is set to infinity.

9.4.8 Results and interpretation

Recall that in Figures 2 and 3 of Chapter 1, we gave plots of the posterior distribution of the date of context 358, θ_{358}, computed with and without the archaeological information. In Figure 9.9, we give plots of the posterior distributions of all the θ_i obtained when the radiocarbon determinations are calibrated separately, without the a priori information provided by the known order of the events. In contrast, Figure 9.10 shows the posteriors which arise when the calibration process includes archaeological constraints as given above. The 95% highest posterior density regions with and without the archaeological prior information are given in Table 9.2. In general we can see that the intervals calculated when the archaeological prior knowledge is included in the analysis are shorter than when it is not. For instance, in the case of context 1235 the interval is some 200 years shorter — quite a dramatic improvement!

Because of its sloping location on a hill spur, St Veit-Klinglberg exhibited complex depositional patterns so that five of the archaeological deposits from which radiocarbon samples were obtained could not be related stratigraphically to the other ten. Within the Bayesian framework, particularly when employing the Gibbs sampler for the computations, it is quite simple to include in the analysis the unstratified contexts. As a result we can make inferences about how they relate to the rest of the contexts. The results are given in Table 9.3. Care must be taken not to regard these results as being extremely accurate. For example in Table 9.3 the posterior probability that context 460 is in Phase I is given as 0.66. This should, we believe, be interpreted as saying that, a posteriori, there is a 60–70% chance

Table 9.3 Posterior probability of assigning the unstratified contexts to the phases at St Veit-Klinglberg.

Sample from context	Radiocarbon determination	Laboratory identifier	Before Phase I	Phase I	Between the phases	Phase II	After Phase II
460	3390±80	OxA-3905	0.27	0.66	0.03	0.04	0.00
45	3480±75	OxA-3906	0.62	0.38	0.00	0.00	0.00
119	3250±75	OxA-3907	0.01	0.51	0.10	0.28	0.10
1308	3115±70	OxA-3908	0.00	0.04	0.03	0.27	0.65
1319	3460±70	OxA-3909	0.55	0.45	0.00	0.00	0.00

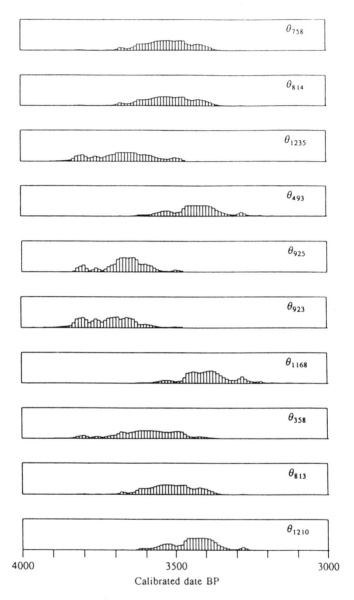

Figure 9.9   Posteriors of the St Veit-Klinglberg dates without the inclusion of prior stratigraphic information. (After Buck *et al.*, 1994a, p. 435, fig. 3.)

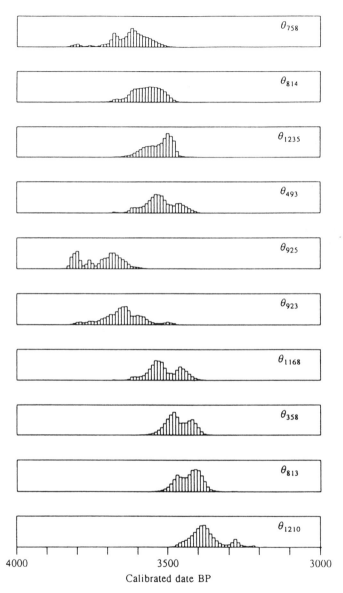

Figure 9.10 Posterior distributions of St Veit-Klinglberg calendar dates with prior stratigraphic information. (After Buck *et al.*, 1994a, p. 436, fig. 4.)

that context 460 belongs to Phase I. The chance that it is before Phase I is 20–30% and that it is later than Phase I is less than 10%.

Such information offers the archaeologists greatly enhanced insight into the chronological relationships of several parts of the site that were not directly linked by archaeological stratigraphy. See Buck *et al.* (1994a) for a more detailed archaeological interpretation of the Bayesian analysis of the radiocarbon determinations from St Veit-Klinglberg.

9.5 Case Study II — Jama River Valley, Ecuador — multiphase

9.5.1 Background

For our second case study we describe the use of the Bayesian paradigm to date the development of the Jama-Coaque culture in the northern Manabí (see Zeidler *et al.*, forthcoming). Despite its importance in the pre-Columbian history of Ecuador as the territory of the Jama-Coaque culture (a series of complex chiefdoms with an elaborate ceramic tradition spanning some 4000 years), temporal information about the Jama River Valley has, until recently, been limited and poorly understood. In a recent archaeological survey in the Jama River Valley covering sixteen different sites, seven major archaeological phases were defined. Initially, one major site was excavated in the upper valley which, through its deep stratigraphy, led to a master ceramic sequence. Over several subsequent seasons of study, the archaeologists undertook further excavations and integrated the information obtained about the sites studied.

Thus a ceramic sequence was established using the methods of relative dating, stratigraphy and typology, but the sequence had no absolute chronology of its own. Rather it was assumed, without any totally reliable evidence, to share the temporal placement of comparable sequences from elsewhere in Ecuador. In order to date the phases, some thirty-seven radiocarbon determinations were made on samples taken from contexts at the sixteen different archaeological sites, whose occupation covered the whole period of interest. Each sample was clearly and unambiguously assigned to one and only one of the seven phases. The primary use of the radiocarbon determinations was to estimate the beginning and ending dates for each of the seven phases as defined by the ceramics.

9.5.2 The model

We need to model the seven phases or time periods that have been identified by the archaeologists. Let α_j and β_j represent the beginning and ending dates (measured BP) of phase j ($j = 1, 2, \ldots, 7$). Thus α_1 represents the beginning of phase 1 and β_1 its ending. Note that, since the dates are measured in years BP, we have $\alpha_1 > \beta_1$. This is true for all phases so we have $\alpha_j > \beta_j$ for $j = 1, 2, \ldots, 7$.

Let n_j be the number of samples assigned to the jth phase and $x_{i,j}$ be the ith radiocarbon determination obtained from phase j with $\sigma_{i,j}$ being the corresponding standard deviation. Let $\theta_{i,j}$ represent the calendar date of the ith radiocarbon sample in the jth phase. Since nothing is known about the date of sample i within phase j, we assume that each date within the phase is equally likely. Therefore we assume that

$$\theta_{i,j}|\beta_j, \alpha_j \sim U(\beta_j, \alpha_j)$$

where we recall that $U(a, b)$ denotes a Uniform distribution on the interval a to b (see Section 5.10.1).

9.5.3 The data

Details of the radiocarbon determinations and their assignment to the phases are given in Table 9.4. A plot of the data with the relevant section of the calibration curve is given in Figure 9.11.

9.5.4 The prior

As mentioned earlier, current understanding of the archaeological phasing of sites in northern Manabí has been established as a result of many years of field work on sixteen different sites. The seven phases are all characterized in different ways and the nature of evidence for the phase boundaries is varied. The phases have all been labelled on the basis of the types of pottery associated with them and are known (in chronological order) as Piquigua, Tabuchila, Muchique 1, Muchique 2, Muchique 3, Muchique 4, and Muchique 5.

Several methods were used to establish the relationship between the archaeological phases. The most secure of these was the identification of tephra (volcanic ash) on several sites which could be linked directly to occupational breaks associated with the end of one phase and the start of another. Such tephra levels form the major part of the evidence for the end of the Tabuchila phase and the start of Muchique 1, and for the end of Muchique 1 and the start of Muchique 2. In addition, some phase boundaries were associated with abrupt changes in ceramic tradition, in particular the end of Piquigua and the start of Tabuchila, and the end of Tabuchila and the start of Muchique 1.

Other phase boundaries are marked by more subtle or gradual changes in ceramic tradition which were studied by seriation. Such evidence is available for the end of Muchique 2 and the start of Muchique 3, and for the end of

Table 9.4   The radiocarbon determinations from the Jama River Valley shown with their associated phase information. (See Zeidler *et al.*, forthcoming for more details.)

Site number	Site name	Ceramic phase	Laboratory identifier	Radiocarbon determination	Phase number
M3D2-001	San Isidro	Piquigua	ISGS-1221	3630±70	I
M3D2-001	San Isidro	Piquigua	ISGS-1222	3620±70	I
M3D2-001	San Isidro	Piquigua	1SGS-1223	3560±70	I
M3D2-001	San Isidro	Piquigua	PITT-426	3545±135	I
M3D2-001	San Isidro	Piquigua	ISGS-1220	3500±70	I
M3B3-001	La Mina	Tabuchila	ISGS-2366	3030±80	II
M3D2-001	San Isidro	Tabuchila	AA-4140	2845±95	II
———	Véliz	Tabuchila	HO-1307	2800±115	II
M3B4-031	El Mocorral	Tabuchila	ISGS-2377	2500±160	II
M3B2-001	Don Juan	Muchique 1	PITT-1128	2430±170	III
M3B3-009	La Cabuya	Muchique 1	PITT-1125	2170±40	III
M3D2-065	Capaperro	Muchique 1	PITT-870/871	2125±300	III
M3B4-037	(unnamed)	Muchique 1	ISGS-2372	1990±100	III
M3B2-001	Don Juan	Muchique 1	ISGS-2376	1980±70	III
M3B2-001	Don Juan	Muchique 1	AA-4138	1960±90	III
M3B4-043	(unnamed)	Muchique 1	ISGS-2373	1950±70	III
M3B3-002	El Tape	Muchique 2	ISGS-2367	1610±70	IV
M3B3-029	La Ladrillera	Muchique 2	ISGS-2370	1590±80	IV
M3B3-003	Jama	Muchique 2	ISGS-2368	1540±70	IV
M3B3-029	La Ladrillera	Muchique 2	ISGS-2371	1520±80	IV
M3B4-011	Pechichal	Muchique 2	PITT-710	1480±75	IV
M3B4-001	Pechichal	Muchique 2	PITT-709	1330±70	IV
M3B3-002	El Tape	Muchique 2	PITT-410	1260±30	IV
M3B4-011	Pechichal	Muchique 2	PITT-708	1240±40	IV
M3D2-065	Capaperro	Muchique 3	PITT-869	1195±85	V
M3D2-065	Capaperro	Muchique 3	PITT-890	1170±340	V
M3B3-001	La Mina	Muchique 3	PITT-417	1120±30	V
M3B3-001	La Mina	Muchique 3	AA-4137	1120±90	V
M3B3-002	El Tape	Muchique 3	AA-4136	1030±90	V
M3B2-001	Don Juan	Muchique 3	PITT-1127	960±35	V
M3B4-022	P Muñoz	Muchique 3	PITT-711	880±70	V
M3B2-001	Don Juan	Muchique 3	PITT-1126	870±45	V
M3B3-045	Pasaborracho	Muchique 3	ISGS-2375	820±70	V
M3B3-012	El Acrópolis	Muchique 3	PITT-1129	800±40	V
M3B4-001	Moncayo	Muchique 4	PITT-707	630±30	VI
M3B3-001	La Mina	Muchique 4	PITT-415	515±40	VI
M3B3-001	La Mina	Muchique 5	PITT-414	305±35	VII

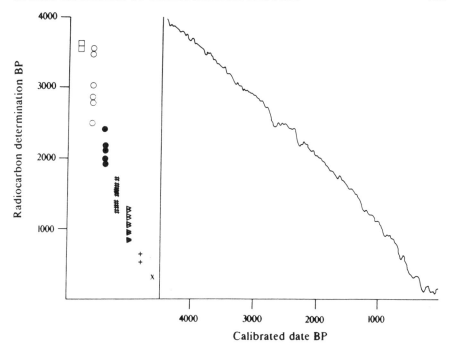

Figure 9.11   Plot of the radiocarbon determinations from Phases I–VII in the Jama River Valley, Ecuador, together with the relevant section of the calibration curve. (After Zeidler *et al.*, forthcoming.)

Muchique 3 and the start of Muchique 4. The boundary between Muchique 4 and Muchique 5 is different from the others in that it is marked by a historical event, the arrival of the Spanish in the area. This is associated with general continuity of ceramic tradition, but with considerable impoverishment of the ceramic arts.

Naturally much discussion was needed to clarify the beliefs that the archaeologists had about the phase relationships before the above understanding was established. This process was helped by the use of simple sketches which allowed representation of possible relative chronological relationships between the various boundaries. That finally agreed and adopted for the published analysis is given in Figure 9.12.

In mathematical terms, this information was expressed in the form of the following relationships between the beginning and ending dates of the phases.

$$\alpha_1 > \beta_1 \geq \alpha_2 > \beta_2 \geq \alpha_3 > \beta_3 \geq \alpha_4 > \beta_4$$

$$\alpha_4 > \alpha_5 > \alpha_6$$

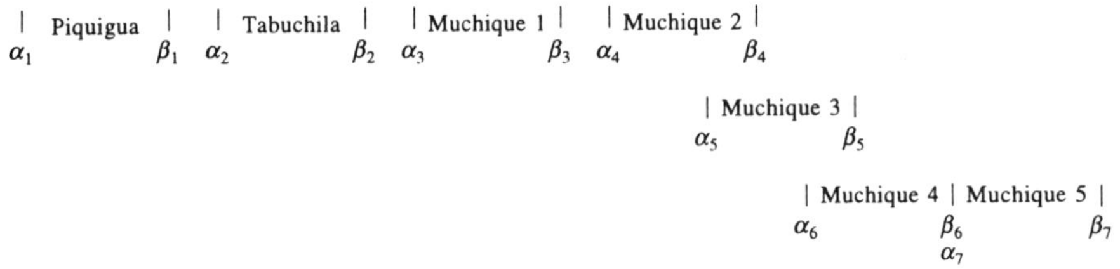

| Piquigua | | Tabuchila | | Muchique 1 | | Muchique 2 |
α_1 β_1 α_2 β_2 α_3 β_3 α_4 β_4

| Muchique 3 |
α_5 β_5

| Muchique 4 | Muchique 5 |
α_6 β_6 β_7
                    α_7

Figure 9.12 Diagram of phases representing the beginning, α_j, and the ending, β_j, dates of Phases I–VII from the Jama River Valley, Ecuador. (After Zeidler *et al.*, forthcoming.)

$$\beta_4 > \beta_5 > \beta_6$$

and

$$\beta_6 = \alpha_7.$$

The relationships between the other α_j and β_j were unknown *a priori* (except that $\alpha_j > \beta_j$). Part of the purpose of the statistical investigation was precisely to provide information about them *a posteriori*. In terms of absolute dates nothing is known so that we will assume that all values of the α_j and β_j satisfying the above constraints are equally likely. That is the prior for α and β is given by

$$p(\alpha, \beta) \propto I_A(\alpha, \beta) \tag{9.9}$$

where

$$I_A(\alpha, \beta) = \begin{cases} 1 & \alpha, \beta \in A \\ 0 & \text{otherwise} \end{cases}$$

where A is the set of values of $\alpha = (\alpha_1, \ldots, \alpha_7)$ and $\beta = (\beta_1, \ldots, \beta_7)$ satisfying the above constraints.

9.5.5   The likelihood

The likelihood is developed in three stages. Firstly, given the calendar date $\theta_{i,j}$, $X_{i,j}$ has a Normal distribution with mean $\mu(\theta_{i,j})$ and variance $\sigma^2_{i,j}$. Therefore the probability density function of $X_{i,j}$ is

$$p(x_{i,j}|\theta_{i,j}) = \frac{1}{\sigma_{i,j}\sqrt{2\pi}} \exp\left\{ -\frac{(x_{i,j} - \mu(\theta_{i,j}))^2}{2\sigma^2_{i,j}} \right\}.$$

Secondly, within a phase, any value of $\theta_{i,j}$ is assumed to be as likely as any other so that

$$p(\theta_{i,j}|\alpha_j, \beta_j) = \frac{1}{\alpha_j - \beta_j} \quad \text{for } \beta_j \leq \theta_{i,j} \leq \alpha_j.$$

Now the likelihood based on $x_{i,j}$ may be expressed as

$$l(\theta_{i,j}, \alpha_j, \beta_j; x_{i,j}) \propto p(x_{i,j}|\theta_{i,j})p(\theta_{i,j}|\alpha_j, \beta_j)$$

and therefore is given by

$$l(\theta_{i,j}, \alpha_j, \beta_j; x_{i,j}) \propto \frac{1}{\alpha_j - \beta_j} \exp\left\{-\frac{(x_{i,j} - \mu(\theta_{i,j}))^2}{2\sigma_{i,j}^2}\right\}.$$

Thirdly, as the observations are assumed to be conditionally independent, the likelihood for the whole sample, x, is given by

$$l(\boldsymbol{\theta}, \boldsymbol{\alpha}, \boldsymbol{\beta}; \boldsymbol{x}) = \prod_{j=1}^{7} \prod_{i=1}^{n_j} l(x_{i,j}|\theta_{i,j}, \alpha_j, \beta_j)$$

and so is equal to

$$\prod_{j=1}^{7} \left\{ (\alpha_j - \beta_j)^{-n_j} \prod_{i=1}^{n_j} z_{i,j} \mathrm{I}_{(\beta_j, \alpha_j)}(\theta_{i,j}) \right\} \tag{9.10}$$

where

$$z_{i,j} = \exp\left\{-\frac{(x_{i,j} - \mu(\theta_{i,j}))^2}{2\sigma_{i,j}^2}\right\}$$

and

$$\mathrm{I}_{(\beta_j, \alpha_j)}(\theta_{i,j}) = \begin{cases} 1 & \beta_j \leq \theta_{i,j} \leq \alpha_j \\ 0 & \text{otherwise.} \end{cases}$$

9.5.6   The posterior

By Bayes' theorem, the joint posterior density of $\boldsymbol{\theta}$, $\boldsymbol{\alpha}$ and $\boldsymbol{\beta}$ is found by multiplying the prior (9.9) and the likelihood (9.10) together to obtain

$$p(\boldsymbol{\theta}, \boldsymbol{\alpha}, \boldsymbol{\beta}|\boldsymbol{x}) \propto \mathrm{I}_A(\boldsymbol{\alpha}, \boldsymbol{\beta}) \prod_{j=1}^{7} \left\{ (\alpha_j - \beta_j)^{-n_j} \prod_{i=1}^{n_j} z_{i,j} \mathrm{I}_{(\beta_j, \alpha_j)}(\theta_{i,j}) \right\}.$$

9.5.7   The calculations

As with the St Veit-Klinglberg example, it is possible to evaluate the posterior densities of the α_j and β_j, using quadrature methods although sophisticated software would be essential. In fact Naylor and Smith (1988) did just this for a similar but slightly simpler example which involved dating the boundaries between four ceramic phases from the Iron Age hill-fort at Danebury, England. However, Markov chain Monte Carlo methods allow a less complex method of computation (Buck *et al.*, 1992). We note that the partial ordering information offers no real difficulty since the $\theta_{i,j}$ can be sampled from a restricted range as in the previous example. The full conditionals of the α_j and β_j are fairly easy to write down and thus to sample.

For instance, since Phases IV and V may or may not overlap but Phase III must precede Phase IV, the full conditional density for α_4 is proportional to $(\alpha_4 - \beta_4)^{-n_4} I_{(c,d)}(\alpha_4)$ where c is the largest of $\alpha_5, \theta_{1,4}, \ldots, \theta_{n_4,4}$ and $d = \beta_3$. Thus, α_4 can be sampled using an inversion method (see Section 8.7.2).

The sampling of β_6 is slightly more complicated as Phases VI and VII abut one another so we have $\beta_6 = \alpha_7$. Furthermore Phase V is known to have ended before Phase VI ended so $\beta_5 > \beta_6$. The full conditional for β_6 is, therefore, proportional to

$$(\alpha_6 - \beta_6)^{-n_6}(\beta_6 - \beta_7)^{-n_7} I_{(c,d)}(\beta_6)$$

where

$$c \text{ is the largest of } \theta_{1,7}, \ldots, \theta_{n_7,7}$$

and

$$d \text{ is the smallest of } \beta_5, \theta_{1,6}, \ldots, \theta_{n_7,7}.$$

One possible method of sampling β_6 is to use rejection (see Section 8.7.3).

9.5.8   *Results and interpretation*

Of archaeological interest are the dates of the beginning and end of each of the phases. Posterior densities are given in Figure 9.13 in the form of histogram-like plots in the same way as the posteriors in the St Veit-Klinglberg case study. The calendar dates provided for the phase boundaries offer a reliable and specific calendar-based chronology for the Jama River Valley. This can now profitably be compared with other, more generalized chronologies established by other workers for different parts of Ecuador. This, in turn, will permit greater understanding of the development of ancient cultures in Ecuador and further afield. For a detailed discussion of the results, the interpretation and the implications for the understanding of the Jama-Coaque culture, we refer the reader to Zeidler *et al.* (forthcoming).

9.6   Case Study III — Stolford, England — wiggle-matching

9.6.1   *Background*

An obvious extension to the type of problem illustrated in Case Study I is to include some archaeological information in the analysis about the time

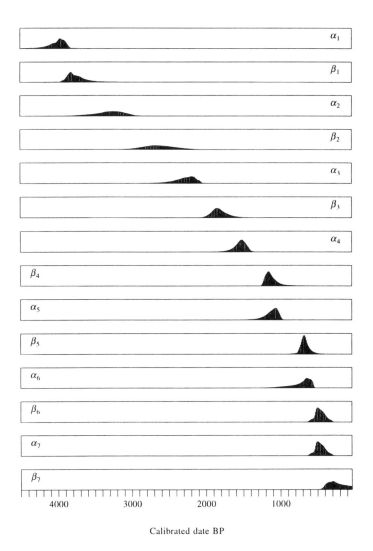

Calibrated date BP

Figure 9.13 Posterior densities of the beginning and ending dates of Phases I–VII of the Jama River Valley, Ecuador. (After Zeidler *et al.*, forthcoming.)

interval between events. Such a situation is that of "wiggle-matching". Wiggle-matching is a method for dating a sequence of related samples of known deposition rate using the high-precision radiocarbon calibration curve. In broad terms, samples are taken for radiocarbon analysis from archaeological events of known relative calendar years (although their absolute calendar dates are unknown). The corresponding radiocarbon results are then formed into a floating curve which is subsequently compared with the high-precision calibration curve to give a precise calendar date for the events. Examples of wiggle-matching are found in Pearson (1986), Baillie and Pilcher (1988), Baillie (1990), Clymo *et al.* (1990), Weninger (1986) and Manning and Weninger (1992). However it is clear from the literature that there is no common approach to carrying out the necessary comparisons. Some workers use highly subjective visual matching of graphs, while others use *ad hoc* statistical methods based on least squares.

The question is how to identify, quantify and incorporate into the statistical analysis prior information occurring in various situations that are suitable for "wiggle-matching". In this case study and the subsequent one we will demonstrate how the problem can be formulated, modelled and subsequently analysed using the Bayesian paradigm. We follow the approach adopted by Christen (1994a) and Christen and Litton (1995).

To illustrate the Bayesian approach to wiggle-matching we will use as an example a floating (that is undated) tree-ring chronology from Stolford, England (see Baillie and Pilcher, 1988). As subsequent dendrochronological work resulted in the precise dating of this chronology, it makes an ideal example to illustrate the methodology.

9.6.2   The model

In general, suppose we have a sequence of n samples; in the Stolford case we have $n = 8$. Let the calendar dates of the samples be represented by $\theta_1, \theta_2, \ldots, \theta_n$. We also assume that the dates are ordered so that θ_i is after θ_{i+1}. In other words, we have $\theta_1 < \theta_2 < \cdots < \theta_n$ (in years BP). Let x_i be the estimated radiocarbon age of the ith sample and σ_i the corresponding standard deviation.

In the case of Stolford, the samples were taken from wood and so the time interval between successive samples could easily be measured. Let ϕ_j be the known interval between the dates of sample j and sample $j - 1$. That is

$$\theta_j - \theta_{j-1} = \phi_j \quad \text{where the } \phi_j \text{ are known, } j = 2, 3, \ldots, n.$$

Hence interest focuses on estimating just one parameter, θ_1, since the others may be easily calculated once the value of θ_1 has been established.

Table 9.5 Radiocarbon determinations and associated relative calendar years for the eight samples from Stolford. Following Baillie (1990) the standard deviation of each average radiocarbon determination is taken as 10.

Sample number (j)	Average radiocarbon determination	Relative calendar year interval (ϕ_j)
1	5030	–
2	5025	29
3	5031	30
4	5070	28
5	5094	30
6	5173	29
7	5196	27
8	5168	30

9.6.3 The data

Samples were taken for radiocarbon dating at fairly regular intervals from the 272-year long tree-ring chronology and each sample was dated by up to twenty laboratories. The results from the different laboratories were combined to form a weighted average. These weighted averages all had standard deviations of less than 10 years and therefore could be considered to be high-precision dates. Following Baillie (1990), we take the standard deviation of each mean determination to be 10 years. The weighted means and their associated relative calendar year intervals (the ϕ_j) are given in Table 9.5.

In Figures 9.14 and 9.15 are plotted the high-precision calibration curve and the floating radiocarbon curve composed of the Stolford samples. The problem is to "fit" the floating curve against the calibration curve in the best possible manner. That is, we move the floating curve against the calibration curve until we find the position of "best" fit, a highly subjective process if done by eye.

9.6.4 Remodelling the calibration curve

We note that the standard deviations in Table 9.5 are much smaller than those we have been dealing with up to now in our case studies. In fact a standard deviation of 10 is roughly of the same size as that of the calibration curve itself at each knot. As a consequence of this we need to include in the analysis some allowance for the fact that the curve itself is uncertain. Christen (1994a) discusses this in some depth and we follow his suggestion. Christen shows that when the standard deviation of the radiocarbon determination is less than about 30, the uncertainty we have about the precise form of the

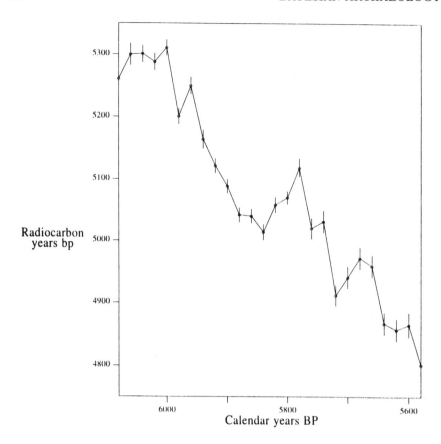

Figure 9.14 Plot of the piece-wise linear calibration curve from 6080 to 5580 BP.
The vertical bars indicate the 1σ ranges where σ is the standard deviation
reported by the laboratory.

curve should be taken into account. He suggests that in these circumstances
X should be modelled as having a Normal distribution with mean $\mu(\theta)$ and
variance given by $\omega^2(\theta)$ where

$$\omega^2(\theta) = \sigma^2 + \gamma^2(\theta).$$

The first term on the right-hand side is the reported laboratory standard
deviation. The second term reflects the uncertainty we have in our knowledge
of the curve itself. Christen shows that a reasonable estimate of $\gamma^2(\theta)$ for θ

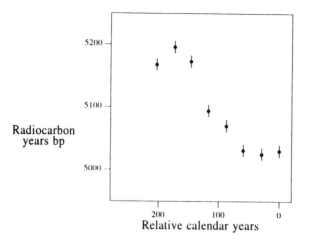

Radiocarbon
years bp

Relative calendar years

Figure 9.15 Plot of the eight radiocarbon results from the Stolford floating tree-ring chronology. The vertical bars indicate the 1σ ranges where σ is the standard deviation reported by the laboratory.

between the $(k-1)$st and kth knots is

$$\gamma^2(\theta) = \left(\frac{\theta - t_{k-1}}{t_k - t_{k-1}}\right)^2 \tau_k^2 + \left(\frac{t_k - \theta}{t_k - t_{k-1}}\right)^2 \tau_{k-1}^2 + \nu^2 \frac{(\theta - t_{k-1})(t_k - \theta)}{(t_k - t_{k-1})}$$

where t_k is the calendar date of the kth knot and τ_k the standard deviation of the calibration curve at the kth knot. Using modern data, Christen estimates ν to be about 20.

9.6.5 The prior

With no prior knowledge about the date θ_1, we take its prior to be

$$p(\theta_1) \propto \begin{cases} 1 & 0 < \theta_1 < \infty \\ 0 & \text{otherwise.} \end{cases}$$

9.6.6 The likelihood

In this example the likelihood may be expressed as

$$l(\theta_1; \boldsymbol{x}) \propto \prod_{j=1}^{8} \frac{1}{w_j(\theta_j)} \exp\left\{-\frac{(x_j - \mu(\theta_j))^2}{2w_j^2(\theta_j)}\right\}$$

where $\theta_j = \theta_1 + \sum_{i=1}^{j} \phi_i$, ϕ_1 is defined to be zero and

$$\omega_j^2(\theta_j) = \sigma_j^2 + \gamma^2(\theta_j).$$

9.6.7　The posterior

Using the prior given above, the posterior density of θ_1 is expressed as

$$p(\theta_1|\boldsymbol{x}) \propto \prod_{j=1}^{8} \frac{1}{\omega_j(\theta_j)} \exp\left\{ -\frac{(x_j - \mu(\theta_j))^2}{2\omega_j^2(\theta_j)} \right\} \quad \text{for } 0 < \theta_1 < \infty.$$

9.6.8　Results

As in the previous two case studies, the nature of the high-precision calibration curve, $\mu(\theta)$, means that no simple analytical form exists for this posterior density and it has to be evaluated numerically. The resulting 95% highest posterior density region consists of the two disjoint intervals 5810–5805 and 5795–5765 BP. That is, based on the prior information and the radiocarbon determinations, we are 95% sure that the date of the tree ring corresponding to θ_1 lies in the intervals 5810–5805 and 5795–5765 BP.

Incidentally, as we remarked earlier, the Stolford tree-ring chronology has now been dated using dendrochronology (Hillam et al., 1990) with the dates of the first and last rings being 6000 BP and 5729 BP respectively. The samples of wood taken from the chronology for radiocarbon dating were subsections spanning about 10 years with the piece corresponding to θ_1 being centred on 5786 BP, that is 3837 BC, which is within the 95% highest posterior density region reported above (see Figure 9.16). This is an outstanding example of radiocarbon dating techniques and Bayesian analysis working so well together. The HPD region is some 35 years long and the true date is within it despite the events being dated occurring about 6000 years ago!

9.7　Case Study IV — Kastanas, Greece — wiggle-matching

9.7.1　Background

In the previous case study the time interval between successive θ_i was known accurately. We now consider a problem where the information about the intervals is less precise. For example, an archaeologist may know that the samples used for radiocarbon dating come from stratified contexts from an excavation, and moreover may be willing to express opinions, based on other information, as to the time periods, ϕ_j, between successive events. Thus although ϕ_j may not be known exactly, the uncertainty may be expressed

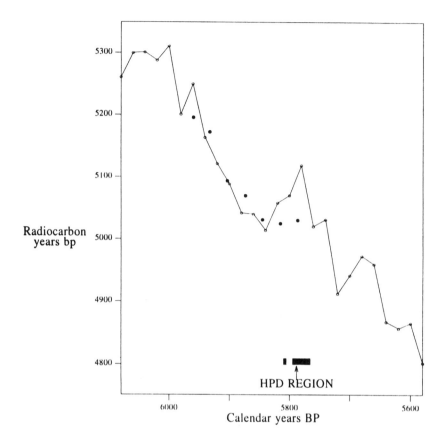

Figure 9.16 Highest posterior density region for the date, θ_1, of sample 1 from the floating tree-ring chronology from Stolford. The radiocarbon determinations for the chronology are plotted at their correct date as determined by dendrochronology.

in terms of a prior density.

We take as our example forty-six radiocarbon determinations from a stratified sequence of samples from Kastanas, Greece. This was first reported by Manning and Weninger (1992) who, on archaeological grounds, combined the determinations into eleven weighted averages, each average corresponding to a stratigraphic level of interest. (Essentially this is assuming that all the radiocarbon determinations within a level correspond to the same calendar year rather than to an interval. This assumption was made by the original researchers and we follow their example in order to illustrate our methodology.) One average was rejected because there was reason to believe it was an outlier. The radiocarbon determinations were wiggle-matched using a heuristic method based on least squares.

However, as the time intervals between the levels are not known exactly, Manning and Weninger experimented by taking values of 10 to 60 in 5-year increments, therefore assuming, perhaps unrealistically, that the intervals were always the same for each pair of consecutively numbered levels. Since no radiocarbon determinations were available for levels 7 and 15, the intervals between levels 6 and 8, and 14 and 16 were assumed to be double that of consecutively numbered levels. It was decided that the interval which gave the "best" fit was 35 years. An "average radiocarbon result" for each level was calculated with the average being formed over the results for intervals equal to $10, 15, \ldots, 60$ years. These average results, together with estimated standard deviations, are given in the third column of Table 9.6. Of course, in calculating average results in this way, it is tacitly assumed that the lengths of the time intervals are equally likely.

9.7.2 The model

Let us now try to formulate the above explanation in terms of a model and prior information. Let θ_i be the date of level i. Then the model is

$$
\begin{aligned}
\theta_8 - \theta_6 &= 2\phi \\
\theta_j - \theta_{j-1} &= \phi \qquad \text{for } j = 9, 10, 11, 12, 13, 14, 17 \\
\theta_{16} - \theta_{14} &= 2\phi
\end{aligned}
$$

where ϕ is the common time interval between consecutive levels.

9.7.3 The data

The weighted averages of the radiocarbon determinations for each level is given in Table 9.6 together with an estimated standard deviation for each average.

9.7.4 The prior

In this example we have two unknown parameters ϕ and θ_6. Without any evidence to the contrary it seems reasonable to assume that the ϕ and θ_6 are a priori independent so that their joint prior may be expressed as

$$
p(\phi, \ \theta_6) = p(\phi)p(\theta_6).
$$

For the moment we accept the above interpretation that each time interval must be the same, must be a multiple of 5 between 10 and 60 years inclusive

Table 9.6 The weighted radiocarbon determinations for the ten levels from Kastanas, Greece.

Cal-endar date	Level number	Radiocarbon result (weighted average)	Number of radiocarbon deter-minations
θ_6	6	2825±67	3
θ_8	8	2952±33	3
θ_9	9	2980±50	1
θ_{10}	10	2899±37	2
θ_{11}	11	2965±45	2
θ_{12}	12	2982±15	10
θ_{13}	13	2949±29	3
θ_{14}	14	3121±24	5
θ_{16}	16	3136±17	9
θ_{17}	17	3180±55	1

and that within that range no interval length is more likely than any other. Thus we assume

$$p(\phi = a) = \frac{1}{11} \quad \text{for } a = 10, 15, \ldots, 55, 60,$$

where ϕ represents the common interval. As nothing is known about θ_6, we take

$$p(\theta_6) \propto \begin{cases} 1 & 0 < \theta_6 < \infty \\ 0 & \text{otherwise.} \end{cases}$$

9.7.5 The likelihood

The likelihood is given by

$$l(\theta_6, \phi; \boldsymbol{x}) \propto \prod_{\substack{j=6 \\ j \neq 7, 15}}^{17} \frac{1}{\sigma_j} \exp \left\{ -\frac{(x_j - \mu(\theta_j))^2}{2\sigma_j^2} \right\}$$

where $\theta_j = \theta_6 + (j-6)\phi$ and ϕ is the common time interval between successive levels.

9.7.6 The posterior

The joint posterior density of ϕ and θ_6 is given by

$$p(\theta_6, \phi | x) \propto \prod_{\substack{j=6 \\ j \neq 7,15}}^{17} \frac{1}{\sigma_j} \exp\left\{ -\frac{(x_j - \mu(\theta_j))^2}{2\sigma_j^2} \right\}$$

for $0 < \theta_6 < \infty$ and $\phi = 10, \ldots, 60$ where $\theta_j = \theta_6 + (j - 6)\phi$.

9.7.7 Results and interpretation

The marginal posterior density of the date of level 6, θ_6, is given in Figure 9.17 and its 95% highest posterior density region is 3040 BP to 2860 BP. Manning and Weninger (1992) quoted 2995±75 as their estimate of θ_6, thereby implying that a 95% confidence interval is 3145 BP to 2845 BP which is considerably wider than the interval obtained through the Bayesian approach.

The marginal posterior probability mass function of ϕ is given in Figure 9.18; the values of ϕ included in the 95% highest posterior probability region are 30, 35, 40, 45 and 50. We note that the posterior probabilities of $\phi = 40$ and $\phi = 45$ are both 0.30 and that of $\phi = 35$ is 0.20. That is, *a posteriori*, a common interval of either 40 or 45 years is most likely with 35 years being the third most likely. This is in contrast to the choice of Manning and Weninger (1992, p. 64) who decided that 35 years gave the *best* fit. Notice that the marginal posterior density of θ_6 is very spiky, a feature that is a direct result of the discrete form (5-year intervals) of the prior distribution specified for ϕ.

9.7.8 Remodelling

In the above analysis of the data from Kastanas we have followed as closely as possible the assumptions made by Manning and Weninger (1992), two of which are as follows.

(a) The time intervals are the same for consecutively numbered levels.
(b) The time intervals are multiples of 5 between 10 and 60 inclusive, that is $a = 10, 15, \ldots, 60$.

Both of these assumptions are very restrictive and therefore perhaps unrealistic. One of the advantages of the Bayesian framework is that it easily permits us to relax such assumptions, thus making the model more realistic but without making the analysis much more difficult, even if greater computational time is involved.

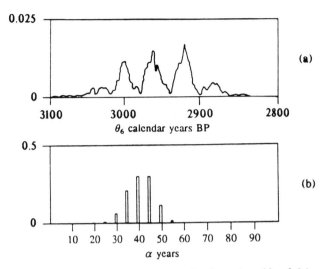

Figure 9.17 (a) Posterior probabilities for the date, θ_6, of level 6 in calendar years BP, at Kastanas, Greece. (b) The marginal posterior probability mass function of the common time interval, ϕ, between successive levels; assuming 5-year intervals. (After Christen and Litton, 1995.)

9.7.9 Sensitivity analysis

Firstly we repeat the analysis but assume that the time interval between successive levels is equally likely to be any year between 10 and 60 inclusive. Thus we have

$$\phi \sim U(10, 60).$$

Thus we are not restricting the intervals to be a multiple of 5 but we are still assuming that all the intervals do have the same value. The marginal posterior functions of θ_6 and ϕ are given in Figure 9.18. Notice that as a consequence, the density of θ_6 is more widespread, smoother and with less pronounced peaks than previously. However the highest posterior density regions for θ_6 and ϕ are not too dissimilar to those given above, 3040 to 2865 BP and 30 to 52 respectively. (See column (ii) of Table 9.7 for the highest posterior density regions for the θ_i.)

Secondly we repeat the previous analysis but now remove the restriction that the intervals between successive levels must be the same. Instead, we assume that each interval can take any value between 10 and 60 years independently of any other interval. In other words if $\theta_j - \theta_{j-1} = \phi_j$ then we are no longer assuming that the ϕ_j have the same value so that

$$\phi_j \sim U(10, 60)$$

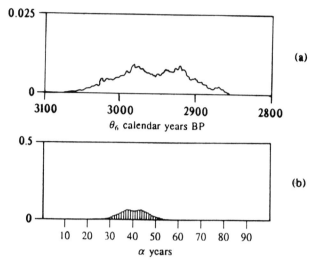

Figure 9.18 (a) Posterior probabilities for the date, θ_6, of level 6 in years BP, at Kastanas, Greece, (b) The marginal posterior probability mass function of the common time interval, ϕ, between successive levels; assuming any value between 10 and 60 years for the common interval is equally likely. (After Christen and Litton, 1995.)

independently of the other ϕ_i.

The marginal posterior densities for $\theta_6, \theta_8, \ldots, \theta_{17}$ are given in Figure 9.19 and the corresponding 95% highest posterior density regions in column (iii) of Table 9.7. We note that the highest posterior density regions given in columns (ii) and (iii) are in broad agreement.

Finally, we remove the restriction that the time interval, ϕ, cannot be less than 10 years or greater than 60 years. To allow both these possibilities we now model ϕ_j as having a Gamma distribution (see Section 5.10.5) with parameters a_0 and b_0. Choosing $a_0 = 5$ and $b_0 = 7$ gives the density shown in Figure 9.20. Using this model, the time interval between successive levels has a mean of 35 years, the same as for both models used above, but in addition there is about a 1% chance of an interval of less than 10 years and about a 7% chance of an interval greater than 60 years.

The 95% highest posterior density regions for $\theta_6, \theta_8, \ldots, \theta_{17}$ for this model are given in column (iv) of Table 9.7. The marginal posterior probabilities are very similar to those of Figure 9.19.

The results in columns (ii), (iii) and (iv) are very similar, thus giving us confidence that our conclusions do not drastically alter if the model or prior is changed slightly. However we must stress the difference between the results in column (i) and those in columns (ii) to (iv). In the former not all information

Table 9.7 Results from Kastanas. Column (i): 95% intervals reported by Manning and Weninger (1992). Columns (ii), (iii) and (iv): 95% highest posterior density regions for the date of each level with $\phi \sim U(10, 60)$, $\phi_j \sim U(10, 60)$ and $\phi_j \sim Ga(5, 7)$ respectively. (* The 95% HPD region also includes 3280–3270 BP.)

Calendar date	(i) BP	(ii) BP	(iii) BP	(iv) BP
θ_6	3145–2845	3040–2860	3050–2890	3040–2900
θ_8	3160–2960	3100–2960	3100–2980	3100–2980
θ_9	3170–3010	3130–3020	3130–3010	3130–3020
θ_{10}	3175–3075	3170–3070	3160–3050	3160–3050
θ_{11}	3200–3120	3200–3110	3200–3090	3200–3100
θ_{12}	3240–3140	3240–3160	3230–3130	3230–3140
θ_{13}	3285–3165	3290–3210	3260–3170	3260–3180
θ_{14}	3325–3185	3310–3250	3300–3210	3300–3220
θ_{16}	3420–3220	3410–3320	3400–3320*	3400–3320
θ_{17}	3480–3230	3460–3360	3440–3320	3440–3340

is included in the analysis whereas the latter are calculated using the Bayesian framework and so incorporate all the relevant knowledge.

9.8   Case Study V — the Chancay culture of Peru — outlier detection

9.8.1   Background

One drawback of radiocarbon dating techniques is that they are liable to produce erroneous or misleading results. Baillie (1990) reported that, for determinations made in the 1970's and early 1980's, about 30% of the results had true dates outside the ranges predicted by the 95% ranges based on the laboratories' standard deviations. Since then the radiocarbon community have improved the accuracy of their methods but, as with any complex scientific process, erroneous results will still arise from time to time. There is a variety of reasons, some arising as a consequence of sampling problems of the archaeologists and some caused by the laboratories themselves. For example, a 3000-year-old sample (relative to AD 1950) if contaminated with organic material of a later period will not be dated to 1050 BC.

How does one detect such misleading results? If we take only one sample and that becomes contaminated then we will inevitably make erroneous inferences. To avoid this it is prudent to take several samples in the hope that no more than one will be affected. Moreover, if the relationship between

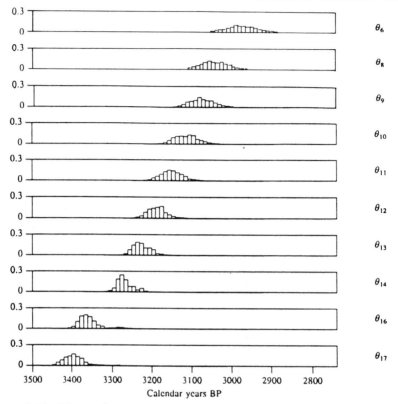

Figure 9.19 Marginal posterior probability density functions of dates of levels 6, 8–14, 16 and 17 at Kastanas, Greece. (After Christen and Litton, 1995.)

the corresponding calendar dates can be modelled, the results can be tested against the model. Consequently, an important requirement of any statistical analysis of radiocarbon data is to be able to identify an erroneous or misleading observation in a set of results. That is, we wish to be able to identify a result that is inconsistent with the others, the model or the prior information. In statistical jargon, such an observation would be called an *outlier*. A Bayesian approach, which we will now outline, to the modelling and identification of outliers in groups of related radiocarbon determinations is given by Christen (1994a, b) and Buck *et al.* (1994b).

9.8.2 Archaeological background

We use as an example the problem of estimating the calendar dates and likely duration of the Chancay culture, which existed in the area close to the

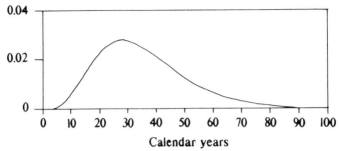

Figure 9.20   The time interval, ϕ_j, between successive levels j and $j + 1$ is assumed to have a Gamma distribution with parameters 5 and 7. Presented is a plot of the probability density function of ϕ_j. (After Christen and Litton, 1995.)

Table 9.8   The radiocarbon determinations relating to the Chancay culture.

Laboratory identifier	Radiocarbon determination	Laboratory identifier	Radiocarbon determination
Gd-2819	520±60	Gd-5672	830±50
Gd-3396	430±30	Gd-5823	670±40
Gd-5304	460±50	Gd-5824	1140±50
Gd-5307	970±50	Gd-6189	1070±60
Gd-5309	910±35	Gd-6196	810±70
Gd-5310	1000±50	Gd-6197	900±70
Gd-5312	390±45		

central Peruvian coast in pre-Hispanic times; the radiocarbon determinations used here were first reported by Pazdur and Krzanowski (1991). Wood from thirteen different tombs at six sites, all associated with the Chancay culture, was sampled and the resulting radiocarbon determinations are given in Table 9.8. In their analysis Pazdur and Krzanowski discarded the earliest (Gd-5824) and latest (Gd-5312) determinations because the associated tombs had been disturbed by robbers. In other words, there was some concern as to whether or not the timber involved did relate to the Chancay culture. If this were the case, then both determinations could be outliers. Christen (1994a, b) reanalysed the data from a Bayesian viewpoint and here we describe his methodology and findings.

9.8.3   The model

The question of archaeological interest is to estimate the most probable period during which the tombs were in use, and hence the dates of the Chancay

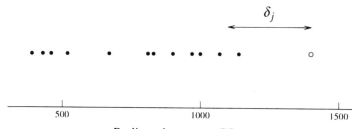

Radiocarbon years BP

Figure 9.21   Diagram showing the radiocarbon determinations, represented by
•, associated with the Chancay culture and another "fake" observation represented
by ○. The fake determination requires a shift of δ_j to be consistent with other
determinations.

culture. However, it is clear that the presence of outliers will grossly affect
any estimates. Thus the methodology employed needs to be able to detect
any outliers.

We represent the date of the start of the Chancay culture by α and its end
by β. Suppose that we have n determinations with the jth determination,
X_j, corresponding to an unknown calendar date θ_j and having a reported
standard deviation of σ_j. A priori we have no information relating to the
calendar dates of the tombs, therefore we assume that their dates are equally
likely to lie anywhere within the time period α to β, and that the dates of the
tombs are independent of each other. Hence we suppose that independently
of each other

$$\theta_j | \alpha, \beta \sim U(\beta, \alpha)$$

where α and β are unknown. In other similar situations, we may have different
a priori information suggesting perhaps that the timber from certain tombs
is more likely to be from the end of the period than the beginning.

To model the possibility of having an outlier, the jth determination is said
to be an outlier if it needs a shift of δ_j radiocarbon years to be consistent with
the rest of the sample (see Figure 9.21). Formally we have

$$X_j | \theta_j, \phi_j, \delta_j \sim N\left(\mu(\theta_j) + \phi_j \delta_j, \sigma_j^2\right),$$

where
$$\phi_j = \begin{cases} 1 & \text{if the } j\text{th determination needs a shift} \\ 0 & \text{otherwise.} \end{cases}$$

Thus an error in the radiocarbon dating process of the jth determination
will be modelled by a shift, δ_j, in the observed value from the expected true
radiocarbon age $\mu(\theta_j)$. The uncertainty concerning whether or not a shift is
needed is measured by $P(\phi_j = 1)$ and $P(\phi_j = 0)$ respectively.

9.8.4   The data

Details of the thirteen radiocarbon determinations associated with the Chancay Culture are given in Table 9.8.

9.8.5   The prior

We assume that whether or not the jth determination needs a shift (and the size of the shift if it is required) is independent of the other determinations and of α and β. Therefore the joint prior of $\boldsymbol{\delta}, \boldsymbol{\phi}, \alpha$ and β, denoted by $p(\boldsymbol{\delta}, \boldsymbol{\phi}, \alpha, \beta)$, may be expressed

$$p(\alpha, \beta) \prod_{j=1}^{n} p(\delta_j) p(\phi_j)$$

and we assess these terms separately as follows.

(i) The Chancay culture is exclusively pre-Hispanic, consequently β must be greater than 400 BP, approximately the date of the invasion of Peru by Pizarro in the mid-sixteenth century. Thus we have

$$\beta > 400.$$

There is, however, no archaeological information to provide an equivalent constraint on α, though, of course, α must be before β $(\alpha > \beta)$. To model this prior knowledge about α and β, we take their joint prior to be

$$p(\alpha, \beta) \propto 1 \quad \text{for } 400 < \beta < \alpha.$$

(ii) As nothing is known, a priori, about the size of any shift, it is reasonable to assume a vague prior for δ_j over some wide range.

(iii) We let the prior probability that the jth determination is an outlier be q_j. That is

$$P(\phi_j = 1) = q_j.$$

In other words our prior assessment that the jth determination needs a shift is q_j and that it does not need a shift is $1 - q_j$. In practice, since accuracy of radiocarbon laboratories has recently improved, taking $q_j = 0.1$ seems not unreasonable although (hopefully) pessimistic.

9.8.6   The likelihood

The likelihood can be expressed as

$$l(\boldsymbol{\theta}, \boldsymbol{\delta}, \boldsymbol{\phi}, \alpha, \beta; \boldsymbol{x}) = \prod_{j=1}^{n} l(\theta_j, \delta_j, \phi_j, \alpha, \beta; x_j)$$

$$= \prod_{j=1}^{n} p(x_j|\theta_j, \delta_j, \phi_j) p(\theta_j|\alpha, \beta).$$

As θ_j has a Uniform distribution between β and α we have

$$p(\theta_j|\alpha, \beta) = \frac{1}{\alpha - \beta} \quad \text{for } \beta \leq \theta_j \leq \alpha$$

and as $X_j|\theta_j, \delta_j, \phi_j \sim N(\mu(\theta_j) + \delta_j\phi_j, \sigma_j^2)$ we have

$$p(x_j|\theta_j, \delta_j, \phi_j) \propto \exp\left\{ -\frac{(x_j - \delta_j\phi_j - \mu(\theta_j))^2}{2\sigma_j^2} \right\}.$$

For ease of notation we let

$$z_j = \exp\left\{ -\frac{(x_j - \delta_j\phi_j - \mu(\theta_j))^2}{2\sigma_j^2} \right\}.$$

Hence the likelihood is given by

$$l(\boldsymbol{\theta}, \boldsymbol{\delta}, \boldsymbol{\phi}, \alpha, \beta; \boldsymbol{x}) \propto (\alpha - \beta)^{-n} \prod_{j=1}^{n} z_j.$$

9.8.7 The posterior

The joint posterior will be given by

$$p(\boldsymbol{\theta}, \boldsymbol{\delta}, \boldsymbol{\phi}, \alpha, \beta|\boldsymbol{x}) \propto (\alpha - \beta)^{-n} \prod_{j=1}^{n} \left\{ z_j q_j^{\phi_j} (1 - q_j)^{1-\phi_j} \right\}$$

over the appropriate ranges for the parameters.

9.8.8 Implementation

The full conditionals for use with the Gibbs sampler are

$$p(\theta_j|\boldsymbol{x}, \boldsymbol{\theta}_{-j}, \alpha, \beta, \boldsymbol{\delta}, \boldsymbol{\phi}) \propto z_j I_{(\beta,\alpha)}(\theta_j),$$

$$p(\alpha|\boldsymbol{x}, \boldsymbol{\theta}, \beta, \boldsymbol{\delta}, \boldsymbol{\phi}) \propto (\alpha - \beta)^{-n} I_{(d,\infty)}(\alpha),$$

$$p(\beta|\boldsymbol{x}, \boldsymbol{\theta}, \alpha, \boldsymbol{\delta}, \boldsymbol{\phi}) \propto (\alpha - \beta)^{-n} I_{(400,c)}(\beta),$$

Table 9.9 Posterior probabilities of inconsistency of the thirteen Chancay culture determinations.

Laboratory sample number	Posterior probability	Laboratory sample number	Posterior probability
Gd-2819	0.09	Gd-5672	0.09
Gd-3396	0.10	Gd-5823	0.13
Gd-5304	0.09	Gd-5824	0.17
Gd-5307	0.10	Gd-6189	0.10
Gd-5309	0.09	Gd-6196	0.09
Gd-5310	0.11	Gd-6197	0.08
Gd-5312	0.09		

$$p(\delta_j | \boldsymbol{x}, \boldsymbol{\theta}, \boldsymbol{\delta}_{-j}, \boldsymbol{\phi}, \alpha, \beta) \propto z_j \quad \text{for } \phi_j = 1,$$

$$p(\phi_j | \boldsymbol{x}, \boldsymbol{\theta}, \boldsymbol{\delta}, \boldsymbol{\phi}_{-j}, \alpha, \beta) \propto z_j q_j^{\phi_j} (1 - q_j)^{1-\phi_j}$$

where $c = \min(\theta_i; i = 1, \ldots, n)$ and where $d = \max(\theta_i; i = 1, \ldots, n)$. Here $\boldsymbol{\theta}_{-j}$ denotes the vector of θ_i but omitting θ_j, that is $\boldsymbol{\theta}_{-j} = (\theta_1, \ldots, \theta_{j-1}, \theta_{j+1}, \ldots, \theta_n)$; $\boldsymbol{\delta}_{-j}$ and $\boldsymbol{\phi}_{-j}$ are defined in a similar fashion. Note that if $\phi_j = 0$ then X_j does not need a shift and so δ_j is not sampled, in effect its value in this case is 0.

The full conditionals for θ_j, α and β given above can be readily sampled as in earlier examples. The sampling of δ_j and ϕ_j is quite straightforward; δ_j has a Normal distribution, perhaps within some restricted range, and ϕ_j has a Bernoulli distribution. Thus, despite the complicated mathematics involved, the Gibbs sampler is quite simple to implement.

9.8.9 Results and interpretation

Given that we have assumed the *a priori* probability of any particular sample being inconsistent is 0.1, the probabilities of each of the current samples being inconsistent in the light of the other data and the *a priori* information are given in Table 9.9. We note that most of these probabilities are very close to the *a priori* value; that is, the data do not suggest that the samples are inconsistent with each other, the model or the prior information.

Sample Gd-5824 has a sizeable increase in its probability of inconsistency; from 0.1 to 0.17 which is still much less than the value of 0.3 observed by Baillie (1990). This sample has the earliest calendar date range in the group and was one of the two which, in the original publication, was discarded as not truly representative of the period of Chancay cultural activity.

For the second of the two samples which were previously discarded as being of dubious cultural association, we note that the probability of inconsistency with the other samples drops from 0.1 to 0.09. In other words, given no strong *a priori* archaeological evidence to the contrary, sample Gd-5312 should not be discarded on the basis of inconsistency with the remaining samples.

Therefore, there is little support for samples Gd-5824 and Gd-5312 to be discarded as inconsistent with the remaining information. On this basis, the best estimates for the dates of the start and end of the Chancay culture are in the range AD 730 to 1000 (mode $\approx$ 930) and AD 1470 to 1550 (mode $\approx$ 1500) respectively. The corresponding estimate for the duration of the period is 470 to 780 years (mode $\approx$ 570). We refer the reader to Buck *et al.* (1994b) for further discussion of the results.

9.9   Discussion of radiocarbon dating

It is clear that there is a wide range of archaeological and statistical problems associated with radiocarbon dating, many of which can only realistically be approached from a Bayesian perspective. Even in the case of a single determination, the multi-modality of the likelihood function makes the use of classical approaches such as maximum-likelihood-based techniques inappropriate. Having said this, due to the nature of the available prior information, evaluating posteriors of interest using quadrature methods requires specialist software not widely available and often not easy to use. The advantage of Markov chain Monte Carlo methods is their ease of implementation and their intuitive appeal — the simple iterative nature of the algorithms is easily understood by archaeologists and, consequently, they are well received.

It is now possible to address using the Bayesian paradigm in a range of interesting and challenging problems associated with the calibration of radiocarbon determinations that were, before the advent of the Gibbs sampler and related methods, relatively difficult and required highly specialized and sophisticated software. We should, however, consider one or two points relating to the practicalities of these approaches. For event-based problems such as that in Case Study I, convergence of the Gibbs sampler is usually rapid and run times are commonly only a few minutes for 10 000 iterations. For more complex models involving ordered phases, such as Case Study II, the parameters are highly correlated, convergence is slower and typically the run times are much longer — convergence may take up to 100 000 iterations. Generally, outlier analysis takes a matter of minutes. These times are insignificant compared to the time taken in excavation, obtaining radiocarbon determinations, post-excavation interpretation and publication of final reports. For larger sites this process commonly takes several years!

10

Spatial analysis

10.1 Introduction

10.1.1 Spatial analysis in archaeology

Spatial analysis lies at the heart of many archaeological problems because the material finds occur in a spatial as well as a chronological setting. Having considered the time dimension in the previous chapter, therefore, we shall now look at some examples of spatial analysis. Artefacts are located at different points in a level, in different levels and contexts within a site and at different sites across a region. Analysis of their locations, whether within a site or distributed over whole regions, can be used to address many different kinds of archaeological questions. These might range from the examination of functional areas within sites to patterns of trade and exchange in the past, and from landscape archaeology to the spatial patterning of ritual. In the analysis of spatial effects, the discipline comes close to geography, and traditionally many of the models adopted in the one subject have grown out of applications in the other. The rich variety of spatial problems can be appreciated from the seminal overview of this subject by Hodder and Orton (1976), and Kintigh (1990) has written a recent critical review.

An important point that Hodder and Orton stress is how easy it is to "interpret" scatters of points on a map or on a plan, which are in fact purely randomly spaced (see Figure 10.1); there is an innate human tendency to find patterns even where none exists. How are we to avoid purely arbitrary interpretations in spatial analysis? The answer must lie in modelling the underlying situation, and Hodder and Orton give many examples of mathematical techniques applied to questions involving the distribution of artefacts, trading patterns (see Figure 10.2), site hierarchies and central places, defining the boundaries of archaeological cultures and so forth. Problems in spatial analysis have not always been so carefully modelled; in Kintigh's words (1990, p. 197),

> too often our methods have been allowed to dictate our questions rather than the other way round.

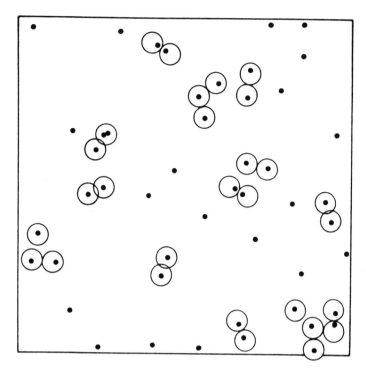

Figure 10.1 Points placed at random in a bounded area. Circles suggest spheres of influence around "sites". (After Hodder and Orton, 1976, p. 5, fig. 1.2.)

and as Baxter (1994, p. 170) has observed of the *k-means* method for spatial clustering (which is a method for distinguishing areas of high density in a two-dimensional scatter of points),

> The need for this, given that the data are readily inspected by eye, may be queried.

Spatial problems, such as the analysis of settlement patterns or the distinction of different activity areas within a site, are complex and require a combination of statistical skill and archaeological judgement. Indeed their very complexity makes them especially suitable for Bayesian treatment. We cannot, within the scope of just one chapter, hope to treat the many different kinds of problem posed by this area of analysis, so the case studies presented here draw only on site survey and prospecting.

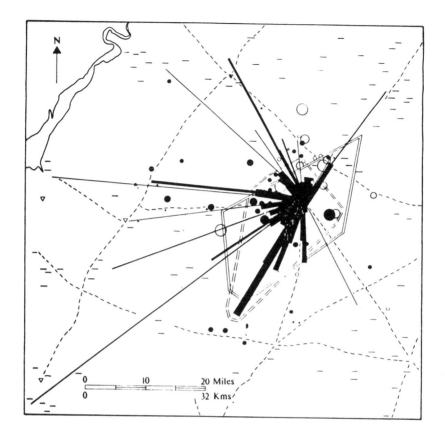

Figure 10.2 "Gravity model" applied to the marketing of pottery from the Romano-British centre at Mildenhall. (After Hodder and Orton, 1976, p. 189, fig. 5.75.) (Reproduced by permission of Cambridge University Press.)

10.1.2 Archaeological survey

Techniques of archaeological survey vary widely in method and purpose, but all, naturally, are concerned with the spatial distribution of archaeological remains. We might mention, in summary, the following.

(i) *Field walking* is used by archaeologists to record surface remains as a means of locating artefact scatters diagnostic of human occupation and other activities. As a result of such work it is possible to trace the development of man's occupation of whole landscapes over long spans of time.

(ii) *Geophysical methods* exploit variations in the soil's properties, which can be measured (usually without the need for excavation), and which shed light on features (archaeological or other) buried beneath the surface. Properties investigated include soil resistance, magnetic properties and the response to ground-penetrating radar signals. Thus resistivity reveals features of low resistance, such as buried pits and ditches (low resistance because they tend to be saturated with more moisture than the surrounding soil), and features of high resistance such as walls and metalled roads (with little moisture). These geophysical techniques have succeeded, for example, in revealing the whole plan of buried settlements, and can also be of value in guiding decisions about where to excavate by indicating the outlines of buried features.

(iii) *Chemical analysis* of elements in the soil is also commonly used to reveal information about the spread and disposition of archaeological traces. Heavy elements, such as lead or copper, or chemicals such as phosphates, concentrate at sites through the human activities focused there. Thus analysis of the phosphate levels from the Shaugh Moor project on Dartmoor, U.K., helped in the interpretation of huts and enclosures as habitation areas (Balaam *et al.*, 1982). A combination of phosphate survey (indicating human and animal waste) and magnetic susceptibility (arising from the burning usually associated with human occupation) was applied to great effect in interpreting the survey of the Iron Age enclosure at Tadworth, Surrey (Clark, 1977) (see Figure 10.3). The main habitation area was distinguished from another area probably used to pen animals.

These types of archaeological data cannot be recorded directly on a photograph or a site drawing in quite the same way as the stone walls of a house or the outline of a grain storage pit. Frequently they need statistical analysis before any interpretation can be made.

10.1.3 *Traditional methods of presenting survey data*

From the archaeological viewpoint, the primary aim is frequently to identify zones corresponding to different types or intensities of human occupation or activity. So at Tadworth different areas of high phosphate and high magnetic susceptibility were distinguished. Visual summaries have traditionally been used to interpret such data. They can range from relatively simple methods such as dot-density plots, grey-scales and colour scales through simple averaging or median smoothing, to the mathematically more sophisticated techniques of contouring and image processing.

In any situation, the choice of technique must depend upon the nature of the data and the problem posed by the researcher. Certain options, such as

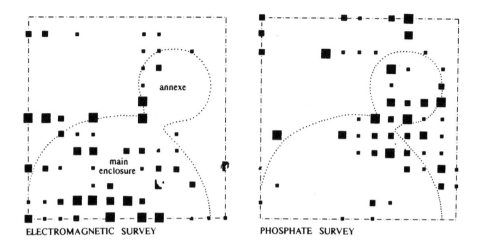

ELECTROMAGNETIC SURVEY   PHOSPHATE SURVEY

Figure 10.3 Magnetic susceptibility and phosphate surveys of the enclosure and annexe at Tadworth, Surrey, U.K. Samples were taken at 5 m intervals. (After Clark, 1977, p. 190, fig. 4.)

simple contouring and dot-density plots, are not always appropriate, and their results may be unreliable or misleading.

In the past the main protection against spurious interpretation has lain in the expertise of the researchers involved: archaeologist and scientific expert have inspected the (computer) output and exercised their judgement developed through experience. With the Bayesian approach, of course, their expertise should be incorporated, ideally as prior knowledge, into the model. In developing an approach the advantages and pitfalls of one technique or another need to be taken into account; contouring is sensitive to anomalous data points and may not be suitable if fairly abrupt changes in level are to be expected; dot-density plots and grey-scales work best if the error in the data is slight, and where any observation is an accurate representation of the underlying archaeological reality.

These types of investigation also call on the resources and instrumentation of scientists such as chemists or physicists, and it is important to allow for information arising from their domains of expertise to be incorporated. In the first place there are many aspects of archaeological soil science where understanding of the underlying processes is far from complete; as far as possible we need to allow for these uncertainties. Secondly there are imprecisions which arise through the instrumentation (let it be acknowledged that archaeologists cannot normally buy access to the very latest hardware, least of all when many samples need to be analysed). In general terms whichever survey method is selected, the data will be affected by inaccuracies and imprecisions, arising either from the instruments or techniques used, or from natural variation in the background level. Statistical methods used to analyse such data must take account of any error.

10.1.4   Overview of the chapter

In what follows we shall examine three case studies, all of which use data derived from the chemical analysis of soil phosphate. This is convenient in allowing us to concentrate on the statistical detail, with a broadly similar background in all three cases for the type of data and method of chemical analysis. In fact the types of approach developed here can be and have been applied to other types of data such as resistivity data and magnetic susceptibility readings (Buck *et al.*, 1990), and have a very general applicability.

High concentrations of phosphate develop on settlement sites through the accumulation and concentration of organic remains which arise, one way or another, through human activities. Thus organic refuse can pile up in kitchen middens, usually sited close to the living quarters, dung heaps can amass close to byres or animal pens. Generally vegetable refuse, animal ordure and all organic remains have high concentrations of phosphorus. Indeed the value of manure for muck-spreading on fields arises in part from its high phosphate content.

The first case study will consider soil phosphate readings taken along a transect. This type of analysis can be used, for example, in prospecting for archaeological sites or broad areas within larger sites, with high concentrations of phosphate (or, indeed, other indicators of archaeological activity). The second case study looks at data collected over a grid and distinguishes zones of high phosphate concentration (referred to as "on-site") or low phosphate concentration ("off-site"), taking into account not only readings along a single transect, but the more general scene across a whole area. Finally the third case study extends the methodology of the second by looking at the situation where there are more than two levels of concentration.

10.2   Case Study I — analysis of phosphate data along a transect — change-point analysis

10.2.1   Archaeological background

As part of a project aimed at increasing our understanding of the rural economy in the past, in the summer of 1987 a phosphate survey was carried out at a number of locations, including the area around the site at present identified as LS N418, of the Laconia Survey in Greece (Buck *et al.*, 1988, 1990). The aim of the survey was to investigate the relationship between phosphate concentrations, which might coincide with the buildings, or might indicate other areas associated with a site such as middens or animal pens, and surface artefacts, most notably roof tiles, which may indicate the location of roofed buildings. Apart from the surface remains there were no other visible indications as to the extent of the site. Soil samples were collected at regular intervals along a series of transects through the site.

We think of an "off-site" or "background" level of phosphate concentration, which represents the level of phosphate naturally present in the soil; that is to say given the particular geology, ecology, cultivation of the area and a host of other local conditions, there will be a particular level of phosphate without the enhancement due to an added archaeological effect. "On-site" concentration is the local natural phosphate level enhanced due to archaeological effects, such as those outlined in the previous section. Essentially, therefore, the aim of the analysis is to recognize and delimit broad zones or contiguous areas of high and low phosphate concentration. In other circumstances one might wish to distinguish more than just two levels of phosphate, indeed we shall consider just this situation in Case Study III, but given the sampling in this case, the distinction of two levels seemed best.

10.2.2   Sampling and chemical analysis

We preface the statistical analysis with a summary discussion of sample collection and chemical analysis. This will help underline those specific practical circumstances which so often affect the quality of the data and have an influence on the statistical treatment.

The soil samples were gathered over a grid at 10 m intervals. The soil was scooped from the top 5 cm of the ground surface; this is far from ideal not least because the uppermost soil surface is unstable and subject to movement. This constraint, a condition of the research permit, nevertheless made sampling rapid. The "total" phosphate was extracted from the soil sample and assayed for phosphate concentration (see Craddock *et al.*, 1985). The chemical analysis is crude and therefore rather imprecise, but has the advantage that it can be carried out in a field laboratory, relatively rapidly. More expensive, laboratory-based methods (see the review with references in Proudfoot, 1976, p. 94–98) would give more precise results, but are more time-consuming and could not be carried out in the field.

Exactly how high concentrations of organic phosphorus become more permanently bound into archaeological sediments is not fully understood, though calcium, aluminium and iron compounds in the clay fraction are believed to be crucial (see Proudfoot, 1976). Research is still needed into how

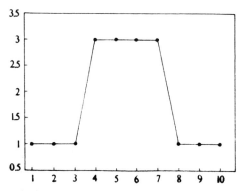

Figure 10.4   Idealized phosphate readings on a transect across an archaeological
site.

phosphates move within soils, and the effects of change over time in different
soil environments. In geochemical processes such as these it is not unusual for
concentrations of elements to follow a log-normal distribution. Indeed it has
been shown, on the basis of samples collected from the area, that the natural
logarithm of the soil phosphate concentration has, approximately, a Normal
distribution (Cavanagh *et al.*, 1988).

10.2.3   The model

Let us consider the phosphate concentrations along a single transect through
the site. We aim to distinguish "high (archaeological or on-site) phosphate"
from "low (non-archaeological or off-site) phosphate". Of course in practice
statistical definitions of "high" and "low" are required, but for the moment
we shall take these as given. In theory we expect a graph looking something
like Figure 10.4: the readings in the middle representing high phosphate, the
readings at the beginning and end representing low phosphate. Essentially we
wish to detect changes in the levels of the underlying process; change-point
analysis is a statistical technique that does precisely this (see Hinkley, 1970;
Smith, 1975).

   We describe the phosphate model in two stages: the first is concerned with
the number and the position of the change-point(s), and the second with
modelling the levels of phosphate concentration and the natural variation
about those levels.

Modelling the change-points

We can imagine moving along the transect starting outside the on-site zone
of high phosphate concentration, passing through it and coming out at the

other side and continuing some way into the low phosphate background. In this ideal case the readings from position 1 up to position r would be at low level, the readings from positions $r + 1$ to s would be at high level and the readings from positions $s + 1$ to n would be at low level again.

However that is the theory and, as field archaeologists are well aware, practice does not always match theory. In this example when collecting the soil samples in the field, the artefact scatter is only a *guide* to where the on-site phosphate might extend. In fact the "phosphate site" may not coincide exactly with the "artefact" site. Indeed different configurations of the artefact and phosphate concentrations would be of particular interest. The high phosphate might not, therefore, sit squarely central to the sampling grid. In practice we need six possible variations of the model which we now describe.

(i) We find no change-point as all the readings are low or "non-archaeological" — we call this model $M_{n,n}$ (see Figure 10.5(a)).

(ii) We find no change as all the readings are high or "archaeological" — denoted by model $M_{1,1}$ (see Figure 10.5(b)).

(iii) There is one change-point — from high to low — between positions r and $r + 1$ (for $r = 1, \ldots, n - 1$). This we call model $M_{r,n}$ (see Figure 10.6).

(iv) If there is one change between positions r and $r + 1$ (for $r = 1, \ldots, n - 1$) from low to high, then the model is denoted by $M_{n,r}$.

(v) There are two change-points — from low to high to low — the changes occurring between positions r and $r + 1$ and between positions s and $s + 1$ for $r = 1, \ldots, s - 1$; $s = 2, \ldots, n - 1$). This we call model $M_{r,s}$ (see Figure 10.7).

(vi) There are two changes — from high to low to high — the changes occurring between positions r and $r + 1$ and between positions s and $s + 1$ for $r = 1, \ldots, s - 1$; $s = 2, \ldots, n - 1$). This model is denoted by $M_{s,r}$.

Note that models (iii)–(vi) include a large number of different variants, depending on the position of the change-point. Indeed (iii) and (iv) each have $n - 1$ variants and (v) and (vi) each have $(n - 1)(n - 2)/2$ variants. If we count all the possible models (including (i) and (ii)) there is a total of $n^2 - n + 2$. Where we wish to refer to the general model which might take on any one of these forms we shall denote it by $M_{i,j}$.

In the above we have made several modelling assumptions. In the first place it is assumed that there are only two levels of phosphate concentration: high and low, or archaeological and non-archaeological. This is justified in terms of the data set examined here, but in other cases more levels may better fit the situation under scrutiny (see Section 10.4). Secondly it is assumed that soil samples with the same basic phosphate concentration will be found in a contiguous block one beside another: high beside high and low beside low. Thirdly it is assumed that there will be an abrupt change from one level to the

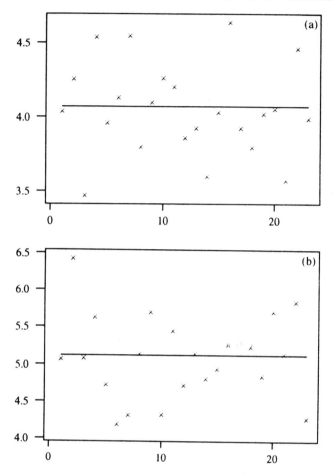

Figure 10.5 (a) Illustration of model $M_{n,n}$ with all the observations at the low level. (b) Illustration of model $M_{1,1}$ with all the observations at the high level.

other. Finally it is assumed that the distinction between high phosphate and low phosphate is likely to be due to archaeological rather than other causes. It is for the archaeologist to be conscious of these modelling assumptions, and to justify them. Do they adequately cover the situation, without either over-simplifying or over-refining?

Modelling the phosphate concentration levels

Consider a single transect and suppose that we collect soil samples at n different equidistant positions along the transect. Let X_i represent the

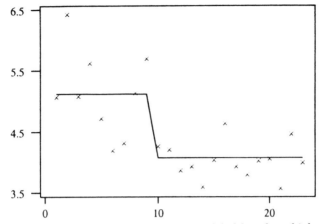

Figure 10.6   Illustration of one-change-point model, $M_{r,n}$, from high to low.

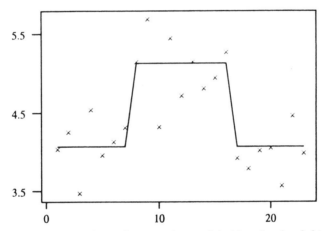

Figure 10.7   Illustration of two-change-point model, $M_{r,s}$: low level, high level, low level.

phosphate concentration at the ith position and let $Y_i = \ln X_i$. The model states that the expected value of the logarithm of phosphate concentration is constant at a value μ_1 within a site and then drops abruptly to the background value of μ_2 outside the boundary of the site. This is a crude model but not totally unrealistic given the scale (samples taken at 10 m intervals) and the limited number of samples taken (perhaps 20 to 30 per transect depending upon the terrain).

As we have mentioned above it is reasonable to suppose that within an area of high, archaeological, phosphate the Y_i have a Normal distribution, and within an area of low, non-archaeological phosphate, the Y_i have a different Normal distribution. For high phosphate we assume that the Y_i form a sample from a Normal distribution with mean μ_1 and variance σ_1^2 and for low phosphate we have mean μ_2 ($\mu_1 > \mu_2$) and variance σ_2^2.

It is convenient for what follows to model Y_i not in terms of its variance but in terms of its *precision*, where the precision of Y_i is the reciprocal of its variance. Thus the greater the precision the smaller the variance and the smaller the precision the larger the variance. We denote the precision of the on- and off-site distributions of Y_i by τ_1 and τ_2 respectively so that $\tau_1 = 1/\sigma_1^2$ and $\tau_2 = 1/\sigma_2^2$. Hence if Y_i is from the high phosphate area we have

$$Y_i \sim N(\mu_1, \tau_1^{-1}),$$

and if Y_i is from the low phosphate area

$$Y_i \sim N(\mu_2, \tau_2^{-1}).$$

10.2.4   The data

Soil samples were taken, at LS N418, along a series of E–W transects each with 23 sampling positions. We shall here consider three such transects labelled Row 2, Row 3 and Row 9. The data are given in Table 10.1.

10.2.5   The prior

We have to express our prior beliefs about

(i) the model, denoted by $M_{i,j}$, which indicates the number of changes along a transect, their positions and the direction of the changes (if any), and
(ii) the means, μ_1 and μ_2, and precisions, τ_1 and τ_2, of the on-site and off-site distributions.

Firstly, experience has shown that it is reasonable to assume that the on-site and off-site precisions are equal, that is $\tau_1 = \tau_2$. We let τ denote the common precision. Secondly, we assume that the number of changes of level and the whereabouts of any changes are *a priori* independent of on- and off-site distributions of the natural logarithm of the phosphate concentration. Therefore we can write the joint prior of $M_{i,j}, \mu_1, \mu_2$ and τ as

$$p(M_{i,j}, \mu_1, \mu_2, \tau) = p(M_{i,j})p(\mu_1, \mu_2, \tau)$$

where $p(M_{i,j})$ is the prior for model $M_{i,j}$ and $p(\mu_1, \mu_2, \tau)$ the joint prior for μ_1, μ_2 and τ.

Table 10.1 Phosphate concentration (mg phosphate per 100 g soil) from LS N418 (* denotes missing data).

Row																							
1	71	57	52	59	*	*	88	53	40	50	77	*	*	*	*	60	61	62	55	57	60	45	59
2	*	55	37	33	*	64	112	48	60	22	62	48	48	60	55	66	83	47	38	50	55	42	104
3	50	97	41	62	*	64	55	55	48	45	33	34	66	64	116	50	55	40	21	45	37	24	57
4	53	62	47	43	23	83	60	48	76	64	71	62	60	57	62	60	60	73	68	104	68	62	44
5	47	52	41	37	52	75	57	83	91	71	80	68	60	77	94	66	203	88	83	59	41	53	45
6	52	60	57	66	60	59	143	290	295	203	146	116	188	125	68	188	71	104	66	*	47	18	33
7	91	53	53	55	66	85	222	275	280	285	176	125	203	101	64	85	97	55	*	44	80	80	60
8	85	59	112	104	166	75	131	203	260	265	203	188	157	125	101	143	270	66	51	45	43	80	59
9	62	57	64	24	77	71	75	80	157	131	143	85	157	157	94	64	73	76	52	60	50	50	57
10	53	80	59	50	40	59	60	62	19	44	52	66	77	91	57	39	71	27	49	55	24	52	*
11	45	48	68	64	71	34	62	108	116	83	80	83	69	80	55	73	93	60	48	*	*	*	*

Prior for the position of the change-points

Let the prior probabilities of no change, one change and two changes be denoted by p_0, p_1 and p_2 respectively. Let us take, as an example, the case that there is one change, but we have no *a priori* information as to its position. Suppose that, conditional on there being exactly one change, the position of the change has a discrete Uniform distribution over the positions $1, 2, \ldots, n - 1$. Also we have no information about whether the readings go from high to low (model $M_{r,n}$) or from low to high (model $M_{n,r}$) — so we assume them to be equally likely. Therefore the prior probability of model $M_{r,n}$ (for $r = 1, 2, \ldots, n - 1$) is given by

$$p_{r,n} = \frac{p_1}{2(n - 1)},$$

and the prior probability of model $M_{n,r}$ (for $r = 1, 2, \ldots, n - 1$) will be

$$p_{n,r} = \frac{p_1}{2(n - 1)}.$$

Similarly for two changes, with no prior information about their positions, nor about whether the model is low-high-low or high-low-high, we assume that each of the $(n - 1)(n - 2)$ pairs of positions is equally likely. Therefore the prior probability of model $M_{r,s}$ (for $r = 1, 2, \ldots, n - 1$; $s = 1, 2, \ldots, n - 1$; where $r \neq s$) is

$$p_{r,s} = \frac{p_2}{(n - 1)(n - 2)}.$$

Prior for the phosphate levels

To model the prior information for μ_1, μ_2 and τ we assume that their joint prior may be expressed in the form

$$p(\mu_1, \mu_2 | \tau) = p(\mu_1, \mu_2 | \tau) p(\tau)$$

where (μ_1, μ_2) conditional on τ have a Bivariate Normal distribution (see Section 6.3.6) and τ has a Gamma distribution (see Section 5.10.5) with parameters α and δ.

In more detail, conditional on τ, we are assuming that

(i) μ_1 and μ_2 have prior means given by λ_1 and λ_2 respectively,
(ii) their prior variances are given by $\nu_{2,2}/\omega\tau$ and $\nu_{1,1}/\omega\tau$ respectively, where $\omega = \nu_{1,1}\nu_{2,2} - \nu_{1,2}^2$, and
(iii) the covariance between μ_1 and μ_2 is equal to $-\nu_{1,2}/\omega\tau$.

In this case the joint prior probability density function of μ_1 and μ_2, conditional on τ, may be expressed in the form

$$p(\mu_1, \mu_2 | \tau) \propto \tau \exp\left\{-\frac{\tau\xi}{2}\right\}$$

where

$$\xi = (\mu_1 - \lambda_1)^2 \nu_{1,1} + (\mu_1 - \lambda_1)(\mu_2 - \lambda_2)\nu_{1,2} + (\mu_2 - \lambda_2)^2 \nu_{2,2}.$$

Since the prior for τ is modelled as a Gamma distribution with parameters α and δ the prior probability density function of τ is given by

$$p(\tau) = \frac{\delta^\alpha}{\Gamma(\alpha)} \tau^{\alpha-1} \exp\left\{-\delta\tau\right\}.$$

From Section 5.10.5 we note that the prior mean, mode and variance of τ are given by

$$\frac{\alpha}{\delta}, \quad \frac{\alpha-1}{\delta} \quad \text{and} \quad \frac{\alpha}{\delta^2}$$

respectively.

Furthermore it can be shown (see Bernardo and Smith, 1994, p. 435) that the unconditional prior means of μ_1 and μ_2 are λ_1 and λ_2 respectively, and the unconditional prior variances of μ_1 and μ_2 are

$$\left(\frac{\delta-1}{\alpha}\right) \frac{\nu_{2,2}}{\nu_{1,1}\nu_{2,2} - \nu_{1,2}^2} \quad \text{and} \quad \left(\frac{\delta-1}{\alpha}\right) \frac{\nu_{1,1}}{\nu_{1,1}\nu_{2,2} - \nu_{1,2}^2} \qquad (10.1)$$

respectively. Finally the prior covariance between μ_1 and μ_2 is given by

$$-\left(\frac{\delta-1}{\alpha}\right) \frac{\nu_{1,2}}{\nu_{1,1}\nu_{2,2} - \nu_{1,2}^2}. \qquad (10.2)$$

In any particular situation prior values for $\alpha, \delta, \lambda_1, \lambda_2, \nu_{1,1}, \nu_{1,2}$ and $\nu_{2,2}$ need to be specified. It is difficult to offer general advice on the choice of prior values for the parameters, as these will depend on the local geology of the area; certainly it can be recommended that off-site or background soil samples are taken from the vicinity of the site, but, as far as the soil scientist and archaeologist can judge, away from any archaeological remains and these samples can help in deciding plausible prior values for the parameters. In Section 10.2.8 we shall give the rationale for our choices for site LS N418.

10.2.6   The likelihood

The model tells us that for Y_i within a high-level zone, Y_i has a Normal distribution with mean μ_1 and precision τ. That is the probability density function of Y_i is

$$p(y_i|\mu_1,\tau) = \left(\frac{\tau}{2\pi}\right)^{\frac{1}{2}} \exp\left\{-\frac{\tau}{2}(y_i - \mu_1)^2\right\}.$$

If Y_i is from a low-level zone, Y_i has a Normal distribution with mean μ_2 and precision τ. That is the probability density function of Y_i is

$$p(y_i|\mu_2,\tau) = \left(\frac{\tau}{2\pi}\right)^{\frac{1}{2}} \exp\left\{-\frac{\tau}{2}(y_i - \mu_2)^2\right\}.$$

For the two-change model $M_{r,s}$ (with $r < s$) — from low to high to low — the likelihood is given by

$$
\begin{aligned}
l(M_{r,s},\mu_1,\mu_2,\tau;\boldsymbol{y}) &= \prod_{i=1}^{r} p(y_i|\mu_2,\tau) \prod_{i=r+1}^{s} p(y_i|\mu_1,\tau) \prod_{i=s+1}^{n} p(y_i|\mu_2,\tau) \\
&= \left(\frac{\tau}{2\pi}\right)^{n/2} \exp\left\{-\frac{\tau}{2}(S_1 + S_2)\right\}
\end{aligned}
$$

where

$$S_1 = \sum_{i=1}^{r}(y_i - \mu_2)^2 + \sum_{i=s+1}^{n}(y_i - \mu_2)^2 \text{ and } S_2 = \sum_{i=r+1}^{s}(y_i - \mu_1)^2.$$

For the two-change model $M_{s,r}$ (with $r < s$) — from high to low to high — the likelihood is given by

$$
\begin{aligned}
l(M_{s,r},\mu_1,\mu_2,\tau;\boldsymbol{y}) &= \prod_{i=1}^{r} p(y_i|\mu_1,\tau) \prod_{i=r+1}^{s} p(y_i|\mu_2,\tau) \prod_{i=s+1}^{n} p(y_i|\mu_1,\tau) \\
&= \left(\frac{\tau}{2\pi}\right)^{n/2} \exp\left\{-\frac{\tau}{2}(S_3 + S_4)\right\}
\end{aligned}
$$

where

$$S_3 = \sum_{i=1}^{r}(y_i - \mu_1)^2 + \sum_{i=s+1}^{n}(y_i - \mu_1)^2 \text{ and } S_4 = \sum_{i=r+1}^{s}(y_i - \mu_2)^2.$$

By a similar argument the likelihood for the one-change model from high to low, $M_{r,n}$, is

$$l(M_{r,n}, \mu_1, \mu_2, \tau; \boldsymbol{y}) = \prod_{i=1}^{r} p(y_i|\mu_1, \tau) \prod_{i=r+1}^{n} p(y_i|\mu_2, \tau)$$

$$= \left(\frac{\tau}{2\pi}\right)^{n/2} \exp\left\{-\frac{\tau}{2}(S_5 + S_6)\right\}$$

where

$$S_5 = \sum_{i=1}^{r}(y_i - \mu_1)^2 \text{ and } S_6 = \sum_{i=r+1}^{n}(y_i - \mu_2)^2.$$

For the one-change model from low to high, $M_{n,r}$, the likelihood is

$$l(M_{n,r}, \mu_1, \mu_2, \tau; \boldsymbol{y}) = \left(\frac{\tau}{2\pi}\right)^{n/2} \exp\left\{-\frac{\tau}{2}(S_7 + S_8)\right\}$$

where

$$S_7 = \sum_{i=1}^{r}(y_i - \mu_2)^2 \text{ and } S_8 = \sum_{i=r+1}^{n}(y_i - \mu_1)^2.$$

For the no-change model all at the low level, $M_{n,n}$, the likelihood is

$$l(M_{n,n}, \mu_1, \mu_2, \tau; \boldsymbol{y}) = \left(\frac{\tau}{2\pi}\right)^{n/2} \exp\left\{-\frac{\tau}{2}S_9\right\}$$

where

$$S_9 = \sum_{i=1}^{n}(y_i - \mu_2)^2.$$

For the no-change model all at the high level, $M_{1,1}$, the likelihood is

$$l(M_{1,1}, \mu_1, \mu_2, \tau; \boldsymbol{y}) = \left(\frac{\tau}{2\pi}\right)^{n/2} \exp\left\{-\frac{\tau}{2}S_{10}\right\}$$

where

$$S_{10} = \sum_{i=1}^{n}(y_i - \mu_1)^2.$$

10.2.7   The posterior

Of primary interest to the archaeologist are the number and positions of the change-points along a transect as these provide information about the areas of archaeological phosphate, indicating where within a site organic remains were concentrated. Making inferences about the means and precisions of the high and low phosphate concentrations are of secondary importance, although they will be of use in developing our understanding of the distinction between archaeological and geological phosphate. Thus the statistical interest focuses upon determining the posterior probabilities of the number of changes and the positions of those changes.

Let us consider calculating the two-change-point model from low to high to low, with changes at r and s — this model is denoted by $M_{r,s}$. Using Bayes' theorem, the joint posterior probability of $M_{r,s}$ (for $r = 1, 2, \ldots, s - 1; s = 2, \ldots, n$), μ_1, μ_2 and τ is given by

$$p(M_{r,s}, \mu_1, \mu_2, \tau | \boldsymbol{y}) \propto l(M_{r,s}, \mu_1, \mu_2, \tau; \boldsymbol{y}) p(M_{r,s}) p(\mu_1, \mu_2 | \tau) p(\tau).$$

To make inferences about $M_{r,s}$ we need to obtain the marginal posterior probability, denoted by $q_{r,s}$, of $M_{r,s}$. We can do this by integrating over μ_1, μ_2 and τ so that

$$q_{r,s} \propto p_{r,s} \int_{-\infty}^{\infty} \int_{-\infty}^{\infty} \int_{0}^{\infty} p(M_{r,s}, \mu_1, \mu_2, \tau | \boldsymbol{y}) \, d\tau \, d\mu_2 \, d\mu_1.$$

This is an intricate exercise in mathematics — the enthusiastic reader may like to refer to Broemeling and Tsurumi (1987, p. 55–58) for more details. The result is a fairly complicated mathematical expression which may be written in the form

$$q_{r,s} \propto p_{r,s} \times f(\boldsymbol{y}, \lambda_1, \lambda_2, \nu_{1,1}, \nu_{1,2}, \nu_{2,2}, \alpha, \beta)$$

where $f(\boldsymbol{y}, \lambda_1, \lambda_2, \nu_{1,1}, \nu_{1,2}, \nu_{2,2}, \alpha, \beta)$ is a function of the data, $\boldsymbol{y}$, and the parameters of the prior for μ_1, μ_2 and τ. If we take the special case of $\nu_{1,1} = \nu_{2,2} = \nu$ and $\nu_{1,2} = 0$ then we obtain

$$q_{r,s} \propto \frac{p_{r,s}}{(a_1 a_2)^{0.5} \left(2\beta + d + \nu(\lambda_1^2 + \lambda_2^2) - \left(a_1^{-1} b_1^2 + a_2^{-1} b_2^2\right)\right)^{0.5n + \alpha}}$$

where

$$a_1 = \nu + n + r - s,$$
$$a_2 = \nu + s - r,$$
$$b_1 = \nu \lambda_1 + \sum_{i=1}^{r} y_i + \sum_{i=s+1}^{n} y_i,$$

$$b_2 = \nu\lambda_2 + \sum_{i=r+1}^{s} y_i,$$

and

$$d = \sum_{i=1}^{n} y_i^2.$$

Even in this special case it is difficult to interpret $q_{r,s}$ without resort to numerical calculations.

Finally we let q_j $(j = 0, 1, 2)$ be the posterior probability of j changes in a transect. Then we have

$$q_0 = q_{1,1} + q_{n,n},$$

$$q_1 = \sum_{r=1}^{n-1} q_{r,n} + \sum_{r=2}^{n} q_{n,r},$$

$$q_2 = \sum_{1 \le r < s < n} q_{r,s} + \sum_{1 \le r < s < n} q_{s,r}.$$

The question is how to interpret these posterior probabilities. Of course it depends upon the problem in hand. Really the posterior probabilities of the number of changes need to be compared with the prior probabilities. If the posterior probability of no change is much larger than the prior probability, then this is telling us that the data tend to support the no-change model. If, on the other hand, the posterior probability of two changes is high compared with the corresponding prior, then the data support the two-change model.

One possible way of proceeding with the interpretation is to say that if q_0 is greater than the sum of q_1 and q_2, then, a posteriori, we infer that there are no changes present in the transect, in which case all the soil samples along the transect are at the low level or at the high level. On the other hand if q_0 is less than the sum of q_1 and q_2, then at least one change is more likely than no change. In this case the number of changes is indicated by the larger of q_1 and q_2. In the two-change case the positions of the changes may be inferred from the mode of $q_{r,s}$. A similar argument may be applied to the one-change case.

Once a decision has been reached whether there is no change, one change or two changes, it is possible to decide the location of the change(s), by examining a histogram of the posterior probabilities of change(s) at each of the positions along the transect. The modal value will give the position with the highest posterior probability.

10.2.8   Example of the change-point methodology

We turn now to analysis of the data from three transects through site LS N418. We select Rows 5, 8 and 9 of the data given in Table 10.1. The natural logarithm of the phosphate concentration is plotted against each position along the transects, in Figures 10.9–11. As in Section 10.2.5 first we consider specifying the prior probabilities for the number of changes, secondly specifying the positions of the changes, and thirdly specifying the prior probabilities for the levels of phosphate concentration and their variability.

Specifying the priors for transects through LS N418

(a) We thought one or two changes to be equally likely, and no change as likely as a change of one sort or the other. These are conservative assumptions, given that the presence of archaeological remains should, *a priori*, increase the probability of the presence of archaeological or high phosphate. Therefore the prior probability of no change was assessed to be 0.5; and the prior probability of a change of some sort, whether with one change-point or two change-points, was also assessed at 0.5. The priors of one or two changes, p_1 and p_2, were both judged to be 0.25.

(b) The prior probability, conditional on only one change-point, that the change-point is at position r along the transect, is assumed to have a discrete Uniform distribution, over the values $1, 2, \ldots, n - 1$. In Rows 5, 8 and 9 $n = 23$, therefore the prior probability that the model is $M_{r,n}$ is given by $p_{r,n} = 0.25/(2(n - 1)) = 1/176$, for $r = 1, \ldots, 22$, and similarly for model $M_{n,r}$. The prior probability of two changes at r and at s is also uniform over all the possible pairs of positions in the transect, $(n - 1)(n - 2) = 22 \times 21$ in all; thus the prior probability that the model is $M_{r,s}$ is given by $p_{r,s} = 0.25/(22 \times 21) \approx 0.0005$.

(c) In the light of other related work carried out in Greece prior values were assigned to the remaining parameters as follows.

   (i) First we need a prior estimate for the variability in the natural logarithm of the phosphate concentration about the true high and low levels, that is to say the variation due to imprecisions in the chemical analysis, the local natural variability in phosphate concentration, and related factors. The standard deviation, σ, which represents the variability, was thought to be about 0.4 and was judged to lie between about 0.3 and about 0.6. Therefore, *a priori* the precision $\tau = 1/\sigma^2$ was thought to have a modal value of about 6, and to lie in the interval from about 3 to about 11. To model the prior information values for α and δ were chosen to be 7 and 1 respectively. Using this density the prior mode was at 6; the prior probability that the precision is smaller than 3 is about 0.03, that

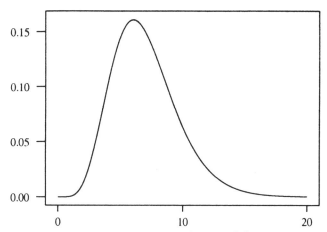

Figure 10.8   Plot of the prior probability density of the common precisions τ, where the parameters of the Gamma distribution are $\alpha = 7$ and $\delta = 1$; the modal value is 6.

is $P(\tau < 3) \approx 0.03$; the prior probability that it is greater than 11 is about 0.08, that is $P(\tau > 11) \approx 0.08$. This specification was believed to provide adequate representation of the variability of the phosphate concentration about the means. A plot of the prior probability density for τ is shown in Figure 10.8.

(ii) The mean of the low (or background) level of the natural logarithm of the phosphate was thought to be 4.0, but it was judged that it could vary considerably from area to area, perhaps between 3.5 and 4.5. Therefore we take λ_2 to be 4.0. Using the property that approximately 95% of the probability of a Normal distribution lies within ± 2 standard deviations of the mean (see Section 5.10.3) we can say that the prior variance of μ_2 is 1/16.

(iii) It was believed that the difference between the high and low levels of the natural logarithm phosphate concentration was about 1.0, that is $\lambda_1 - \lambda_2 \approx 1.0$, and so we set $\lambda_1 = 5.0$. Moreover it was thought that this difference between levels was independent of μ_2 and might vary between 0.5 and 1.5. This we interpret to imply that the variance of $(\mu_2 - \mu_1)$ is 1/16.

We can use (10.1) and (10.2) to determine suitable values for $\nu_{1,1}$, $\nu_{1,2}$ and $\nu_{2,2}$ as follows. We recall that $\omega = \nu_{1,1}\nu_{2,2} - \nu_{1,2}^2$ and note that the mode of a Gamma distribution is $(\alpha - 1)/\delta$, so that in this case $(\alpha - 1)/\delta = 6$ and hence we have

$$Var(\mu_1) = \frac{1}{8} = \frac{1}{6} \times \frac{\nu_{2,2}}{\omega}$$

Table 10.2 Posterior probabilities of no change, one change and two changes for Rows 5, 8 and 9 of LS N418. In each case the prior probability of no change, q_0, was set at 0.5, of one change, q_1, at 0.25 and of two changes, q_2, at 0.25.

Row	q_0	q_1	q_2
5	0.65	0.00	0.35
8	0.00	0.17	0.82
9	0.00	0.00	0.99

$$Var(\mu_2) = \frac{1}{16} = \frac{1}{6} \times \frac{\nu_{1,1}}{\omega}$$

and

$$Cov(\mu_1, \mu_2) = \frac{1}{16} = \frac{1}{6} \times \left(\frac{-\nu_{1,2}}{\omega} \right).$$

Solving for $\nu_{1,1}$, $\nu_{2,2}$ and $\nu_{1,2}$ yields

$$\nu_{1,1} = \frac{8}{3}, \quad \nu_{2,2} = \frac{16}{3}, \quad \nu_{1,2} = -\frac{8}{3}.$$

Thus we have completed our specification of the prior.

Interpreting the posteriors for transects through LS N418

Posterior probabilities of no change-point, one change-point and two change-points in Rows 5, 8 and 9 of the data are given in Table 10.2.

(i) In Row 5 it can be seen that the posterior probability of no change, 0.65, is larger than its prior probability, 0.5. Although this is higher than the combined posterior of one and two change-points, in fact the posterior probability of two changes, 0.35, is higher than its prior, 0.25. In this case the preference for the no-change model arises in part from the conservative prior probability favouring no change. Given two changes, the modal posterior value for the positions of the changes is 0.47, for positions 16 and 17. A glance at Figure 10.9 will show that there is a single high reading at position 17. This single, isolated, high reading is interpreted, therefore, on the conservative assumption underlying the prior probability of the no-change model, as an exceptional reading, and not to indicate a true change of level.

(ii) In Row 8 (Figure 10.10) the posterior probability of two changes, 0.82, is considerably higher than its prior, 0.25. The posterior probability of one change-point, 0.17, is lower than its prior. In other words the data have considerably increased our belief in the two-change model. Within Row

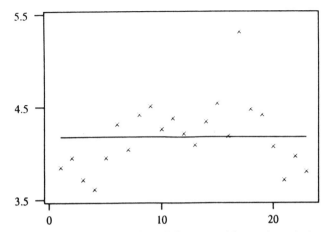

Figure 10.9 LS N418, Row 5: plot of the natural logarithm of phosphate concentrations. The posterior probability for the no-change model is 0.66, whilst for the one-change model it is 0.00 and for the two-change model it is 0.35.

8, *given* that there are two change-points, the modal value is 0.56, for their positions at 2 and 17; the next highest posterior for the positions of the two change-points, at 3 and 17, is 0.15. The natural interpretation of this analysis is that we are confident that there are changes from low to high to low level, and that the centre of the transect is indeed marked by an area of archaeological phosphate.

(iii) In Row 9 (Figure 10.11) the posterior probabilities are overwhelmingly in favour of two changes. Given that there are two change-points, the posterior for $r = 8$ and $s = 15$ is 0.30; the next highest value, for $r = 8$ and $s = 14$, is 0.25. Row 9, like Row 8, can be interpreted as strongly supporting the conclusion that the centre of the transect has high phosphate of archaeological origin.

We hope to have illustrated, in Case Study I, the development of the change-point models, the ascription of prior probability distributions to the parameters of the models, the calculation of the posterior probabilities in the light of the data, and the interpretation of those prior probabilities. All of this, of course, has been performed within the framework of the Bayesian approach to data analysis.

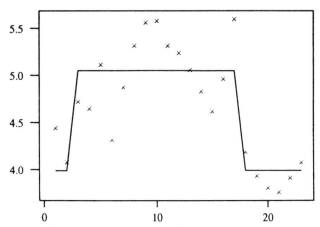

Figure 10.10   LS N418, Row 8: plot of the natural logarithm of phosphate concentrations together with the change-points at positions 2 and 17. The posterior probability for the two-change model is 0.82 and that for the one-change model is 0.17.

10.3   Case Study II — analysis of phosphate data in a grid, with two levels — image segmentation

The change-point methodology described in the previous section is particularly effective for recognizing changes in some underlying process for data collected along a single transect. It has been used quite successfully in the field to provide a quick and easy method of establishing the extent of high phosphate areas in the vicinity of a site. However, the data at site LS N418 were collected over a grid (see Table 10.1), and a methodology which is able to view the whole picture right over the grid, rather than merely along transects, would have the advantage of taking all the data simultaneously into account. In this case the archaeologist is interested in broader zones of high and low phosphate concentration, rather than the linear information available from transects.

Methods which process whole images in this way are referred to as *image segmentation* methods, because they divide up or *partition* a region into separate significant zones or segments. Image segmentation has a long history with much of the development work carried out by physicists, electronic engineers and computer scientists. The introduction of Bayesian methods by statisticians has allowed important improvements, and in particular made possible a unified approach to a greater variety of problems than had previously been considered. Ripley (1988) gives a comprehensive review of the subject. A later review, including the use of Markov chain Monte

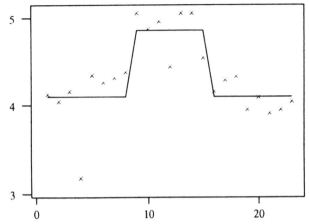

Figure 10.11 LS N418, Row 9: plot of the natural logarithm of phosphate concentrations together with the change-points at positions 8 and 15. The posterior probability for the two-change model is 0.99.

Carlo methods, is given by Besag and Green (1993). In this example the methodology of Besag (1986) is used to illustrate the approach.

10.3.1 Background to the model

We can imagine the grid of phosphate squares to have been photographed from a satellite, so that the definition of the "photograph" is limited to the 10 m squares. The soil phosphate specimens were taken at the centre of each square, and each square is referred to as a cell; each cell contains an observation of the phosphate concentration. The phosphate concentrations are rather like the different shades of grey in a black-and-white photograph. Because of various imprecisions (not least the rather poor definition of the data) the picture appears blurred and smudged, and it is the task of image segmentation to clean up the dirty image. In the case of the LS N418 data, we wish to improve the definition of areas of archaeological occupation, indicated by phosphate but not necessarily evident from artefact remains. That is we wish to divide the grid into zones of high and low phosphate concentration.

The image segmentation method developed by Besag allows for the observation in any particular cell to be ascribed to one level or another, after taking into account the ascription of other cells in its immediate neighbourhood. In our case let us suppose that an observation might belong either to the high phosphate level or to the low phosphate level. The image-processing algorithm will look at the other neighbouring cells immediately surrounding the uncertain observation. If the neighbouring cells are of high level, then our confidence is increased that the unascribed observation should also be high. If, on the other hand, the neighbouring cells are low, then our belief increases that the unascribed observation should belong to the low level.

10.3.2 The model

Consider a continuous two-dimensional region which is divided into an m by n rectangular array of cells labelled (i, j) for $i = 1, 2, \ldots, m; j = 1, 2, \ldots, n$. Suppose that each cell can take one of L levels, labelled $1, 2, \ldots, L$, but the actual level of a particular cell (i, j) is unknown. At LS N418 the grid over the entire area of the site covered 230×110 m, and the sampling points were spaced at 10 m intervals so $m = 23$ and $n = 11$. In this case just two levels are proposed, on-site and off-site, so that $L = 2$.

We do not know which cells are at which level. In fact it is the purpose of this analysis to make inferences about which parts of the region are at which levels.

Let $\theta_{i,j}$ denote the true level of cell (i, j) so that if cell (i, j) is at level l then $\theta_{i,j} = l$. A *scene* of the region will be denoted by $\boldsymbol{\theta} = (\theta_{1,1}, \theta_{1,2}, \ldots, \theta_{m,n})$, where $\theta_{i,j}$ is the corresponding level of cell (i, j). Let $y_{i,j}$ be the observed value of the natural logarithm of the phosphate concentration at cell (i, j) and $\boldsymbol{y} = (y_{1,1}, y_{1,2}, \ldots, y_{m,n})$ be the corresponding vector, which is a realization of the random vector $\boldsymbol{Y} = (Y_{1,1}, Y_{1,2}, \ldots, Y_{m,n})$. Then given $\boldsymbol{y}$, what can be inferred about the true scene $\boldsymbol{\theta}$?

As in the previous section we will develop our model in two stages.

Modelling the variation within a level

The model posits that observations of the natural logarithm of the phosphate concentration coming from cells from the same level, say level l, can be thought of as a sample from a Normal distribution with mean μ_l and and variance σ_l^2. That is to say if $Y_{i,j}$ is the random variable representing the natural logarithm of the phosphate concentration in cell (i, j) and if cell (i, j) has level l so that $\theta_{i,j} = l$ then

$$Y_{i,j} | \mu_l, \sigma_l^2, \theta_{i,j} \sim N(\mu_l, \sigma_l^2).$$

As in the change-point example, it is in this case reasonable to suppose that the variation is the same for all levels, so that $\sigma_l^2 = \sigma^2$ for $l = 1, \ldots, L$, where σ^2 is the common variance. For ease of notation we let $\boldsymbol{\mu} = (\mu_1, \ldots, \mu_L)$ and for simplicity of the argument that follows, we assume that $\boldsymbol{\mu}$ and σ^2 are known. Of course in practice this is not so, but our argument is easily extended to include any uncertainty about these parameters.

Table 10.3 A central cell of level $\theta_{i,j}$ and the surrounding cells in its neighbourhood.

$\theta_{i-1,j-1}$	$\theta_{i-1,j}$	$\theta_{i-1,j+1}$
$\theta_{i,j-1}$	$\theta_{i,j}$	$\theta_{i,j+1}$
$\theta_{i+1,j-1}$	$\theta_{i+1,j}$	$\theta_{i+1,j+1}$

Modelling the relationship between adjacent cells

We are attempting to divide the region into several homogeneous zones. Moreover, we expect that significant concentrations of phosphate will cluster together. Therefore it seems sensible that in deciding the true level $\theta_{i,j}$ of any cell we should take into account not only the observation, $y_{i,j}$, the natural logarithm of the phosphate concentration in that cell, but also the levels assigned to cells in the immediate *neighbourhood*.

Of course we need to define carefully what we mean by neighbourhood. One possibility is to include in the neighbourhood of cell (i,j) all eight adjacent cells, see Table 10.3. Alternatively a neighbourhood could consist of the four cells which share a boundary with cell (i,j), namely cells $(i,j-1)$, $(i-1,j)$, $(i,j+1)$ and $(i+1,j)$ (which in the case of site LS N418 are all 10 m from (i,j)). Or one could also include the next nearest neighbours: $(i-1,j-1)$, $(i-1,j+1)$, $(i+1,j-1)$ and $(i+1,j+1)$ (which in the case of site LS N418 are all about 14 m from (i,j)).

An important consequence of saying that $\theta_{i,j}$ is dependent only upon neighbouring cells is that the conditional probability of cell (i,j) being at level R, i.e. $\theta_{i,j} = k$, given the values of *all* the other $\theta_{r,s}$ in the *whole* region is equal to the conditional probability of $\theta_{i,j} = k$ given the values of the $\theta_{r,s}$ in just the *neighbourhood* of cell (i,j). We let $\boldsymbol{\theta}_{-\{i,j\}}$ denote all the $\theta_{r,s}$ in the whole region but with $\theta_{i,j}$ deleted. We let $\boldsymbol{\theta}_{\partial\{i,j\}}$ denote the $\theta_{r,s}$ in the neighbourhood of cell (i,j). Then mathematically we are saying that

$$P(\theta_{i,j} = k | \boldsymbol{\theta}_{-\{i,j\}}) = P(\theta_{i,j} = k | \boldsymbol{\theta}_{\partial\{i,j\}}). \tag{10.3}$$

10.3.3   The prior

The specification of a prior for $\boldsymbol{\theta}$ is difficult, but Besag (1986) suggests a suitable form, which takes into account the number of pairs of cells, the one of level k and the other of level l. The form suggested is

$$p(\boldsymbol{\theta}) \propto \exp\left\{ -\sum_{k=1}^{L-1} \sum_{l=k+1}^{L} \beta_{k,l} n(k,l) \right\} \tag{10.4}$$

Table 10.4   Eight neighbours at levels 1 and 2 and a central cell whose level is unknown.

1	1	1
1	$\theta_{i,j} = k$	1
2	2	2

where $n(k, l)$ is the number of distinct adjacent pairs of levels k and l in the *whole* scene, and the $\beta_{k,l}$ are parameters, whose significance will be explained below.

An important consequence of this choice of prior is that

$$P(\theta_{i,j} = k|\boldsymbol{\theta}_{-\{i,j\}}) \;=\; P(\theta_{i,j} = k|\boldsymbol{\theta}_{\partial\{i,j\}})$$

$$\propto \; \exp\left\{ -\sum_{\substack{l=1 \\ l\neq k}}^{L} \beta_{k,l} u_{i,j}(k, l) \right\}$$

where $u_{i,j}(k, l)$ is the number of neighbours of cell (i, j) of level l.

For LS N418 $L = 2$, so that k and l can take the values 1 or 2. Let us consider a central cell and its eight neighbours arranged as in Table 10.4. For ease of illustration, we shall assume that the diagonal cells have the same weight as those directly adjacent to cell (i, j). If $k = 1$ then $u_{i,j}(1, 2) = 3$. If $k = 2$ then $u_{i,j}(2, 1) = 5$. Therefore the prior probability that cell (i, j) is at level 1 is proportional to $\exp\{-3\beta_{1,2}\}$ and that the cell is at level 2 is proportional to $\exp\{-5\beta_{1,2}\}$. Since the cell is either at level 1 or at level 2 then

$$P(\theta_{i,j} = 1|\boldsymbol{\theta}_{-\{i,j\}}) = \frac{e^{-3\beta_{1,2}}}{e^{-3\beta_{1,2}} + e^{-5\beta_{1,2}}} = \frac{1}{1 + e^{-2\beta_{1,2}}}$$

and

$$P(\theta_{i,j} = 2|\boldsymbol{\theta}_{-\{i,j\}}) = \frac{e^{-5\beta_{1,2}}}{e^{-3\beta_{1,2}} + e^{-5\beta_{1,2}}} = \frac{e^{-2\beta_{1,2}}}{1 + e^{-2\beta_{1,2}}}.$$

To pursue the example let us look at the consequences of taking different values of $\beta_{1,2}$.

(i) If $\beta_{1,2} = 0$ then $P(\theta_{i,j} = 1|\boldsymbol{\theta}_{-\{i,j\}}) = P(\theta_{i,j} = 2|\boldsymbol{\theta}_{-\{i,j\}}) = 0.5$ and so the surrounding pair-wise comparisons have no effect on the ascription of the central cell.

(ii) If $\beta_{1,2} = 0.1$ then the probability that $\theta_{i,j} = 1$ is 0.55 and that $\theta_{i,j} = 2$ is 0.45, that is to say the surrounding cells have a moderate effect on assigning the central cell to level 2.

(iii) If $\beta_{1,2} = 0.5$ then the probabilities that the central cell belongs to level 1 or level 2 are 0.73 and 0.27 respectively, that is the effect of the surrounding cells is very much stronger.

To summarize: the higher the value of $\beta_{1,2}$ the more likely it is that $\theta_{i,j}$ will be ascribed (in this example) to level 1. The result of this is to smooth out differences of level between cells in the same neighbourhood. If, on the other hand, $\beta_{1,2}$ is negative, the more likely it becomes that the central cell will be not of level 1 but of level 2.

Thus $\beta_{k,l}$ can be thought of as representing our prior belief about how smooth the scene should be. In the above example a small value of $\beta_{1,2}$ indicates our prior belief that small isolated patches of high phosphate concentration are likely. On the other hand taking a high value of $\beta_{1,2}$ indicates that we believe that the region should be divided into broad contiguous zones of either high or low phosphate concentration.

10.3.4 The likelihood

The $Y_{i,j}$ are assumed to be conditionally independent and as $Y_{i,j}$ depends only upon $\theta_{i,j}$, the likelihood can be expressed as

$$l(\boldsymbol{\theta}; \boldsymbol{y}) = \prod_{i=1}^{m} \prod_{j=1}^{n} l(\theta_{i,j}; y_{i,j}). \tag{10.5}$$

Furthermore, since $Y_{i,j}$ conditional on $\theta_{i,j} = k$ has a Normal distribution with mean μ_k and variance σ^2, $l(\theta_{i,j}; y_{i,j})$ is proportional to the probability density function of a Normal random variable with mean μ_k and variance σ^2. Therefore if $\theta_{i,j} = k$ we have

$$l(\theta_{i,j}; y_{i,j}) \propto \exp\left\{-\frac{(y_{i,j} - \mu_k)^2}{2\sigma^2}\right\}.$$

10.3.5 The posterior

Since we are assuming that $\boldsymbol{\mu}$ and σ^2 are known, we have (by Bayes' theorem) that

$$p(\boldsymbol{\theta}|\boldsymbol{y}) \propto l(y; \theta) p(\boldsymbol{\theta})$$

and as a consequence of (10.3), (10.4) and (10.5) we have, given the rest of the scene and the data, that the posterior probability of cell (i, j) belonging to level k is

$$P(\theta_{i,j} = k|\boldsymbol{\theta}_{-\{i,j\}}, \boldsymbol{y}) \quad \propto \quad l(\theta_{i,j}; y_{i,j})P(\theta_{i,j} = k|\boldsymbol{\theta}_{-\{i,j\}})$$

$$\propto \quad \exp\left\{-\frac{(y_{i,j} - \mu_k)^2}{2\sigma^2}\right\}\exp\left\{-\sum_{\substack{l=1 \\ l \neq k}}^{L}\beta_{k,l}u_{i,j}(k, l)\right\}.$$

10.3.6   Implementation

There are two ways of proceeding from here. Firstly one can use Markov chain Monte Carlo methods (Besag and Green, 1993) where each $\theta_{i,j}$ is sampled in turn and the process continues until convergence is achieved. Alternatively there is Besag's method (1986), in which the strategy for cell (i, j) is to choose the level k that maximizes the posterior probability that the level is k, given the rest of the scene; that is we find k which maximizes $P(\theta_{i,j} = k|\boldsymbol{\theta}_{-\{i,j\}}, \boldsymbol{y})$. This is equivalent, for each cell (i, j) of the grid, to determining the level k that *minimizes* the expression

$$\frac{(y_{i,j} - \mu_k)^2}{2\sigma^2} + \sum_{\substack{l=1 \\ l \neq k}}^{L}\beta_{k,l}u_{i,j}(k, l) \tag{10.6}$$

where μ_k is the mean value for level k, $u_{i,j}(k, l)$ is the number of neighbours of cell (i, j) of level l and $\beta_{k,l}$ is the weighting or smoothing factor. Here we use Besag's method.

Let us examine this in more detail for part of the grid of the LS N418 data. The raw data and the natural logarithm of the data are given in Table 10.5 (a) and (b), together with the allocation of the neighbouring cells to levels 1 and 2 (Table 10.5 (c)).

Suppose that the mean of the high phosphate level for the site is set at 5.10, and the mean of the low phosphate level is 4.05, and their common standard deviation is 0.4. That is $\mu_1 = 5.10$, $\mu_2 = 4.05$ and $\sigma = 0.4$. Suppose we take $\beta_{1,2}$ as equal to 0.5. Such a choice indicates that, a priori, we believe that the region should be divided into broad zones exclusively of high phosphate or of low phosphate.

We note that $u_{i,j}(1, 2) = 3$ and $u_{i,j}(2, 1) = 5$. If the central cell were assigned to level 1, then (10.6) becomes

$$\frac{(4.44 - 5.10)^2}{2 \times 0.4^2} + 0.5 \times 3 = 1.36 + 1.5 = 2.86$$

Table 10.5 Part of the grid of data from LS N418: (a) raw data, (b) natural logarithm of raw data, and (c) levels to which the observations have been assigned.

(a)

208	188	157
142	85	157
51	66	77

(b)

5.33	5.24	5.06
4.96	4.44	5.06
3.93	4.19	4.34

(c)

1	1	1
1	$\theta_{i,j} = k$	1
2	2	2

whereas if it were assigned to level 2 then we have

$$\frac{(4.44 - 4.05)^2}{2 \times 0.4^2} + 0.5 \times 5 = 0.48 + 2.5 = 2.98.$$

Thus with $\beta_{1,2}$ set to 0.5 the central cell should be assigned to level 1, as 2.86 is less than 2.98 and so minimizes (10.6). If, on the other hand, $\beta_{1,2} = 0.3$, then the first sum would come to 2.26 and the second would be 1.98, so that with a lower smoothing factor the central cell would be ascribed to level 2.

The procedure then advances by considering each cell in turn. Initially a cell (i, j) is allocated to that level k for which its reading $y_{i,j}$ is closest to the mean, μ_k, of level k. Thus in the example of Table 10.5, 4.44 is closer to 4.05 than to 5.10 and so initially the central cell would be ascribed to level 2. Then the algorithm is applied using (10.6) until convergence is reached. For cells near the edge of the region some suitable adjustment must be made in the definition of the neighbourhood.

10.3.7 Detailed methodology for LS N418

For the reasons outlined in Section 10.2.8 it was believed that, in the case of LS N418, the prior mean value for the high phosphate zone was about 0.5, that is $\mu_1 = 5.0$, the mean of the low phosphate zone was 4.0, that is $\mu_2 = 4.0$, and the prior variance was 0.4^2, that is $\sigma^2 = 0.16$. Figure 10.12 shows a grey-scale representation of the data with no attempt at image segmentation. The numbers in the cells indicate the artefact counts at the point where the soil sample was taken.

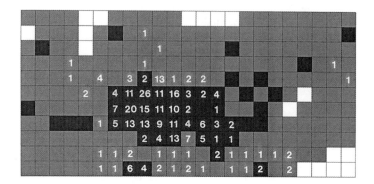

Figure 10.12   Grey-scale image of the natural logarithm of the phosphate observations from LS N418 (see Table 10.2). The different levels of phosphate are indicated by shading, the numbers in the cells indicate artefact counts.

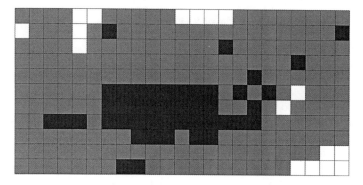

Figure 10.13   Image segmentation analysis of the phosphate observations from LS N418. The smoothing factor $\beta_{1,2}$ was set at 0.05.

To achieve the image shown in Figure 10.13, the image-processing methodology was applied using a value of $\beta_{1,2}$ of 0.05; as the discussion in Section 10.3.3 has indicated, this choice will smooth the image only slightly. Thus if there were five out of eight adjacent cells at level 1, the consequent probability that $\theta_{i,j} = 1$ would be just 0.52. All the same it can be seen that a number of the more isolated cells, visible in Figure 10.12, have been smoothed away. Most of the high phosphate is concentrated at the centre of the grid.

Figure 10.14 shows the scene which results when $\beta_{1,2}$ is set at the extremely high value of 2.88. In this case if there were five out of eight adjacent cells at level 1, the consequent probability that $\theta_{i,j} = 1$ would be 0.99. This can be interpreted as imposing an extremely strict rule for ascribing a cell to a level.

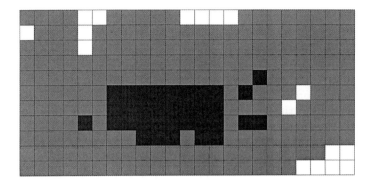

Figure 10.14 Image segmentation analysis of the phosphate observations from LS N418. The smoothing factor $\beta_{1,2}$ was set at 2.88.

The surrounding cells have a very strong influence, as we are convinced that a cell really is of a different value from its neighbours only if it is very high indeed (in a low neighbourhood) or very low indeed (in a high neighbourhood).

10.3.8 Interpretation of the results

The raw phosphate data from site LS N418 are given in Table 10.1, and the results of the image processing of the data are illustrated in Figures 10.12–10.14. There is a distinct area of high phosphate concentrated in the centre of the grid in cells which also have high sherd counts.

This contiguous zone, precisely because the readings are consistently of high level, survives when $\beta_{1,2}$ is very high. The isolated high spots, even when very rich in phosphate, ultimately will succumb when the probabilities are stacked in favour of their low neighbours. How is this to be understood archaeologically? The natural interpretation is that the building remains and the "phosphate site" occupied the same area; contrast this with the situation recognized at Tadworth (see Figure 10.3) where the inhabited area was different from that with high phosphate. In more detail there are some parts of the site with high phosphate and low artefact counts, and others with low phosphate and high sherd counts, which suggest distinct functional areas within the site. The isolated high readings for example in cell $(7, 2)$ or in cell $(15, 3)$ are more difficult to interpret; it is safest to assume, in the absence of other indications, that they are not archaeologically significant. There are, indeed, limits to the interpretation of surface remains, and more refined conclusions would require confirmation from further research, through excavation or the application of more detailed prospecting techniques.

10.4   Case Study III — Manor Farm, Lancashire, U.K. — image segmentation with three levels

In this case study we examine data from a soil phosphate survey carried out in the course of the excavation of the Bronze Age cairn at Manor Farm near Borwick, north Lancashire. We refer the reader to Olivier (1987) for a full discussion of the site, its excavation and its interpretation. In brief we mention that once the mound was removed, inhumation and cremation graves were revealed; at that stage the excavation team collected soil samples from the ancient surface. In the acid soil of the site it was thought quite possible that the bones of some of the burials might have disappeared altogether. It was hoped, therefore, that phosphate analysis of the samples might reveal the locations of graves not recognized through the bone remains, but still traceable through enhanced phosphate in the soil, as bone has a high concentration of phosphate. With this aim in mind a survey interval of 0.5 m was selected and soil samples were taken over the entire burial area of approximately 25 × 25 m.

Although applied to phosphate data the analysis required here was different from that in the case of Laconia survey site LS N418. Rather than continuous blocks of phosphate delimited by one change of level, a number of discrete patches, a few metres across, were sought. Note that the sampling interval of 0.5 m is very much smaller than the 10 m of LS N418. It was not clear that two basic levels of concentration, distinguished in the model for surface survey data, were appropriate to this investigation. After discussion with the excavator it was decided that an image segmentation approach, with more than just two levels, offered the best chance of interpretable results.

10.4.1   The model

We think of the phosphate concentration in any one of the cells as belonging to one of a number, L, of levels; in the previous case study two basic levels were selected, since we were attempting to delimit broad zones, namely off-site and on-site. How many levels should be selected in this instance? As indicated in the previous section, in this rather more refined analysis there may be more than one level of "archaeological" intensity, or more than one level of "background" intensity. Thus there might be a high level of phosphate concentration at the centre of a burial, a halo of intermediate level on a burial's periphery, receding to a low level beyond that. When the Manor Farm data were first published, four levels were chosen (Olivier, 1987, p. 182) to display the raw data. However, the histogram of the natural logarithm of phosphate concentration (see Figure 10.15) suggested three levels, and three were chosen as the basis for the analysis carried out here. That is, we take L to be 3.

The basic principles of the methodology have been explained in the previous section and there is no need to expand them further.

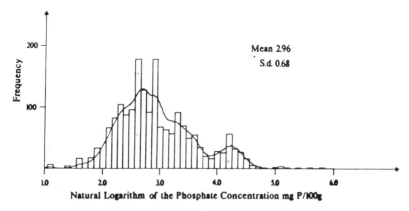

Figure 10.15   Histogram of the natural logarithm of phosphate concentration of soil samples taken from the Bronze Age cairn at Manor Farm, Lancashire. The continuous curve represents a "smoothed" version of the histogram.

10.4.2   Detailed methodology

Having decided that three levels were appropriate, estimates of their means were taken to be 4.2 for the highest level, 3.3 for the level below that, and 2.7 as the mean for the lowest level. That is $\mu_1 = 4.2$, $\mu_2 = 3.3$ and $\mu_3 = 2.7$. Our experience in dealing with phosphate data has persuaded us that the assumption of equal variances is reasonable, and so the common standard deviation σ was taken as 0.5. A grey-scale image is given in Figure 10.16, based simply on the natural logarithm of the observations, where each cell is ascribed to the level to whose mean its value is closest. The grey-scale plot is difficult to interpret archaeologically.

For the image processing, the eight adjacent cells were chosen as the neighbourhood, because we expected cremation burials to extend over several cells of the grid, and the cells on the diagonals were given the same weighting as those with a common side. We recall from Section 10.3.3 that the smoothing of the image takes place by looking at the number of distinct adjacent pairs of levels k and l, where $k \neq l$. With three levels we can have the pairs 1 and 2 (or 2 and 1), 2 and 3 (or 3 and 2), and 1 and 3 (or 3 and 1). For the smoothing parameters, therefore, we need to express our beliefs about suitable values: $\beta_{1,2}$ and $\beta_{2,3}$ were given the same value, as adjacent levels were judged equally probable to occur in adjoining cells; on the other hand $\beta_{1,3}$ was set at a higher value because an abrupt jump from level 1 to level 3 or *vice versa* was thought less likely. The effect of a higher value of $\beta_{1,3}$ is to increase the probability that, for example, a single cell initially at level 1 be raised to level 3, if all the cells in its neighbourhood are of level 3.

Figures 10.17 and 10.18 when contrasted with Figure 10.16 show the much

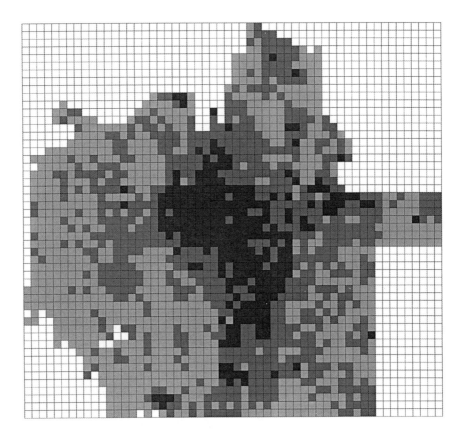

Figure 10.16   Grey-scale image of the natural logarithm of the phosphate
observations from Manor Farm, Lancashire, U.K. The three levels of phosphate are
indicated by shading; their means were judged to be 4.2, 3.3 and 2.7.

clearer images which result from the image segmentation procedure. In Figure
10.17 the smoothing parameters for the pair of levels 1 and 2, and the pair
2 and 3, were both 0.25, whilst that for the pair 1 and 3 was 0.5; that is
$\beta_{1,2} = \beta_{2,3} = 0.25$ and $\beta_{1,3} = 0.5$. Here a central area of high phosphate can
be observed together with a circle of satellite patches, each perhaps a metre
or two across, surrounding it.

  If the smoothing parameters are increased an image like that in Figure 10.18
is produced. Here the smoothing parameters for the pair of levels 1 and 2, and
the pair 2 and 3, were both 0.50, whilst that for the pair 1 and 3 was 1.0;
that is $\beta_{1,2} = \beta_{2,3} = 0.50$ and $\beta_{1,3} = 1.0$. The basic image remains the same,
and it is interesting to note that most of the satellite patches survive even
this extreme degree of smoothing, which in turn suggests that they conform

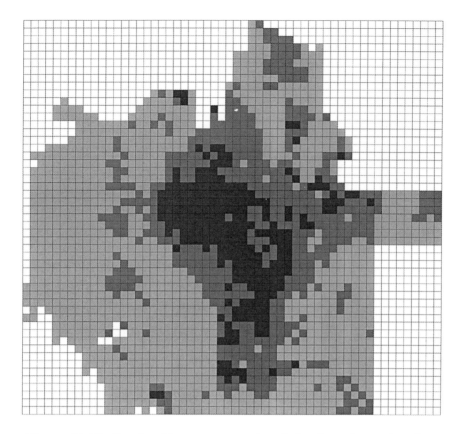

Figure 10.17 Results of image segmentation of phosphate observations from Manor Farm, Lancashire. The smoothing parameters $\beta_{1,2}$ and $\beta_{2,3}$ were both 0.25 and that for $\beta_{1,3}$ was 0.5.

to the original model very well.

10.4.3 Interpretation

An interesting picture is produced by the analysis. It is clear that the highest intensity of phosphate concentration is at the centre of the cairn. Curiously enough the majority of the burials located in excavation by the presence of bone were not found within this core of very high phosphate, though a primary burial was. It is only possible to speculate that this central area was the focus of some other activity, perhaps some funerary rite or ritual immolation, that would not result in the deposition of bone remains, but which involved some

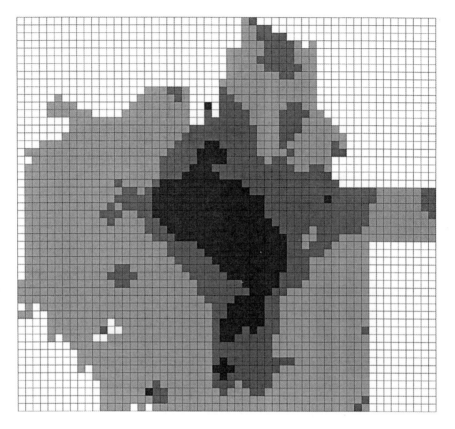

Figure 10.18   Results of image segmentation of phosphate observations from Manor Farm, Lancashire. The smoothing parameters $\beta_{1,2}$ and $\beta_{2,3}$ were both 0.50 and that for $\beta_{1,3}$ was 1.0.

organic residue. On the whole, however, burials did coincide with some of the patches of high phosphate distinguished through the image segmentation analysis. In the light of this it is plausible to suggest that similar such blocks of higher phosphate in the western and northern sectors of the cairn were likewise the sites of burials whose bones have entirely decayed. This possibility is given weight by the fact that it is not uncommon for late Bronze Age cairns to contain satellite burials arranged in just such a circle. This representation of the results reflects back on the selection of three fundamental levels of phosphate concentration, now interpreted as the very high levels at the core, the medium levels associated with burials, and the low level associated with background. In this case the analysis is sensitive to the choice of parameter values, and any interpretation is best corroborated by further evidence, such as reanalysing the relevant soil samples for other elements which might be associated with bone.

10.5 Conclusion

In these case studies we have been concerned with data arising from soil phosphate concentrations, but these techniques of spatial analysis have also been applied successfully to other types of data, such as magnetic susceptibility and resistivity (Buck *et al.*, 1990). The models have a general applicability, therefore, but they may need adapting or fine-tuning to the particular problem or the particular type of data. The Bayesian approach, by insisting on explicit specification of these values, forces us to acknowledge their influence on the final result. Change-point analyses, at least in the case of LS N418, are robust; the prior values of the parameters need to be altered quite considerably before substantially different interpretations emerge. In the case of the image segmentation analysis of the Manor Farm data, on the other hand, the results could be rather sensitive to the choice of values for the various parameters.

We have used Besag's image segmentation method for Case Studies II and III. This was for ease of explanation rather than a preference for that methodology over the Markov chain Monte Carlo approach. In fact for the majority of applications we would expect to use the latter, as it is more versatile, and only a little more difficult to implement. The Bayesian approach to spatial analysis is continually being refined and improved, particularly since the advent of MCMC methods. The reviews by Besag and Green (1993) and Ripley (1988) provide a useful starting point for those wishing to follow up our brief exposition.

11

Sourcing and provenancing

11.1 Introduction

One of the fundamental questions posed by archaeologists about the artefacts they find is "where do they come from?" The answer can reveal information about the society in which the artefact was used. For example, are such items found elsewhere? Is there evidence for formalized trade routes? Is there evidence of itinerant craftsmen (such as similar artefacts being made at different sites)? Is it possible that similar items on different sites represent the spread of the knowledge of manufacture rather than of the goods themselves? Such questions are many and varied. Our ideas about the internal and external relationships between peoples in the past develop in the light of such investigations.

Since the late nineteenth century archaeologists and scientists have recognized that a fruitful field of joint investigation is the chemical or geological analysis of archaeological artefacts (Richards, 1895). The basic premise for this collaboration is that chemical (and physical) properties of raw materials vary from deposit to deposit and that, with careful analysis of both geological beds and the artefacts themselves, it ought to be possible to match the material from which the artefact was manufactured to its *source*.

Over the last century, very many such investigations have been completed. In particular, materials such as obsidian (Renfrew *et al.*, 1966, 1968), jade (Hammond *et al.*, 1977), steatite (Allen *et al.*, 1975), flint (Craddock *et al.*, 1983), and lead (Sayre *et al.*, 1992; Pernicka, 1992; Gale and Stos-Gale, 1992; Budd *et al.*, 1993) have been investigated. For each of these studies samples were available both from archaeological artefacts and from likely geological source material.

What of situations where the locations (and hence nature) of the raw materials are unknown? In such cases we still require answers to the question "where do they come from?", but archaeologists tend to talk about artefact *provenance* rather than source. Different authors ascribe different meanings when they use the term provenance, but here we simply mean:

the location at which the raw materials were brought together and manufactured into the artefact in question.

So, for example, the provenance of a pot is usually the site of the potter's workshop. The provenance of a bronze brooch is usually the site of the foundry. The provenance of a fragment of glass is the site at which the components of its manufacture were brought together, melted and formed into the object of which the fragment was a part (though there is evidence that raw glass was transported in antiquity). Excavation of a production site allows us to identify one of the possible provenances for the type of artefact manufactured there. Identification of workshop and foundry sites is, however, quite rare and so commonly we must undertake provenance studies without much information about the location of (and hence chemical composition of materials from) workshop sites.

The technique most commonly used to aid in provenance studies is that of chemical compositional analysis and over the last century it has been applied to almost all archaeological artefacts whose constituent parts are of geological origin. These include glass (Christie *et al.*, 1979), metals (Pernicka *et al.*, 1990) and the most commonly and routinely analysed material type: archaeological ceramics (for a general review of ceramic analysis see Rice, 1987 and for a review of the techniques as applied to Greek ceramics see Jones, 1986).

In summary then, this chapter focuses on the use of chemical compositional data to aid investigations of source and provenance. The important difference between the two is that in order to source artefacts we require information about the composition of all possible source materials, whereas to provenance we can talk in rather more abstract terms and still gain useful information about the relationships between specimens. The data available from known sources are described by statisticians as *training data* and it is these that make the crucial difference between the statistical approaches required. This means that when training data from production sites are available, problems which archaeologists would normally think of as questions of *provenance* may be tackled using methodologies developed to solve problems of *source*.

11.2   Plan of the chapter

As will become clear in what follows, sourcing and provenancing are particularly challenging areas for statistical investigation and interpretation and we must stress that our methodology is at an early stage. The approach outlined here is not intended as the definitive guide to the interpretation of chemical compositional data for archaeological sourcing and provenancing. Instead we intend to give the reader an appreciation of how the Bayesian method may be applied in this area of research and to give an indication of its potential for the future.

Before we describe our methodology, however, there is an important point which must be highlighted. For sourcing and provenancing problems the data for each specimen consist of observations of the concentrations of several chemical elements all of which are inter-related. Due to this dependency, the statistical distributions are complicated and consequently difficult to explain in an intuitive manner. Furthermore, since archaeological sourcing requires a rather different approach from that when training data are absent, the main part of this chapter will be split into two sections. Nevertheless a single statistical model can be used as the foundation for both methodologies; so we have chosen to describe the basic model before embarking on the case studies.

11.3   A basic model for sourcing and provenancing studies

A wide range of chemical analysis techniques are currently used to obtain compositional data from archaeological artefacts. These include: neutron activation analysis (NAA), x-ray fluorescence analysis (XRF) and inductively coupled plasma spectroscopy (ICPS). Full descriptions and applications of these can be found in Tite (1972), Parkes (1986) and Slater and Tate (1988). It is sufficient to note that selection of the most appropriate technique is usually made as the result of discussion between the chemists and archaeologists involved. The outcome of this discussion will depend upon many factors including the nature of the material to be analysed, the amount of archaeological specimen available, the speed at which results are required, the budget available for this particular piece of work and the range of chemical elements or isotopes for which information is required.

We should perhaps note here that, although there have been marked improvements in the precision of the measurement of elemental concentrations over the years, there are still unavoidable sources of imprecision in the data, whatever the analytical technique. Specimens are, in any case, unlikely to be completely homogeneous and so even two specimens taken from the same object will not give rise to identical chemical compositions.

11.3.1   Notation

The data arising from compositional analyses take the form of collections of elemental concentrations. Such collections, one per specimen analysed, are known as *data vectors*. Throughout this chapter, we will represent such data vectors by x_j where $j = 1, 2, \ldots, n$ and n is the number of archaeological specimens whose provenance is to be determined. Each x_j is composed of q observations, one for each chemical element analysed in specimen j. Thus, in the illustrative extract of data from a known pottery kiln site given in

Table 11.1 we see that $n = 10$. Throughout this chapter we will use q to represent the number of chemical elements under investigation, so that for the data in Table 11.1, $q = 5$. We can then refer to the concentration of elements $r = 1, 2, \ldots, q$ in specimen j using notation of the form $x_{j,r}$. Referring to our example data again, we see that $x_{1,1} = 10.8$, $x_{1,2} = 15.3$, $x_{1,3} = 6.84$, $x_{1,4} = 2.36$ and $x_{1,5} = 0.96$ are the five concentrations found in the first specimen.

The elemental concentrations in Table 11.1 are in parts per million. The values for each specimen do not add up to $1\,000\,000$, however, since we are only looking at a selection of the enormous number of chemical elements present in each of the objects under study. It has been shown that many archaeological minor and trace element data are well modelled by a log-normal distribution (Bieber *et al.*, 1976). (Recall that the probability density function of the log-normal distribution was given in Section 5.10.6.) Therefore, it is conventional to take logarithms of the raw data. In mathematical notation, we let $y_{j,r} = \ln(x_{j,r})$ and it is these transformed data values rather than the raw data values, the $x_{j,r}$, which are used in the subsequent analyses.

If we perform this transformation on our example data, we obtain the values shown in Table 11.2. Thus, we see that $y_{1,1} = 2.38$, $y_{1,2} = 2.73$, $y_{1,3} = 1.92$, $y_{1,4} = 0.86$ and $y_{1,5} = -0.04$. The covariance matrix for the transformed data is given in Table 11.3 and the corresponding correlation matrix in Table 11.4. The covariance matrix is square and symmetric about its diagonal and has the same number of rows and columns as the data vectors have entries (see Section 6.2.8). The correlation matrix is also square and symmetric about the diagonal and its diagonal cells all contain ones (Section 6.2.8).

Looking at the correlation matrix in a little more detail, we can see that Table 11.4 actually contains a great deal of useful and easily interpretable information. High positive correlation indicates that as the concentration of one element increases there is a tendency for the concentration of the other element to do so also. This relationship can be observed between elements 1 and 2 in Table 11.4. Note that we have arranged the data in Tables 11.1 and 11.2 so that the concentrations of element 1 are in increasing order. It can, therefore, easily be seen in the data that specimens with higher concentrations of element 1 tend also to have higher concentrations of element 2.

By contrast, high negative correlation indicates that as the concentration of one element rises there is a tendency for the concentration of the other element to fall. This relationship can be observed between elements 1 and 3 in Table 11.4. With the concentrations of element 1 in ascending order it can easily be seen that specimens with higher concentrations of element 1 tend, broadly speaking, to have lower concentrations of element 3, although this is not always the case.

Table 11.1 An illustrative extract of trace element data from a kiln site in Spain (data are in parts per million).

Object	Trace element 1	2	3	4	5
1	10.8	15.3	6.84	2.36	0.96
2	11.3	15.6	7.99	2.59	0.98
3	11.4	20.4	6.98	2.74	0.89
4	11.5	16.2	6.63	2.27	0.81
5	14.0	19.2	7.26	3.15	1.04
6	15.7	19.1	6.90	2.38	0.91
7	15.7	19.1	6.74	2.43	0.92
8	16.0	19.1	5.75	2.25	2.22
9	16.5	18.7	6.65	2.54	0.87
10	18.8	21.3	6.15	2.71	0.77

Table 11.2 The result of taking natural logarithms of the data in Table 11.1.

Object	Trace element 1	2	3	4	5
1	2.38	2.73	1.92	0.86	−0.04
2	2.42	2.75	2.08	0.95	−0.02
3	2.43	3.02	1.94	1.01	−0.12
4	2.44	2.79	1.89	0.82	−0.21
5	2.64	2.95	1.98	1.15	0.04
6	2.75	2.95	1.93	0.87	−0.09
7	2.75	2.95	1.91	0.89	−0.08
8	2.77	2.95	1.75	0.81	0.80
9	2.80	2.93	1.89	0.93	−0.14
10	2.93	3.06	1.82	1.00	−0.26

Table 11.3 The covariance matrix for the natural logarithms of the trace element data in Table 11.2.

Trace element	Trace element 1	2	3	4	5
1	0.039	0.016	−0.010	0.001	0.007
2	0.016	0.013	−0.004	0.005	0.001
3	−0.010	−0.004	0.008	0.004	−0.012
4	0.001	0.005	0.004	0.011	−0.009
5	0.007	0.001	−0.012	−0.009	0.089

Table 11.4   The correlation matrix for the natural logarithms of the trace element data in Table 11.2.

| Trace element | Trace element | | | | |
	1	2	3	4	5
1	1.000	0.725	−0.583	0.063	0.124
2	0.725	1.000	−0.448	0.403	0.030
3	−0.583	−0.448	1.000	0.423	−0.440
4	0.063	0.403	0.423	1.000	−0.284
5	0.124	0.030	−0.440	−0.284	1.000

11.3.2   The basic model

We suppose that each specimen under investigation must come from one and only one of k possible sources or provenances. Each source has a different "mix" of elemental concentrations. In addition, suppose that specimen j comes from the ith source so that y_j is assumed to have a Multivariate Normal distribution (see Section 6.5.2) with mean vector μ_i and covariance matrix Ω_i. This we write as

$$y_j | \mu_i, \Omega_i \sim MVN(\mu_i, \Omega_i).$$

11.3.3   Discriminant analysis and cluster analysis

Given the basic model (and data of the form described), we need a method of identifying the source or provenance of archaeological artefacts on the basis of their chemical composition. There is, however, one further complication and that relates to knowledge about the number of likely sources represented within any given data set. In situations where we have training data (see Section 11.1) from known sources or production sites, the value of k is fixed *a priori*; what we need to do is allocate each of the specimens of unknown origin to exactly one of these predefined sites. In situations where we have no training data, however, the number of sources or provenances is not known *a priori* and so uncertainty about k must be included in the data analysis process.

   In other words, when we have training data, we require a statistical methodology which will allow us to characterize the nature of data from known sources. Then, on the basis of this information, we need to assign other specimens of unknown origin to those sources in such a way that specimens ascribed to each source are as similar to the training data as possible. The statistical approach to this problem is known as *discriminant analysis*.

   Situations where we have no training data are statistically rather more difficult to address. In such cases, we must rely upon the model, the prior and the data to allow us to define the characteristics which distinguish specimens from different provenances, and thus obtain information about the likely number of provenances represented. We refer to statistical approaches to this problem under the general title of *cluster analysis*.

11.4   Some important issues

As with many applications of scientific methods to archaeology, chemical compositional analysis was initially seen as a reliable solution to the problem of assigning artefacts to their source or provenance. In recent years, however, various complications have come to light which mean that chemical compositional studies are treated with a little more scepticism.

Although the basic archaeological premises and the statistical model thus far described are very straightforward, this is all rather deceptive. Throughout this book we have stressed the scientific importance of learning and developing ideas on the basis of experience. This is exactly what the archaeological science community has done in the case of the study of provenance. Some examples will help us to illustrate just how difficult sourcing and provenancing studies are and will indicate how, in the light of experience, understanding has improved.

At this point we must make it quite clear that we in no way wish to detract from the importance and value of earlier pioneering work. Indeed quite the contrary is the case. We hope that the examples we have chosen illustrate the processes by which the understanding of problems evolves as new data are collected and ideas refined.

11.4.1   Provenancing pottery

Consider first the provenancing of pottery. Here we focus on the work of Catling and colleagues who in the 1960's began investigation into the provenance of Bronze Age pottery from the Mediterranean. One particular study investigated coarse jars which were used to transport olive oil. Evidence on the sources of these *stirrup jars* derives not only from chemical analysis and thin-sections, but also, rather unusually, from place-names painted on the jars in the *Linear B* script of the time. Work continues on the project but a convenient summary of the earlier stages can be found in Catling *et al.* (1980).

Initial chemical compositional analysis of specimens from the jars concluded that some of the inscribed vessels had their source at two sites in *eastern* Crete. This original conclusion does not, however, agree with more recent archaeological evidence and research on the place-names, both of which point towards a source for inscribed jars in *western* Crete. Over time, a larger number of specimens was analysed and control material from clay deposits was obtained. Eventually, it became clear that there was chemical compositional, as well as archaeological, evidence for the existence of a source in western Crete. This source is now widely accepted to have provided raw materials for one group of the stirrup jars. This example illustrates how, as new data accumulate and understanding improves, the analysis and data interpretation develop. Note also the various areas of expertise which fed into the analysis: archaeology, chemistry and epigraphy as well as statistics.

11.4.2   Provenancing metals

As another example, to bring out the complexities of provenancing and sourcing problems, we consider *lead isotope analysis*. Lead isotope analysis is used to characterize lead-bearing geological deposits (using the ratios of the isotopes lead-204, lead-206, lead-207 and lead-208). Archaeological artefacts are assigned to the sources on the basis of the ratios of their lead isotope concentrations. This technique has received a great deal of attention for two reasons. Firstly, the approach promised accurate definition of source materials and, consequently, unequivocal assignment of archaeological objects to their source. Secondly, many types of artefact contain, at least, traces of lead.

In particular copper-ore lodes frequently also contain lead, which transfers through the processes of smelting and casting into the finished artefact. Copper ore is of particular interest as a vital raw material during the Bronze Age (bronze is an alloy of copper and, usually, tin). Unlike the other examples we consider (obsidian and pottery), provenancing through lead isotopes does not rely on the *concentration* of lead, which of course would alter, perhaps unpredictably, on the way from ore to finished artefact. The *ratios* of the different isotopes, which are thought not to fractionate in the course of production, stay the same from ore to ingot to artefact. Thanks to this direct relationship between artefact and source (ore body), lead isotope was seen as a much more hopeful key to provenancing than trace element analysis, which had been tried before and found less satisfactory.

The ratios effectively reflect the geological age at which a given ore body was laid down. The major copper-ore bodies of the Mediterranean seemed to be of different dates, and therefore to have different lead isotope signatures; though research has (not unexpectedly) indicated sources recognized in the archaeological material but not, as yet, identified geologically. Whilst many of the results of lead isotope analysis have made excellent archaeological sense, some recent findings, not least a report that Sardinia was importing copper from Cyprus, were unexpected; Sardinia has plenty of copper of its own. Once again, as in the case of the stirrup jars, the process of learning from experience is passing through another phase. This controversy is still unresolved, and the recent state of research can be traced through Sayre *et al.* (1992), Pernicka (1992), Pernicka (forthcoming), Gale and Stos-Gale (1992) and Budd *et al.* (1995). The discussion at the moment centres on questions such as how exactly to define ore bodies and their extent, sampling problems, the definition of statistical outliers, the possibility of artefacts containing recycled copper and so on. Even such a brief summary of this controversy should provide some insight into why we believe the Bayesian approach offers a useful framework for dealing with these highly complex problems.

11.4.3   Provenancing obsidian

We consider next a different material widely used in antiquity: obsidian. Obsidian sourcing is a classic application of chemical compositional analysis in archaeology. The chemical composition of obsidian stays the same from source to artefact, whereas pottery and metals are altered by the craftsmen who work them,

Recently Hughes (1994) has investigated more than 200 obsidian specimens taken from outcrops in the Casa Diablo region of California. The results of this research indicate that two (or even three) distinct varieties of obsidian can be distinguished within what was once identified as a single source. These findings suggest that each obsidian flow is chemically distinguishable from others, even those which are geographically close. This is potentially of great significance since it confirms what was already becoming clear from earlier provenance studies: precisely located sources need to be characterized through specimens gathered at source.

11.4.4   Comment

We have seen from the examples just discussed that archaeological ideas are constantly being refined and modified in provenance studies. The use of model-based statistical techniques will allow the statistical methodology to develop hand-in-hand with progress in the other disciplines involved, such as geochemistry, archaeology and materials science.

11.5   Case Study I — sourcing — Bayesian discriminant analysis

Sourcing and provenancing are, in some respects, very similar, and the models for the two have certain characteristics in common. However, sourcing is the easier to model and so we take this problem as our first case study. The initial example is somewhat artificial but helps to convey the fundamental ideas with clarity.

We have available to us a collection of data from tin-glazed ceramics produced at three *known* medieval kiln sites: two in Spain (one at Valencia and the other at Seville) and one in Italy (at Castelli). All three kiln sites produced pottery of the same type which is characterized by being highly decorated. Some of the decoration is in metallic copper lustre and some in a wide range of colours (blues, greens, yellows and purples) on a white background. Ceramics of this type have been admired and discussed for many years (Caiger-Smith, 1973) and they are of particular interest as they are known to have been exported to other European countries and even America (Hurst *et al.*, 1986).

In addition to kiln data, we also have data from a further 94 sherds found in excavation and known to have been manufactured at one or other of the Spanish kiln sites. Since we know that each of the sherds of unknown origin must have come from one or other of the two known kiln sites, this data set is an ideal one with which to demonstrate our Bayesian approach to sourcing. The sherds from the kiln sites have the same role that specimens from geological sources do when we attempt to attribute objects to the site of extraction of their raw materials. We can think of our kiln sites as sources in that we wish to assign each of the sherds (specimens) of unknown origin to one or other of them. For simplicity in this case study we shall refer to specimens rather than sherds.

11.5.1 The model

Recall in our general statement of the model (in Section 11.3) that there were n archaeological specimens under investigation and that for each of these we had compositional information relating to q chemical elements. We then represented the data vector (or some transformation of it, for example its logarithm) for the jth specimen (sherd) by $\boldsymbol{y}_j$ ($j = 1, 2, \ldots, n$) so that $y_{j,r}$ is the concentration of element r (or its transformation) in specimen j where $r = 1, 2, \ldots, q$. Now, when we have training data from known kiln sites, the number of sources, denoted by k, is known exactly and so will not be altered during the analysis.

Specimens of unknown source can come from any one of the k sites. We suppose that a specimen comes from source i with probability θ_i ($i = 1, 2, \ldots, k$) — the θ_i are called the *mixing probabilities*. In situations where all production sites produced equal quantities of pottery all the θ_i would be equal. In other situations where the production sites manufactured different quantities, θ_i would reflect the proportion of the total production believed to have occurred at site i.

Recall also that it is widely accepted by both chemists and geologists that chemical compositions of many geological materials are either normally distributed or can easily be transformed to normality. This means that the data from each kiln site are modelled as having a Multivariate Normal distribution so that if specimen j comes from kiln site i then

$$\boldsymbol{y}_j | \boldsymbol{\mu}_i, \Omega_i, \theta_i \sim MVN(\boldsymbol{\mu}_i, \Omega_i)$$

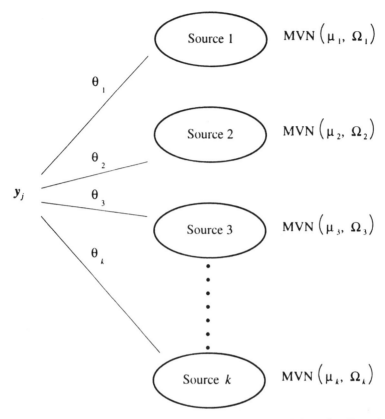

Figure 11.1   Diagrammatic representation of the model used in Bayesian
discriminant analysis.

where μ_i is the mean vector and Ω_i the covariance matrix for kiln site i. In
Figure 11.1 we provide a graphical representation of the links between the
data, the archaeological problem and the parameters of the model.

We then have all the components of a model for adopting a Bayesian
approach to *discriminant analysis* (see Lavine and West, 1992). The approach
is known as discriminant analysis because we wish to establish what
distinguishes or *discriminates* one source from another and then to use this
information to allocate the specimens of unknown origin to one or other of
them.

11.5.2   The data

In this case study, we have 32 training specimens, 16 from Valencia and 16 from Seville; for ease of reference later in the chapter the training specimens from Valencia will be numbered 1 to 16 and those from Seville 17 to 32. We also have access to 24 specimens from Castelli in Italy. The kiln site at Castelli was operating at about the same time as those at Valencia and Seville and produced ceramics of a similar type: We propose, therefore, to use the Castelli data to provide prior information for our analysis of the data from Spain.

In addition to the data known to come from specific kiln sites, we have 94 specimens of unknown origin. For future reference these are numbered 33 to 126.

In all cases, the ceramic specimens were obtained from the main body of the object and care was taken not to include glaze in the specimen since glaze is not usually made from the same materials as the rest of the object. The clay specimens from each sherd were then analysed for fifteen chemical elements using neutron activation analysis (see Tite, 1972).

11.5.3   The prior

For ease of notation, let $\boldsymbol{\mu} = (\boldsymbol{\mu}_1, \ldots, \boldsymbol{\mu}_k)$, $\boldsymbol{\Omega} = (\Omega_1, \ldots, \Omega_k)$ and $\boldsymbol{\theta} = (\theta_1, \ldots, \theta_k)$. We then seek prior information about $\boldsymbol{\mu}$, $\boldsymbol{\Omega}$ and $\boldsymbol{\theta}$.

Since the model and approach suggested in this chapter are in the early stages of development, we have chosen to adopt conjugate priors throughout (see Section 8.3). This is largely for mathematical convenience and ease of exposition — use of non-conjugate priors would make for much more complicated explanations. We do not intend to imply that the distributions used here are the only ones that reasonably capture the prior information. Others can and will no doubt be proposed in subsequent investigations.

Prior for μ_i — $p(\mu_i | \Omega_i)$

The prior for μ_i is assumed to be conditionally independent of θ_i. In other words, a priori, the information we have about the mean for each source, μ_i, is assumed to be independent of the probability that specimens come from that (or any other) source. Moreover, μ_i conditional on Ω_i is assumed to be Multivariate Normal. That is, we assume that

$$\mu_i | \Omega_i \sim MVN(m_i(0), h_i(0)^{-1}\Omega_i)$$

for some mean $m_i(0)$ and parameter $h_i(0)$.

This means that the prior information about μ_i must be elicited in terms of its expectation, $m_i(0)$. The strength of belief in $m_i(0)$ is reflected by the value given to $h_i(0)$. Large values of $h_i(0)$ represent strong belief in the value given to $m_i(0)$, small values of $h_i(0)$ represent little strength of belief in $m_i(0)$, and when $h_i(0) = 0$ the prior for $m_i(0)$ is vague.

When investigating archaeological ceramics, we generally have very little prior information about the μ_i since these are the mean vectors for each kiln site which we will not learn about until our training data are analysed. In future analyses of data of this type, we might expect to have more useful prior information about $m_i(0)$ from earlier analyses. In our current investigation, however, we adopt a vague prior by taking $h_i(0) = 0$.

Prior for Ω_i — $p(\Omega_i)$

Like μ_i, the prior for Ω_i is assumed conditionally independent of θ_i. In addition, Ω_i is assumed to have an inverse-Wishart distribution (see Section 6.5.4) so that

$$\Omega_i \sim W^{-1}(v_i(0), V_i(0))$$

where $v_i(0) > 0$ is the number of degrees of freedom and $V_i(0)$ is a $q \times q$ matrix. Note that the prior expectation of Ω_i is $V_i(0)/(v_i(0) - 2)$.

By defining the prior for Ω_i in this way, we must elicit information about $V_i(0)$ and $v_i(0)$. Here $V_i(0)$ reflects the prior knowledge about the variability of each element within source i and how pairs of elements vary together. We shall refer to these two types of variability as the *within kiln site variability*. The parameter $v_i(0)$ reflects the strength of belief in $V_i(0)$. Large values of $v_i(0)$ reflect high strength of belief in the prior for the variability, while small values of $v_i(0)$ indicate weaker prior knowledge. In practice, of course, prior information about Ω_i will need to be sought from expert chemists who have worked with such data and have knowledge about both the general and specific variability of within kiln site compositions.

As mentioned above, we have data from Castelli in Italy which is a similar production site to those at Seville and Valencia. The 24 specimens from there can be used to provide useful prior information about within kiln site variability for the problem at hand. Since there are 24 specimens from Castelli we set $v_1(0) = v_2(0) = 24$. In the light of no information to the contrary we take $V_1(0) = V_2(0)$ and their common value is estimated from the within kiln site variability of the Castelli data.

Prior for θ — $p(\theta)$

The vector of mixing probabilities, θ, is assumed to be independent of μ and Ω and to have a Dirichlet distribution (see Section 6.5.3). That is, $\theta \sim D(a(0))$ where $a(0) = (a_1(0), \ldots, a_k(0))$.

Prior information about θ may be elicited in terms of the relative production quantities at the kiln sites under study. Such information might come from expert archaeologists. However, for the purposes of this case study, we did not have access to any such information so we used a vague prior for $\boldsymbol{\theta}$ by setting $a_1(0) = a_2(0) = 0.5$.

11.5.4 Analysis of the training data

The analysis of this problem will illustrate how Bayes' theorem permits the sequential updating of information in the light of new knowledge. We commence the analysis by using the training data to arrive at an initial posterior for $\boldsymbol{\theta}$, $\boldsymbol{\mu}$ and $\boldsymbol{\Omega}$. Later, we use this posterior as our prior for investigating the data of unknown source in order to arrive at our new posterior for these parameters and as a means of allocating specimens of unknown origin to kiln sites.

The training data from the known kiln sites consist of trace element analyses for $t = 32$ perfectly classified specimens. Let $\boldsymbol{y}(T)$ denote all the observations from the training data so that $\boldsymbol{y}(T) = (\boldsymbol{y}_1, \boldsymbol{y}_2, \ldots, \boldsymbol{y}_t)$.

We now introduce a new random variable z_j which is used to indicate the kiln site to which specimen j is assigned or *classified*. If specimen j is assigned to kiln site i, then $z_j = i$. In the case of our training data, the origin of all the specimens is known. As the first 16 specimens are from Valencia, which is labelled site 1, we have $z_j = 1$ for $j = 1, \ldots, 16$. The second 16 specimens are from Seville, labelled site 2, so that we have $z_j = 2$ for $j = 17, \ldots, 32$. We let $\boldsymbol{z}(T) = (z_1, z_2, \ldots, z_t)$ be the *classification* vector for the training data.

Finally, let $G_i(T)$ represent the set or collection of all training specimens assigned to kiln site i, and let $g_i(T)$ be the number of specimens assigned to kiln site i. Note that $t = g_1(T) + \cdots + g_k(T)$.

Having set up the notation for the model we now are in a position to proceed with the analysis of the training data. We shall adopt the methodology given in Lavine and West (1992). As the mathematics involved is complicated we will not give all the details — these can be found in the original paper.

Full conditional distribution of $\boldsymbol{\mu}_i$ — $p(\boldsymbol{\mu}_i | \Omega_i, \boldsymbol{y}(T), \boldsymbol{z}(T), \boldsymbol{\theta})$

The full conditional distribution of $\boldsymbol{\mu}_i$ is

$$\boldsymbol{\mu}_i | \Omega_i, \boldsymbol{y}(T), \boldsymbol{z}(T) \sim MVN(\boldsymbol{m}_i(T), h_i^{-1}(T)\Omega_i)$$

where

$$h_i(T) = h_i(0) + g_i(T),$$

$$\boldsymbol{m}_i(T) = h_i^{-1}(T)(h_i(0)\boldsymbol{m}_i(0) + g_i(T)\bar{\boldsymbol{y}}_i(T))$$

and

$$\bar{y}_i(T) = g_i^{-1}(T) \sum y_j(T)$$

with the summation being over those training specimens classified as originating from kiln site i.

Here $m_i(T)$ represents the current information about the expected value of the mean vector at kiln site i based upon both the prior information and the training data. Notice that $m_i(T)$ is the weighted average of the prior mean vector, $m_i(0)$, and the mean vector for the training data for site i. The weights are $h_i(0)/(h_i(0) + g_i(T))$ and $g_i(T)/(h_i(0) + g_i(T))$ respectively. Consequently, as the number of training samples increases, less weight is attached to the prior mean vector. The strength of belief in $m_i(T)$ is represented by $h_i(T)$ whose value is affected by the number of samples from which the prior and training information are obtained.

Full conditional distribution of Ω_i — $p(\Omega_i | y(T), z(T), \theta)$

The full conditional distribution of Ω_i is

$$\Omega_i | y(T), z(T) \sim W^{-1}(v_i(T), V_i(T))$$

where

$$v_i(T) = v_i(0) + g_i(T),$$

$$V_i(T) = V_i(0) + S_i(T) + (\bar{y}_i(T) - m_i)(\bar{y}_i(T) - m_i)' g_i(T) h_i(0) h_i^{-1}(T),$$

and

$$S_i(T) = \sum (y_j(T) - \bar{y}_i(T))(y_j(T) - \bar{y}_i(T))',$$

with the summation being over those training specimens classified as coming from kiln site i. Here $(y_j(T) - \bar{y}_i(T))'$ is the *transpose* of the vector $(y_j(T) - \bar{y}_i(T))$ (see Namboodiri, 1984).

Here $V_i(T)$ represents the information we have about within kiln site variability, based upon both the prior knowledge and the training data. There are three components which contribute to our knowledge about within kiln site variability. The first is the *a priori* knowledge, represented by $V_i(0)$; the second, $S_i(T)$, is the information available from the training data; and the third, $(\bar{y}_i(T) - m_i)(\bar{y}_i(T) - m_i)' g_i(T) h_i(0) h_i^{-1}(T)$, is the variability between the expected value of the chemical compositional vectors and the values obtained from the training data. Note that the degrees of freedom, $v_i(T)$, which can be thought of as a measure of the strength of belief in $V_i(T)$, are obtained from the degrees of freedom of the prior, $v_i(0)$, and the number of training samples, $g_i(T)$.

Full conditional distribution of θ — $p(\theta|y(T), z(T), \mu, \Omega)$

Given $z(T)$, the vector of mixing probabilities, θ, is conditionally independent of $(y(T), \mu, \Omega)$ and has a Dirichlet distribution with parameter a. That is, $\theta|z(T) \sim D(a)$ where $a = (a_1, \ldots, a_k)$ and $a_i = a_i(0) + g_i(T)$. Since for the training data all the z_j are known, this completes the analysis of the training data. Thus we have the full conditional distributions of μ_i, Ω_i and θ.

11.5.5 Analysis of the unsourced data

We now use the posterior from the analysis of the training data as the prior for the analysis of the data from the remaining 94 unsourced specimens. Let $y(U)$ denote the data vectors from the specimens of unknown source so that $y(U) = (y_{t+1}, \ldots, y_{t+u})$ where $u = 94$.

Let $z(U) = (z_{t+1}, \ldots, z_{t+u})$ be the corresponding classification vector. Unlike the training specimens, the value of z_j for $j = t + 1, \ldots, t + u$ is unknown. As a result, we want to be able to make inferences about the value of z_j for $j = t + 1, \ldots, t + u$. Despite the lack of information about $z(U)$, the full conditional distributions are only slightly more complex than for the training data.

If we let $y = (y(T), y(U))$ and $z = (z(T), z(U))$, then the methodology of Lavine and West (1992) gives the following results.

Full conditional distribution of μ_i — $p(\mu_i|\Omega_i, y, z, \theta)$

The full conditional distribution of μ_i is given by

$$\mu_i|\Omega_i, y, z \sim MVN(m_i, h_i^{-1}\Omega_i)$$

where

$$h_i = h_i(T) + g_i(U)$$

$$m_i = h_i^{-1}(h_i(T)m_i(T) + g_i(U)\bar{y}_i(U)),$$

$g_i(U)$ is the number of unsourced specimens currently classified as being from kiln site i and

$$\bar{y}_i(U) = g_i^{-1}(U)\sum y_j(U)$$

with the summation being over those unsourced specimens currently classified as originating from kiln site i.

Here m_i represents the current information about the expected value of the mean vector at kiln site i based upon the prior information, the training data and the information available from the unsourced data. Notice that m_i is the weighted average of $m_i(T)$ and the mean vector for the unsourced data currently ascribed to site i. The weights are $h_i(T)/(h_i(T) + g_i(U))$ and $g_i(U)/(h_i(T) + g_i(U))$ respectively.

Full conditional distribution of Ω_i — $p(\Omega_i|y, z, \theta)$

The full conditional distribution of Ω_i is given by

$$\Omega_i|y, z \sim W^{-1}(v_i, V_i)$$

where

$$v_i = v_i(T) + g_i(U)$$

and

$$V_i = V_i(T) + S_i(U) + (\bar{y}_i(U) - m_i)(\bar{y}_i(U) - m_i)'g_i(U)h_i(T)h_i^{-1}.$$

Since both v_i and V_i have components from the prior, the training data and the data of unknown origin, Ω_i can now be seen to encapsulate all available information about the within kiln site variability at site i.

Full conditional distribution of θ — $p(\theta|\mu, \Omega, y, z)$

The full conditional distribution of θ is a Dirichlet. That is, $\theta|z \sim D(a)$ with $a_i = a_i(0) + g_i(T) + g_i(U)$. In other words, the parameter of the Dirichlet, a_i, corresponding to kiln site i is determined by the total number of specimens classified as belonging to site i (those from the training data and those unsourced specimens currently allocated there), and the original prior value of the parameter which is $a_i(0)$.

Full conditional distribution of z — $p(z|y, \mu, \Omega, \theta)$

Finally, the full conditional distribution for the classification vector, $z(U)$, for the unsourced specimens can be obtained as follows. Given y, μ, Ω and θ, the z_j are conditionally independent. Then, for each $j = t + 1, \ldots, t + u$, the conditional probability mass function of z_j is

$$P(z_j = i|y, \mu, \Omega, \theta) \propto \theta_i p(y_j|\mu_i, \Omega_i, z_j = i) \quad i = 1, \ldots, k.$$

Note that $p(y_j|\mu_i, \Omega_i, z_j = i)$ is the joint probability density function of a Multivariate Normal distribution with mean vector μ_i and covariance matrix Ω_i.

11.5.6   Implementation

Using the conditional distributions given above, there are several ways to implement a Gibbs sampling scheme which allows us to cycle through each of the parameters of the model. In this particular case we chose to initialise z by assigning the unsourced samples arbitrarily to one or other of the two known kiln sites. We then sampled from the full conditional distributions of each of the parameters in the following order:

(i) Ω_1, Ω_2,
(ii) μ_1, μ_2,
(iii) θ, and finally
(iv) z.

As explained in Section 8.6.3, it was then necessary for us to decide whether to base our inferences about the parameter values upon just one long run of this sampling scheme or to use multiple runs and base our inference upon an average. For two reasons we decided to use the latter approach. Firstly, the values taken by the parameters are likely to be highly correlated and this might result in the sampling scheme becoming "locked" into one particular set of allocations of the specimens from which it cannot "escape". Secondly, the Bayesian approach to discriminant analysis is experimental and we felt it better to monitor several runs in order to begin to understand more about the behaviour of the sampling scheme selected. In more detail, we

(i) ran the sampler for a certain number of iterations,
(ii) established convergence of that run, and then
(iii) repeated the whole sampling procedure a number of times, before
(iv) averaging the final results from each of the runs undertaken.

In the case of our analysis of the Spanish ceramics we were able to establish that 500 iterations allowed us to be sure of obtaining convergence at step (ii) and that 500 repetitions (or restarts) at step (iii) produced a large enough sample to allow us to be reasonably confident in the results produced.

11.5.7   Displaying and interpreting the results

The posterior information which is most archaeologically interpretable takes the form of probabilities of allocation of the various archaeological specimens to the known kiln sites. These probabilities can be interpreted as a measure of the strength of relationship between the archaeological specimens and the kiln sites. In other words, a posterior probability of 0.9 that specimen j is from kiln site i with the remainder of the probability spread between several other kiln sites allows us to assess that specimen j is more like the training data (and other specimens) assigned to kiln site i than like the training data (and

Table 11.5 Specimens which could be securely allocated on the basis of sixteen training specimens per kiln site and prior information about Ω based on the data from Castelli.

Valencia					Seville				
33	34	35	36	38	80	81	82	83	84
39	40	41	42	43	85	86	87	88	89
44	45	46	47	48	90	91	92	93	94
49	50	51	52	53	95	96	97	98	99
54	55	56	59	60	100	101	102	103	104
61	62	63	64	65	105	106	107	108	109
66	67	68	69	70	110	111	112	113	114
71	72	73	74	75	115	116	117	118	119
76	77	78	79		120	121	122	123	124
					125	126			

other specimens) assigned to any of the other kiln sites in the current study. On the other had, if specimen j is from kiln site i with probability 0.6 and from kiln site m with probability 0.4, we cannot make such secure assessments. Using a model-based Bayesian approach does not necessarily result in clear-cut sourcing of the specimens, but the probability statements are unequivocal and allow unambiguous statements about the posterior information.

The posterior information is then easily summarized in terms of charts and tables. The interpretation is, perhaps, not quite so straightforward. We cannot offer clear-cut guidelines about levels of probability which correspond to "secure" allocation of specimens to a particular kiln site. The cut-off point between clear allocation and uncertainty as to which kiln site a particular specimen came from will depend on the problem under investigation and upon the confidence of the experts in the data available to them. Nonetheless, in most practical situations, there will be a notional probability threshold between secure allocation and uncertain allocation. Provided this level is stated, all further interpretations can then be based around this value and its level adjusted at a later date should other experts disagree.

For the purposes of this case study, we were very demanding of the methodology and set the threshold posterior probability to 0.99. This means that we can show in Table 11.5 all specimens which were, in this quantitative sense, securely allocated to each of the two sites. In Table 11.6 we show the specimens which could not be so securely allocated and their probability of allocation to each of the two known kiln sites.

We see that 91 of the 94 specimens could be allocated to one or other kiln site with a high posterior probability (that is greater than 0.99). But what are we to make of the three specimens which could not be securely allocated

Table 11.6 Samples which could not be securely allocated to either Valencia or Seville on the basis of sixteen training specimens from each kiln site and prior information about Ω based on the data from Castelli.

Sample number	Probability from Valencia	Probability from Seville	Kiln with highest probability
37	0.51	0.49	Valencia
57	0.54	0.46	Valencia
58	0.52	0.48	Valencia

to either Valencia or Seville? Without any further archaeological information (and unfortunately this is the case here) we cannot state categorically why these specimens were so hard to allocate; we can only make some suggestions about the cause of the posterior uncertainty.

We note that the two kiln sites cannot be completely distinguished from one another, using the training data and prior information provided; there is some overlap of the two distributions. This in itself is sufficient to account for the uncertainty observed. In addition, however, there are other factors that might give rise to uncertainty of this sort. For example, perhaps the three specimens did not in fact come from either of the two known kiln sites, but from one not yet identified, or perhaps the current training data were not fully representative of the sites from which they came. Anyone interpreting such results should bear these possibilities in mind and report the posterior probabilities, so that other workers can reassess the posterior information.

11.6   Case Study II — provenancing obsidian — Bayesian cluster analysis

Situations in which source data are non-existent or poorly understood are very common in archaeology. The lack of training data makes the statistical modelling much more complex and the resulting posterior information harder to interpret. All the more reason, then, to extend the methodology for discriminant analysis (in which the number of sources is known *a priori*) to create a methodology which allows for an unknown number of possible sources, or provenances, which we wish to identify as part of the statistical analysis. Just such a methodology, which we call Bayesian *cluster analysis*, is outlined below. Cluster analysis, it should be noted, is a broad term including many different methods used to partition data into groups or clusters (for a review see Baxter, 1994).

In order to illustrate our Bayesian methodology for provenancing we shall first examine data from the analysis of obsidian, and then move on to pottery.

As pointed out above, the case of obsidian is more straightforward because the artefact is derived directly from the source; the chemical composition of obsidian is not altered in the course of fabrication. Pottery, on the other hand, is manufactured through a series of complex operations which involve selecting clays, often mixing clays, tempering, levigating, adding fillers and firing. Most of these operations will affect the chemical composition. So the link between specimen artefact and original source is much more complex than is the case for obsidian.

We illustrate the Bayesian approach to cluster analysis using data from a study of obsidian in the Western Mediterranean reported by Hallam *et al.* (1976). The data were published, easily available and not contentious. Of course in a real problem, collaboration with various subject experts would be necessary — but, as explained above, our work is still in the developmental stage. For this reason, the analysis serves simply to illustrate the methodology.

11.6.1 The model

Adopting the same basic notation as in Section 11.3.1, let y_j be the q-dimensional vector of chemical compositions, or some suitable transformation thereof, for object j, $j = 1, 2, \ldots, n$. The data, y_j, are assumed to be drawn from some Multivariate Normal distribution with parameters μ_j and Ω_j. (Note that we are not necessarily restricted to using the Multivariate Normal distribution; it is the most appropriate here, but other circumstances may require the use of other distributions.) Let k be the (unknown) number of provenances represented within the current data set. For provenance studies (as opposed to those when we have source data) we must allow for the possibility that each specimen is from a unique provenance, that is k could take the value n. However, in practice the number of provenances, k, is likely to be very much smaller than the number of specimens, n. As a consequence, *a posteriori* some specimens will have the same μ_j and Ω_j. Such specimens are said to form a cluster, with the parameters of the ith cluster being μ_i and Ω_i. That is, all specimens in the ith cluster can be thought of as realizations from the same distribution with parameters μ_i and Ω_i.

To aid understanding, we can relate this to the notation used in the discriminant analysis. Recall that z was used as the allocation vector. Under the current model, z_j $(j = 1, \ldots, n)$ takes any value in the set $\{1, \ldots, n\}$, so that z_i will be equal to z_j if, and only if, y_i and y_j are in the same cluster (that is, $\mu_j = \mu_i$ and $\Omega_j = \Omega_i$). Thus μ_j and Ω_j relate directly to the classification for y_j and so z serves the same purpose it did in the discriminant analysis.

11.6.2   The data

Chemical compositional analysis has been applied to the sourcing and provenancing of obsidian for many years now and continues to be a focus for research (see Section 11.4.3). The data upon which we focus attention here (published by Hallam *et al.*, 1976) are the chemical compositions of objects and geological material from Sardinia and the Aeolian Islands. Hallam *et al.* (1976) provide analyses of 124 specimens from Sardinia and Lipari (one of the Aeolian Islands). For ease of explanation the specimens will be numbered 1–124. In undertaking Bayesian cluster analysis of these data, the hope is to divide the 124 specimens into clusters which relate to the source of their obsidian. These clusters are to be based upon the chemical compositional data and prior information which has a bearing on the problem.

11.6.3   The prior

Consider, first, prior information about the number of groups present. To model this we introduce an additional parameter which we denote by α. The details of the modelling of this parameter are complex and beyond the scope of this book (see Escobar and West, 1995). Here it is enough to state that there is a formal relationship between α and k which can be interpreted as follows. In broad terms, a large value of α corresponds to a prior belief in a large number of groups and a small value of α corresponds to prior belief in a small number of groups.

   Although experts will very rarely want to make categorical statements about the number of provenances present, they can usually give broad limits on the number of groups and some indication of their strength of belief in their own assessment. After all, as highlighted above, chemical compositional data are expensive to obtain and commonly involve destruction of at least a small part of the object of interest. Simply by the fact that they have embarked on compositional analysis for the purposes of provenancing, it is reasonable to assume that the experts expect groupings to be present in the data.

   Apart from the broad interpretation given above, the new parameter, α, is difficult to interpret in a specific sense. Fortunately, however, we can exploit a link between it and the number of groups, k, which allows us to produce graphs of the prior probability mass function of k (see Escobar and West, 1995). Examples of these are given in Figure 11.2. The prior form for k shown in Figure 11.2(a) was adopted in the analysis described below.

   In addition to information about k we need to think about possible prior information available about the intra-provenance variability of the compositional data. It happens that, as well as the data from Sardinia and the Aeolian Islands, Hallam *et al.* (1976) report upon analyses from a number of specimens from other Western Mediterranean locations. One of these is

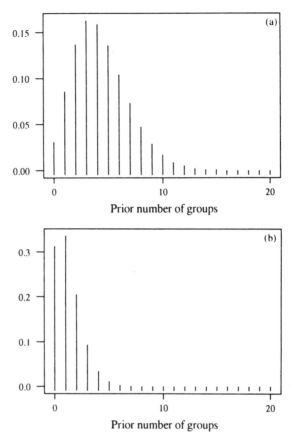

Figure 11.2 Prior probability mass functions of the number, k, of provenances for (a) Case Study II and (b) Case Study III (see Section 11.7).

the small island of Pantelleria. We used the data from Pantelleria to provide prior evidence for the intra-provenance variability for the analysis reported below. (Note that, ideally, specimens should be collected specifically for this purpose.)

Having defined our prior knowledge about the likely number of provenances present and about the intra-provenance variability of the chemical compositional data, the only parameter which remains without prior specification is the μ_j. Since every provenance is expected to have a different mean vector (indeed this assumption is at the core of our model) we cannot make assumptions about the values it is likely to take. As a consequence, we adopt a vague prior for the μ_j.

11.6.4 The posterior

When we undertake a Bayesian cluster analysis, there is a great deal of posterior information available to us. We have information about the posterior mean vector and posterior covariance matrix for each provenance identified, we have a distribution for the posterior number of provenances present and we have posterior information that relates each of the specimens to all of the others in the analysis. We have not included here the full conditionals for all the parameters as we did for the discriminant analysis — they are given in Buck and Litton (forthcoming).

11.6.5 Implementation

As with the Bayesian approach to discriminant analysis, the use of a numerical integration approach to obtain posterior information was clearly out of the question given the complexity of the model. Instead we adopted an MCMC approach. The specific approach we used was to average over replicated runs of the Gibbs sampler (using a different random seed for each run) just as explained in the implementation of the sourcing methodology above (see Section 11.5.6). For each run we used a "burn-in" period of 100 iterations (all of which we discarded); we then took a sample of 500 iterations and repeated this complete process 500 times. Experiments of this type are extremely computer-intensive, but by comparing the results of several such runs, we established that the results obtained had converged and are reproducible.

11.6.6 Displaying and interpreting the results

In examples like this, there are many parameters of interest, including the number of provenances found in the data, the posterior means and covariance matrices for each provenance, the mixing probabilities and the classification vector. However archaeological interest focusses on

(i) the number of provenances, k, and
(ii) which specimens are believed to belong to which provenance.

The posterior information about the number of provenances, k, is provided by its posterior probability mass function which is given in Figure 11.3. We can see that this posterior suggests rather more provenances than we specified in the prior. The mode of the a posteriori distribution of k is at 12 whereas a priori this was at 4. Indeed a glance at Figures 11.2(a) and 11.3 reveals that the prior and posterior mass functions for k are very different. It is clear from this that the evidence from the prior information and that from the data do not completely concur. In the light of the debate about obsidian source definition referred to in Section 11.4.3 these differences might suggest a need for reassessment, both in terms of geological definition and

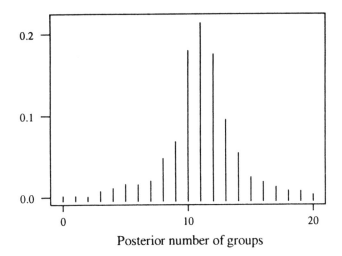

Figure 11.3 The posterior probability mass function of the number, k, of provenances for the obsidian data.

posterior interpretation, of what constitutes an obsidian source. It might be, for example, that the extra clusters relate (as Hughes, 1994 found in California) to individual obsidian flows.

Having considered the posterior number of provenances, we also need to indicate the posterior partitioning of the data: which specimens have been allocated to which provenance. We suggest the adoption of a method already widely used for displaying the results of cluster analysis: that known as the *dendrogram* (or tree diagram). Dendrograms have been used in the past to show links between specimens on the basis of chemical compositional data. To create such tree structures, specimens are linked together in a pair-wise fashion on the basis of some distance or similarity measure. There are many approaches to producing a tree from such measures and we do not propose to discuss these here (the interested reader might consult Shennan, 1988, p. 212–225).

The posterior probability that specimens i and j are in the same group can be calculated during the Gibbs sampling process and used as a posterior measure of similarity between the specimens. A high posterior probability that specimens i and j are in the same group means that they are, in terms of the model, "close" to one another and a low posterior probability that they are "further away". Although the probability dendrogram is similar in appearance to those used in conventional cluster analysis, it should be stressed that the measure used in the Bayesian method is a probability, not one of the more conventional (dis)similarity measures such as Euclidean distance or Mahalanobis distance (see, for example, Baxter, 1994, p. 144 and 156–157).

In Figure 11.4 we provide a posterior probability dendrogram (created using the average linkage method, see Shennan, 1988, p. 215–217) which arises from the analysis of the obsidian data. Observe that the distances between the levels in this tree represent posterior probabilities and thus the larger the distance between specimens the more posterior evidence there is that they are not from the same provenance. Interpretation of such dendrograms is not straightforward, a feel for the probabilistic information is required and some specimens will not be easily assigned to a particular provenance.

Genuine archaeological interpretation of Figure 11.4 is, of course, impossible since both the data and the prior have been selected for convenience. Nonetheless, it is instructive to note that in their original interpretation of these data Hallam *et al.* (1976) identified subdivisions such that specimens 1–29 were grouped together, as were specimens 30–45, specimens 46–63 and specimens 64–124. These subdivisions are broadly supported by the results obtained here, but the Bayesian results suggest that these groups are themselves made up of subgroups. We must stress here that we cannot pretend to make any genuine archaeological interpretation of these observations, we simply suggest that the reader once again considers the findings of Hughes (1994).

11.7 Case Study III — provenancing ceramics — Bayesian cluster analysis

In this section we return to the problems associated with analysis and interpretation of pottery data. For this example we have been able to obtain both data and prior information from subject matter experts and can thus address issues that were considered rather superficially in our example using obsidian data. Readers who would like more detail may consult Buck and Litton (forthcoming).

11.7.1 The model

In this illustration we use the same model as we adopted for the obsidian example. In the future it is likely that, as experience in the application of Bayesian cluster analysis increases, this model will be developed and refined, and, perhaps, new models will be developed.

11.7.2 The data

We use data from the chemical analysis of 150 specimens of medieval tin-glazed ceramics of the type already described in Case Study I (see Section 11.5.2).

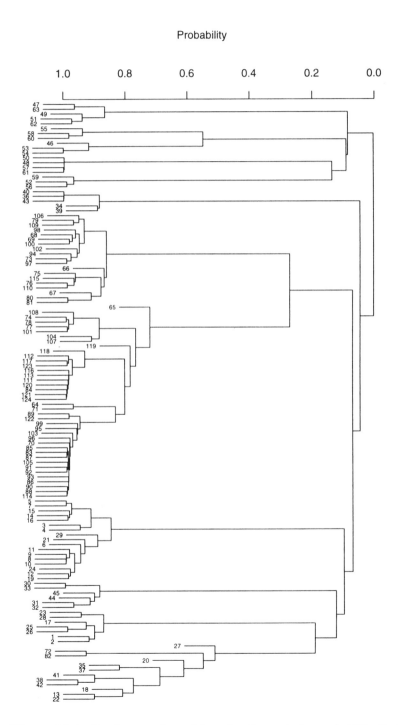

Figure 11.4 Dendrogram formed by using the posterior probability that specimens i and j are from the same provenance as a measure of similarity.

11.7.3   The prior

Consider first the problem of specifying a prior for k. We discussed the prior for the number of provenances with the ceramics experts responsible for these data. In doing so we provided a selection of plots for the prior on k derived by varying the value of the parameter α. After much discussion, it was established that the experts believed that it was unlikely that there would be more than five provenances represented and that in fact it is likely that there are just two or three. As a result, the prior form for k is taken to be that shown in Figure 11.2(b).

In addition to information from ceramics experts, we have prior information available from the chemists. The expert chemists were provided with a simple questionnaire on which they were asked to express their knowledge about inter-element correlations for the particular elements under investigation in this study. The results of using such questionnaires are reported by Buck (1994) from which we give just one expert's opinion in Table 11.7. In this table the code letters refer to correlations as assessed by the expert using the following key HP (highly positive) ≈ 0.7, MP (moderately positive) ≈ 0.3, L (low) ≈ 0.0, MN (moderately negative) ≈ -0.3 and HN (highly negative) ≈ -0.7. In fact, this expert felt that (for these particular elements) none of the elements was negatively correlated with any other. There is no single explanation for this lack of negative correlation between trace elements, but it may well reflect the fact that all the trace elements correlate negatively with silicon (the major elemental constituent of pottery) and as a consequence their proportions tend to correlate positively with each other.

Of course, information about prior correlation does not provide complete prior information for Ω. In order to obtain this it is also necessary to elicit information about within provenance variability for the concentration of each chemical element. Eliciting and using such prior information is somewhat technical and rather beyond the scope of this book. However, the interested reader might consult Buck (1994) where some mechanisms are suggested for eliciting the relevant information.

Having defined the prior knowledge about the likely number of provenances present and about the intra-provenance variability of the data, the only parameter which remains without prior specification is the mean. By virtue of the fact that we are reporting one of the first analyses of this type and that we are attempting to use this methodology to identify groupings in the medieval tin-glazed pottery data for the first time, it is appropriate here (as in our previous analysis) to adopt a vague prior for the μ_js.

Table 11.7 The expert opinion of a chemist regarding inter-element correlations in ceramics from the same provenance.

	Sc	Cr	Co	Rb	Cs	La	Ce	Sm	Eu	Yb	Lu	Hf	Ta	Th	Ba
Sc	-	MP	MP	L	L	HP	HP	HP	HP	HP	HP	MP	MP	MP	L
Cr		-	MP	MP	MP	HP	HP	MP	MP	MP	MP	L	L	MP	L
Co			-	L	L	MP	MP	MP	MP	MP	MP	L	L	L	L
Rb				-	HP	MP	MP	MP	MP	MP	MP	MP	L	MP	L
Cs					-	MP	MP	MP	MP	MP	MP	MP	L	MP	L
La						-	HP	HP	HP	HP	HP	MP	L	MP	L
Ce							-	HP	HP	HP	HP	MP	L	MP	L
Sm								-	HP	HP	HP	MP	L	MP	L
Eu									-	HP	HP	MP	L	MP	L
Yb										-	HP	MP	L	MP	L
Lu											-	MP	L	MP	L
Hf												-	L	MP	L
Ta													-	L	L
Th														-	L
Ba															-

11.7.4 The posterior

The posterior information obtained is of the same type and was computed using the same algorithms as those reported in Case Study II (see Section 11.6).

11.7.5 Displaying and interpreting the results

In Figure 11.5 we provide a plot of the posterior information relating to k. We can see immediately that the posterior information suggests the presence of rather more provenances than was indicated by the prior. The prior mode was at $k = 2$ with the likely range being between 1 and 5 (inclusive). On the other hand, the posterior mode is at $k = 6$ with a likely range of 4 to 8 (inclusive). At first sight, this suggests that there is a lack of concurrence between the prior and the data. In this case, however, it is necessary to consider this possibility in the light of the posterior information about cluster membership.

In Figure 11.6 we provide a dendrogram based upon the posterior probabilities that specimens i and j are from the same provenance. Note first that there is one cluster (that containing specimens 64 to 126) which forms a strong, cohesive group and can reasonably be interpreted as a distinct and clearly defined provenance. From the other specimens it seems that we could also identify two other likely clusters: that associated with specimens in the range 1 to 63 and that associated with specimens 127 to 150. That said, several

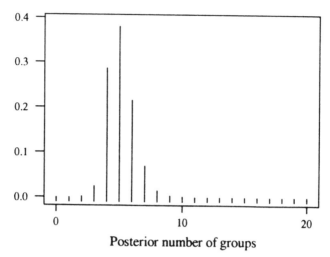

Figure 11.5   The posterior probability mass function of the number, k, of provenances for the tin-glazed pottery data.

specimens (including 21, 23, 41 and 42) can be seen to have a low posterior probability of being from the same provenance as any of the other specimens. This results in several clusters containing only one or two specimens. So, we have three main clusters and about ten specimens that form very small clusters.

In interpreting this information we note that a group consisting of just one specimen is unlikely to be held to constitute a separate provenance and in such circumstances might be thought of as an outlier. Given this observation, and that our prior information is that there are likely to be two or three provenances represented in the data, it seems reasonable to interpret the three largest posterior clusters as representing individual provenances. In doing so, however, we must accept that about ten specimens cannot be reliably ascribed to a provenance. Further work is need to identify whether the data for these specimens are outliers or whether they actually suggest the presence of an, as yet, unidentified provenance for these types of tin-glazed ceramics.

11.8   A useful practical extension to the methodology

In this chapter, we have ignored a complication which often occurs in chemical compositional data and which we must address if the Bayesian discriminant and cluster analysis techniques are to be really useful. This complication is that data vectors often contain one or more missing values. Missing values

Figure 11.6 Dendrogram formed by using the posterior probability that specimens i and j are from the same provenance for European medieval tin-glazed pottery data.

occur when, for various reasons, the chemists are unable to provide a figure: because there is insufficient specimen to analyse for all elements or because some elements are present below (or very occasionally above) the reliable range of the measuring techniques used. Therefore it is highly desirable that our statistical approach be able to cope with occasional missing data values.

The first situation mentioned above is that the chemist is unable to return any value for an element in a specimen. The second situation is that the detection limit of the analytical equipment has been reached and so only limited information about the value can be returned. Consider the first situation, where we have no information about the value which is missing. Without loss of generality, suppose that the vector of chemical compositions, $\boldsymbol{y}_j$, for the jth specimen is partitioned so that $\boldsymbol{y}_{j,r}$ $(r = 1, \ldots, s)$ are all observed and $\boldsymbol{y}_{j,r}$ $(r = s+1, \ldots, q)$ are all missing values. That is,

$$\boldsymbol{y}_j = \begin{pmatrix} \boldsymbol{y}_j^{(o)} \\ \boldsymbol{y}_j^{(m)} \end{pmatrix}$$

where $\boldsymbol{y}_j^{(o)} = (y_{j,1}, \ldots, y_{j,s})'$ are observed and $\boldsymbol{y}_j^{(m)} = (y_{j,(s+1)}, \ldots, y_{j,q})'$ are missing. Now, we have already modelled

$$\boldsymbol{y}_j | \boldsymbol{\mu}_i, \Omega_i, \theta_i \sim MVN(\boldsymbol{\mu}_i, \Omega_i)$$

so that, in the case of missing data,

$$\boldsymbol{\mu}_i = \begin{pmatrix} \boldsymbol{\mu}_i^{(o)} \\ \boldsymbol{\mu}_i^{(m)} \end{pmatrix}$$

and

$$\Omega_i = \begin{pmatrix} \Omega_{11} & \Omega_{12} \\ \Omega_{21} & \Omega_{22} \end{pmatrix}$$

where Ω_{11} is an $s \times s$ matrix, Ω_{12} is $s \times (q - s)$, Ω_{21} is $(q - s) \times s$ and Ω_{22} is $(q - s) \times (q - s)$. In this case the conditional density of $\boldsymbol{y}_j^{(m)}$ given the rest of the data, the assignment of the specimen to kiln site i and the other parameters is given by

$$\boldsymbol{y}_j^{(m)} | \boldsymbol{y}_j^{(o)}, \boldsymbol{\mu}_i, \Omega_i, \theta_i \sim MVN(\boldsymbol{\mu}_i^{(m)} + \Omega_{21}\Omega_{11}^{-1}(\boldsymbol{y}_j^{(o)} - \boldsymbol{\mu}_i^{(o)}), \Omega_{22} - \Omega_{21}\Omega_{11}^{-1}\Omega_{12}).$$

By modelling the problem in this way, we are able to obtain information about the missing values by sampling from the appropriate Multivariate Normal distribution. Clearly, within the MCMC framework, this poses no problem since we simply simulate the missing values as one more step in the iteration process.

This model extension is also suitable for use in situations of missing data of the second type, where we have some information about the processes which gave rise to the missing observations. A common example is when the values are missing because they are below the detection limit of the analytical equipment. Such information will allow us to limit the range of possible values for the missing data and so the previous result holds, but now within a restricted range. This too can easily be incorporated within the MCMC framework.

11.9   Conclusion

Although the development and application of Bayesian techniques for addressing the problems of source and provenance is in its infancy, we believe the approach has much to offer. We recognize that the methodology is difficult mathematically and will raise many challenges in implementation if it is to be used routinely. However, we believe that the case studies described in this chapter have conveyed a basic model, and have given some indication of the detailed model-building, prior specification and posterior computation that are needed for its implementation.

More particularly we believe that the Bayesian approach offers real advantages over the more conventional approaches to problems of this nature:

(i) The Bayesian approach is model-based, so that the methodology relies upon clear statements being made about the problem to be solved, and how the data relate to that problem and the methodology.

(ii) Clear definition of prior information lies at the core of all Bayesian investigations and, through this, we can establish the effect of a range of expert opinion on the results (and hence interpretations) obtained.

(iii) Since all the results obtained are probabilistic, individual specimens can be allocated to particular clusters with a given posterior probability. This allows us to distinguish between stronger and weaker attributions to clusters and hence arrive at better-judged interpretations.

(iv) Adoption of a Bayesian approach also permits us to take account of missing data values, not simply by imputing them or ignoring them, but by making allowance for them in the modelling process.

12

Application to other dating methods

12.1 Introduction

In Chapter 9 we described in some detail the application of the Bayesian paradigm to radiocarbon dating. The question is whether the methodology can be applied to other dating techniques. The answer, we believe, is an emphatic YES, although as with any new area, careful thought needs to be given to developing a suitable model, and to the elicitation and quantification of any prior information. In this chapter we will describe how the Bayesian approach could be applied to two rather specialized but somewhat dissimilar dating techniques, namely *seriation* and *tree-ring dating*. Finally we discuss how it could be used to combine information from two different methods of dating.

Seriation, which was illustrated by an example in Chapter 3, is concerned with determining the chronological order of a collection of archaeological contexts by comparing the relative proportions of certain types of artefacts found in each context. Thus seriation does not produce absolute dates for the contexts nor even relative dates measured in years but just attempts to establish which context is the oldest (or youngest), which is the next oldest (or youngest) and so on until the most recent (or oldest). The method is of some historical interest as its underlying principles can be traced back to Petrie's work on the pre-Dynastic cemeteries of Egypt. More recent examples of seriation and its usefulness in archaeological interpretation can be found in Carver (1985).

In contrast tree-ring dating, or *dendrochronology* to use its formal title, *may* sometimes provide a way of dating to a *precise* year a piece of timber from an archaeological site or, more often, a standing building. As one might expect with a technique that can be so precise, many stringent conditions need to be satisfied before it will work satisfactorily. For example, usually not just one sample of timber is needed but several. Each timber should have a substantial number of annual rings, say at least 80 and preferably 100 or more. Also, there must be a master chronology for the species for the region concerned and covering the period during which the site or building was constructed. In the case of Britain, at least, the best timber for tree-ring dating is oak although tree-ring master chronologies do exist for other species. If any of these conditions are not met, then a British site or building is unlikely to be dated. Moreover, even if these conditions (and some others that we have not mentioned) are met, there is no absolute guarantee that even an experienced dendrochronologist will be able to produce a totally trustworthy date.

12.1.1   Plan of the chapter

As we are discussing the *potential* uses of the Bayesian paradigm to dating techniques, we will in this chapter adopt a slightly different approach. Previously, in the applications chapters, we have tried to let the archaeology drive the statistics needed and the analysis pursued. Now, since we are proposing to illustrate the potential of the Bayesian approach to novel areas of research, our examples will be somewhat less realistic.

We are going to cover three research areas and for each area our approach will follow a similar pattern. We will give some archaeological background, develop a general model and show how the Bayesian method can be applied to it. The examples will be used to give the reader a feel for what is involved and to illustrate the potential of the Bayesian approach. Firstly we will propose a stochastic model of seriation and show how it could be analysed using the Bayesian approach. Secondly we examine the possible uses of the Bayesian paradigm in tree-ring dating. Thirdly we discuss how we can combine evidence from two different sources, namely tree-ring dating and radiocarbon dating, in order to produce evidence for a date when neither technique is successful by itself.

12.2   Seriation

12.2.1   Background

As has been remarked earlier in Section 3.6.3, seriation has developed as a technique for ordering chronologically archaeological contexts, such as graves, by using either the incidence (presence/absence) or relative frequencies of particular artefact types found in them. Robinson (1951) and Kendall (1971a) describe an underlying model of seriation in which the relative popularity of an artefact rises and subsequently falls but never then rises again (see Figure 7 of Chapter 3). Although not always explicitly stated, this model is the only one regularly adopted for general chronological seriation. We will refer to this model as "Kendall's model". Both Kendall (1971b) and Laxton (1976) develop the theoretical consequences of this model and show, by using a similarity measure, how to test whether a data set fits it.

However, many archaeological data sets cannot be seriated using this method because no "perfect" ordering of the data exists by this criterion. There are many reasons for this; sometimes there is some random noise perhaps due to "residual" artefacts left over from earlier periods. As a result archaeologists have turned to other techniques such as correspondence analysis (see Djindjian, 1989; Madsen, 1988) and multi-dimensional scaling (see Boneva, 1971) to interpret their data. Of these two methods, correspondence analysis has proved the most popular; we refer the interested reader to Baxter (1994, p. 100–139) for an explanation of the technique together with a discussion of numerous archaeological examples. However, neither of these methods permits the incorporation into the analysis of any uncertainty caused by the stochastic nature of the data. An alternative method has been proposed by Laxton and Restorick (1989) which does take account of the "noisy" nature of the data but this has yet to gain popularity amongst the archaeological community.

As far as we are aware, one drawback of all the methods of seriation currently is that they produce *one* ordering but give no indication as to other orderings that may fit the model almost as well. That is a method finds the "best" ordering using a particular algorithm, but fails to inform the user of the second, third or fourth "best" orderings. In this section we provide a means of identifying other possible orderings together with some assessment of our relative beliefs in them. Moreover, in some situations there may be other information regarding the possible order; for example, certain orders may be impossible on stratigraphic grounds. None of those currently popular permit the inclusion of "extra" information into the analytical process whereas, of course, the Bayesian approach does.

12.2.2   Stochastic model

We now develop a stochastic model for seriation based upon the work of Buck and Litton (1991). Consider I contexts (or graves) and J artefact types. Let $\theta_j(t)$ be the underlying proportion of artefact j present in a culture at time t. Suppose that context i was laid down at (unknown) time t_i and let for simplicity

$$\theta_{i,j} = \theta_j(t_i).$$

That is $\theta_{i,j}$ is the underlying proportion of artefact type j when context i was deposited. Since for a fixed context, say context i, the $\theta_{i,j}$ are proportions, they must add up to 1. That is we must have for each context i

$$\theta_{i,1} + \theta_{i,2} + \cdots + \theta_{i,J} = 1.$$

12.2.3   The data

Suppose we observe $n_{i,j}$ artefacts of type j in the ith context. We assume that the $n_{i,j}$ $(j = 1, \ldots, J)$ are a sample from the population proportions $\theta_{i,j}$ $(j = 1, \ldots, J)$. If we view the problem in this manner we may assume that $\boldsymbol{n}_i = (n_{i,1}, \ldots, n_{i,J})$ has a Multinomial distribution (see Section 6.5.1) with parameter vector $\boldsymbol{\theta}_i = (\theta_{i,1}, \ldots, \theta_{i,J})$.

12.2.4   The likelihood

As we are assuming that $\boldsymbol{n}_i$ has a Multinomial distribution with parameters $\boldsymbol{\theta}_i$, the likelihood is given by

$$l(\boldsymbol{\theta}_i; \boldsymbol{n}_i) \propto \prod_{j=1}^{J} \theta_{i,j}^{n_{i,j}}. \tag{12.1}$$

12.2.5   The prior

It is difficult to anticipate the prior information and hence choose a prior for this situation. One possibility is to use the Dirichlet distribution (see Section 6.5.3) but we must acknowledge this is for mathematical convenience only — the Dirichlet distribution is the conjugate prior for the Multinomial distribution.

Therefore in order to illustrate our general strategy, we assume that, a priori, $\boldsymbol{\theta}_i$ has a Dirichlet distribution with parameter vector $\boldsymbol{\alpha}_i$, denoted by

$$\boldsymbol{\theta}_i \sim D(\boldsymbol{\alpha}_i).$$

Then the prior density is given by

$$p(\boldsymbol{\theta}_i) \propto \prod_{j=1}^{J} \theta_{i,j}^{\alpha_{i,j}-1}. \tag{12.2}$$

Comparing (12.1) and (12.2) we see that they are very similar except that in (12.2) we have $\alpha_{i,j} - 1$ instead of the $n_{i,j}$ in (12.1). Thus setting $\alpha_{1,1}$ to, say, 4 can be thought of as having prior information $4 - 1 = 3$ artefacts of type 1. Likewise, setting $\alpha_{1,2}$ equal to 7 is equivalent to 6 artefacts of type 2, and so on. This gives us a feel for how the $\alpha_{i,j}$ could possibly be given values in a real situation.

12.2.6   The posterior

Using Bayes' theorem, the posterior density of $\boldsymbol{\theta}_i$ is given by

$$p(\boldsymbol{\theta}_i|\boldsymbol{n}_i) \propto l(\boldsymbol{\theta}_i; \boldsymbol{n}_i)p(\boldsymbol{\theta}_i).$$

Hence, the posterior density of $\boldsymbol{\theta}_i$ is proportional to

$$\prod_{j=1}^{J} \theta_{i,j}^{n_{i,j}+\alpha_{i,j}-1}$$

and so we recognize that, *a posteriori*, $\boldsymbol{\theta}_i$ has a Dirichlet distribution with parameter vector $\boldsymbol{n}_i + \boldsymbol{\alpha}_i$. That is

$$\boldsymbol{\theta}_i|\boldsymbol{n}_i \sim D(\boldsymbol{n}_i + \boldsymbol{\alpha}_i).$$

12.2.7   Implementation

Given this model and the Gibbs sampler methodology (see Section 8.6.1) the analysis is easily implemented as we now describe.

(i) For each context i, simulate $\boldsymbol{\theta}_i$ from a Dirichlet distribution with parameter vector $\boldsymbol{n}_i + \boldsymbol{\alpha}_i$.
(ii) Test whether, in some order, $\boldsymbol{\theta}_1, \boldsymbol{\theta}_2, \ldots, \boldsymbol{\theta}_I$ seriate according to Kendall's model. If so record that order. If not, no order is recorded.
(iii) Repeat (i) and (ii).

12.2.8   Case Study I

As an illustration, we apply our methodology to a highly simplified but, we believe, quite illuminating data set given in Laxton and Restorick (1989) and reproduced in Table 12.1. By reordering the sites as 2, 5, 3, 6, 1, 4 (see Table 12.2) we can easily see that in each column the data values rise and then fall but never then rise again. Thus the data, with the sites in the order 2, 5, 3, 6, 1, 4, satisfy Kendall's model.

As we have said earlier, correspondence analysis is popularly used to interpret such data. However, as Laxton and Restorick (1989) point out, correspondence analysis produces the order 3, 6, 5, 2, 1, 4 which does not satisfy Kendall's model as, in this order, the relative artefact frequencies do fall and then rise again (see Table 12.3). In particular consider artefact type 4; when the sites are ordered 3, 6, 5, 2, 1, 4 the observed frequencies are 0, 3, 0, 0, 42, 13 — they rise, then fall, then rise and then fall again, obviously violating Kendall's model.

Table 12.1   Artefact counts for the six sites.

	Artefact type						
Site	1	2	3	4	5	6	7
1	20	3	4	42	18	0	13
2	85	3	12	0	0	0	0
3	26	40	8	0	0	26	0
4	20	1	4	13	58	0	4
5	67	10	23	0	0	0	0
6	26	29	8	3	0	33	1

Table 12.2   Artefact counts for the six sites when the sites are ordered according to Kendall's model.

	Artefact type						
Site	1	2	3	4	5	6	7
2	85	3	12	0	0	0	0
5	67	10	23	0	0	0	0
3	26	40	8	0	0	26	0
6	26	29	8	3	0	33	1
1	20	3	4	42	18	0	13
4	20	1	4	13	58	0	4

Table 12.3   Artefact counts for the six sites when the sites are ordered according to the results produced by correspondence analysis.

	Artefact type						
Site	1	2	3	4	5	6	7
3	26	40	8	0	0	26	0
6	26	29	8	3	0	33	1
5	67	10	23	0	0	0	0
2	85	3	12	0	0	0	0
1	20	3	4	42	18	0	13
4	20	1	4	13	58	0	4

Table 12.4   Results of the Bayesian analysis.

Order of sites	%
2, 5, 3, 6, 1, 4	80.8
2, 5, 3, 6, 4, 1	14.0
2, 5, 6, 3, 1, 4	3.6
5, 2, 3, 6, 1, 4	1.2
2, 5, 6, 3, 4, 1	0.4

In the light of no information about the $\theta_{i,j}$ we follow conventional Bayesian wisdom by taking $\alpha_{i,j} = 0.5$ to represent vague prior information. So letting $\alpha_{i,j} = 0.5$, we applied the Bayesian methodology to this data set and ran the simulation until 1000 complete orderings of the six sites were obtained. The results are given in Table 12.4. On the basis of the simulations it is clear that the order produced using Kendall's method is the most likely seriation of the data given his model. We note that there are a number of other orders which represent possible seriations of the data but these are less likely. Kendall's ordering occurs 80.8% of the time while the next most likely alternative occurs 14.0% of the time. However, this alternative does not satisfy Kendall's model.

It is interesting that the order 3, 6, 5, 2, 1, 4 — the order obtained using correspondence analysis — is *not* represented in the results from our analysis.

12.2.9   Discussion

In this example we have viewed the observed artefact frequencies as realizations of a Multinomial distribution and then sampled from the posterior distribution of the underlying proportions. Other more justifiable assumptions could no doubt be made, that is a more realistic model and prior could be used. However, once we have sampled the underlying proportions, we then examine them to see whether they satisfy, in this case, Kendall's model. We then repeat this many times and hence assess which ordering is the most likely, which is the second most likely and so on. In doing so we give the archaeologist some idea of what is the most likely answer and what are the other possible orders, together with some indication of their relative importance.

Furthermore, by using this technique it is possible to use any prior information about the order of the contexts that may be available. In our example it is conceivable that some stratigraphic information could provide a partial ordering for the contexts. Alternatively, organic material from a few of the contexts could have been radiocarbon dated. This type of information can easily be incorporated into the algorithm.

The major drawback of the method is that it is extremely time-consuming since many sets of the θ_i are rejected as they do not seriate. Of course, this may be more a reflection of the inadequacy of the model than on the simulation algorithms used. However, if this approach to seriation is to become a routine tool, then more efficient sampling algorithms need to be found.

A final comment is that by using this technique other models of seriation can easily be applied to the data, perhaps allowing the popularity of an artefact to rise slightly after it has fallen, or to incorporate other features in the model. To arrive at alternative models, collaboration between archaeologists and statisticians will be necessary. In particular archaeologists will need to propose different theories/models that better explain the data and statisticians will have to devise efficient ways of fitting these highly complex models.

12.3   Tree-ring dating

12.3.1   Background

The following is not intended as a complete and comprehensive guide to tree-ring dating; readers who want a more exhaustive discussion of the subject may like to consult specialist books such as Baillie (1982) or Laxton and Litton (1988).

Tree growth

Each year an oak tree adds precisely one ring of growth on the outside, just inside its bark, to its trunk and all its branches. (In some other species of trees, a ring may not grow in every year; in other species, two rings or more may sometimes grow in a season.) Figure 12.1 shows a sample of timber which has been cut across the grain and polished to show the ring pattern. In the U.K., a year's growth begins in about April and is quite easily identified by a row of wide cells. These wide cells are followed later in the rest of the growing season, which lasts until about October, by narrow cells (see Figure 12.2). Thus a year's growth is from the start of one set of wide cells to the beginning of the next set. Normally it is easy, particularly in the case of oak, to identify the rings and to measure their widths. The measuring is usually carried out on a travelling microscope linked to a computer for the automatic recording of each width.

The influence of the weather

The weather conditions during the growing season largely determine the amount of growth — the width of these annual rings. Good weather during the season gives rise to a wide ring and poor weather to a narrow ring. So

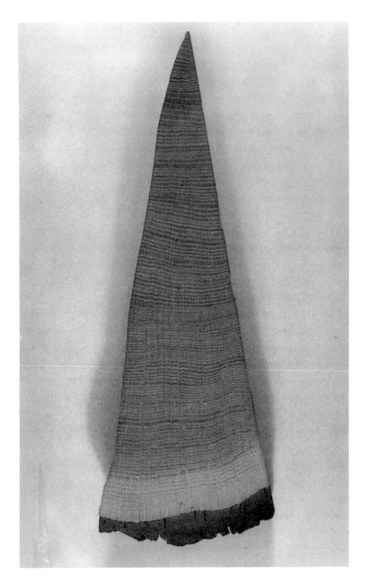

Figure 12.1 Section through the trunk of an oak tree cut across the grain and polished to show the ring pattern. (From Laxton and Litton, 1988, fig. 1.1(a).)

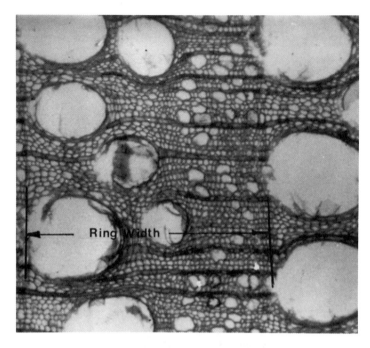

Figure 12.2   An annual tree ring under magnification, showing the wide cells of the spring growth and the narrower cells from later in the growing season. (From Laxton and Litton, 1988, fig. 1.2(b).)

the pattern in the succession of ring widths in a tree reflects, year by year, the pattern of the climate in the corresponding growing seasons. If two oak trees are growing in the same locality, they experience the same climate year by year and so can be expected to show similar (but not exactly the same) sequence of fluctuating ring widths.

Fortunately for archaeologists, the weather from year to year is very variable. If the weather were not so uncertain, then tree-ring dating would not work. In the event it is unlikely that the weather conditions in a sequence of 100 consecutive years, say, is the same, or even nearly the same, as in any other sequence of 100 consecutive years during the last 2000 years. Weather is just too unpredictable for this to happen; this unpredictability will be reflected in the sequences of ring widths of an oak tree. Therefore, it is very unlikely that the ring-width pattern in a sequence of 100 consecutive rings, say, grown in one sequence of 100 consecutive years is the same, or even nearly the same, as in any other sequence of 100 consecutive rings corresponding to a different set of 100 consecutive years. Clearly it becomes even more unlikely the longer the sequence of widths being compared. This principle is the basis of tree-ring dating.

Cross-matching

Suppose timber A was felled in the autumn of 1976 and had 148 rings. Then its outermost ring has a date of 1976 and its innermost 1829. Consider timber B from another oak tree growing in the same locality but which began growing 30 years later than A, in 1859 — *though we do not know this fact to begin with*. Now, provided that factors other than climate are unimportant, we would fully expect to find that the sequence of annual ring widths in B looks like and agrees well ring by ring with that of A, *when and only when the first ring of B is placed exactly 30 rings after the first ring of A*. That is, B is at an offset of +30 rings relative to A. Once done, this will date all the rings of B relative to A and if we knew, as we do here, that the date of the first ring of A is 1829, then the date of the first ring of B would be 1829 + 30 = 1859. By following this process, we obtain a date for B *relative* to A whether we know the date of the first ring of A or not. Relative dating alone can be very useful. For example, A has 148 rings and if B has 118 rings and no rings have been lost from either timber, then we would know that the trees from which the two timbers came were felled in the same year.

This process of comparing one sequence with another at all the different relative positions is called *cross-matching* and is used in an *attempt* to date one sequence of widths relative to another. We deliberately use the word *attempt* because it is important to realize that there is no guarantee that even the most experienced dendrochronologist will be able to date a specific piece of oak timber. For example, the timber may have some abnormal ring-width pattern resulting from insect attack, local weather conditions and the like. As a consequence, its ring-width pattern will not match that of the region.

In Figure 12.3 we illustrate the cross-matching of a tree felled in AD 1979 with the stump of a tree felled in AD 1948, which in turn is cross-matched with a timber removed from a windmill. This process can, if samples can be found, be continued backwards in time — for instance, in the East Midlands of England an 1100-year-long sequence has been developed from AD 1981 back to AD 882.

Removal of non-climatic trends

Earlier we pointed out that the ring-width sequences from different oaks growing at the same time, and even in the same locality, are not identical, though, hopefully, they are not too different. (This can be seen in Figure 12.4 (*a*).) This is to be expected in general since there are several factors which may influence the width of the rings in any particular tree. One tree might be

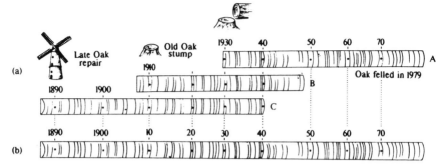

Figure 12.3 Diagram to illustrate the cross-matching of ring widths from a series of overlapping samples. (From Laxton and Litton, 1988, fig. 1.4.)

in an exposed position at the top of a hill, another might be in a valley below, a third might be in competition with a fourth for light and moisture and so on. We do not know in detail how such non-climate factors affect the growth of the rings, but anything that can be done to remove such effects helps in successful cross-matching since, as explained above, this largely depends on the climate common to all the oaks in a region.

One factor which can be removed before cross-matching is known as the *growth trend*. This trend can be seen very clearly in the timbers in Figure 12.4(a) — the ring widths of early growth, the inner rings, tend on the whole to be wider than those of the later growth, the outer rings, quite independently of the rapid year-by-year fluctuations in widths largely due to variable climate. This trend can be and is removed by standard statistical procedures (see Laxton and Litton, 1988). When it is removed, we obtain ring-width *indices*, see Figure 12.4(c), and it is normal practice to cross-match sequences of these rather than the sequences of raw ring widths themselves.

Master chronologies

By sampling timbers from buildings of different periods and using timber from modern trees of known felling date, it is possible to develop a *master tree-ring chronology* for a region. The region may be as large as the whole of the British Isles or as small as Kent (Laxton and Litton, 1989). In broad terms the larger the region, the longer and more replicated the sequences need to be. In contrast, with a chronology for a smaller region, it is possible confidently to date sequences composed of fewer and shorter samples.

Briefly, samples from modern trees whose date of the last ring are known are formed into an initial master. Perhaps if we are lucky this chronology will take us back to, say, 1650. A building constructed in the mid-eighteenth century may have timber spanning the period 1525 to 1725 which matches our

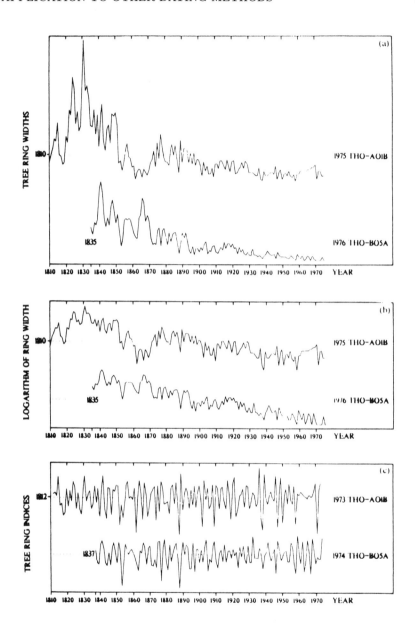

Figure 12.4 Comparison of ways of plotting tree-ring widths of two samples (THO-AO1B and THO-BO5A). (a) Raw ring widths. Note the growth trend: the widths in the earliest years of growth are larger than those in the latest. (b) Taking the logarithm of the widths prevents the widest rings from dominating; the growth trend has been reduced but not eradicated. (c) Baillie-Pilcher indices (see Laxton and Litton, 1988): narrow rings have been enhanced still further and the growth trend removed. (From Laxton and Litton, 1988, fig 1.5.)

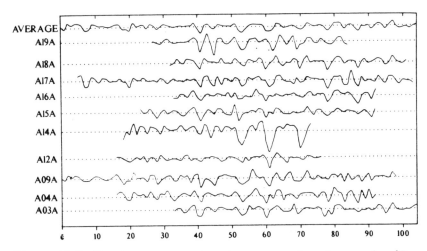

Figure 12.5   Forming an average tree-ring sequence from ten samples. (From Laxton and Litton, 1988, fig. 1.8(a).)

living tree chronology. Thus our chronology goes back to 1550. The search goes on for timber which overlaps this and takes us further back in time. This is repeated until we have covered the time-span of interest. In practice, it is not as easy as we have made it appear, but the general principles are as we have described them.

Dating a building

The more oak samples one can take in a building and the more annual rings they have (preferably 80 or more) the better the chance there will be of dating them. The practice in the Tree Ring Dating Laboratory of the University of Nottingham is to take about eight samples. If several of these eight sequences can be dated with respect to each other by cross-matching, then a good *well-replicated* site sequence can be formed by averaging (see Figure 12.5) and this improves the chances of dating. The dating of an average sequence from Bede House Farm against the East Midlands Master Chronology was illustrated in Figure 12.8 of Chapter 3.

12.3.2   *The traditional approach to tree-ring dating*

Suppose we have a long, well-replicated master tree-ring chronology for a region and we wish to date against it a new sample (or, preferably, an average sequence formed by a collection of samples from the same site that match together). Let the master chronology and the sample (or the average sequence) consist of N and n indices respectively. To simplify our discussion, let us also make two assumptions, namely that the sample is

(i) from the same region as the master chronology and
(ii) from a period completely within the time-span of the master,

so that the date of the first index relative to the master may be any one of $N - n + 1$ years.

For example, the East Midlands Master Chronology for oak spans the 1100-year period AD 882 to AD 1981. Thus in this case $N = 1100$. If the sample to be matched against it has 112 rings corresponding to 112 consecutive years we have $n = 112$. We have $N - n + 1 = 1100 - 112 + 1 = 1089$ comparisons to make. The first comparison will be made when the earliest ring of the sample is at the first ring of the master chronology, that is, at AD 882. The last ring of the sample will be at AD 993. The second comparison will be made when the sample covers the period AD 883 to AD 994, and so on. The last comparison is at AD 1870 to AD 1981.

The comparison between the sample and the master chronology is done visually or by using some statistical means or perhaps by a mixture of these two methods (see Baillie, 1982 or Laxton and Litton, 1988). *A priori*, we have no reason to believe that any one date is more likely than any other. In fact, this is a basic assumption of the "traditional" dendrochronological approach. If we can find a position at which we are sure "beyond all reasonable doubt" that the pattern of the indices of the sample matches that of the master chronology, then the sample is said to be *dated*. If, however, we cannot find such a position, then the sample will be undated, although the dendrochronologist may find one or more positions at which the match is "good" but not "beyond all reasonable doubt". Of course, the dendrochronological community have to agree on what "beyond all reasonable doubt" actually means and convey its meaning to the archaeological community.

To assess and report how good a match is, dendrochronologists in the U.K. use a *t-value*. A *t*-value is a measure of how similar two tree-ring sequences are. We refer the reader to Baillie and Pilcher (1973) or Laxton and Litton (1988) for information about its computation and a discussion of its interpretation.

Conventionally with a single sample of length 112 rings, tree-ring dating laboratories would be happy to see a *t*-value of 6 or more as representing very good evidence in support of a date. In contrast, a *t*-value of 4.5 or less would be unacceptable as evidence for a date unless there is some better information provided by cross-matching with some other chronologies. This raises the question of what a *t*-value between 4.5 and 6 represents. Currently tree-ring dating laboratories err very much on the side of caution and are not prepared to report a date with a *t*-value in this range. However, in some cases they do have some evidence for a date but that evidence is not sufficiently strong. Thus there is, in fact, a "grey" area between the black (no date) and the white (a date about which we are totally certain). The question is how to quantify and report evidence in this grey area. That is what we hope to address in this section.

12.3.3   A statistical model of tree-ring dating

Fritts (1976) proposes a model of tree growth in which he assumes a climatic signal common to all trees growing in a particular region plus some noise term due to non-climatic or localized factors. This model (with extensions) and its properties have been examined by Kronberg (1981), Laxton and Litton (1982), Wigley et al. (1984), Zainodin (1988) and Litton and Zainodin (1991).

Consider a homogeneous region for tree growth, for example, the East Midlands of England. Let $\phi_1, \phi_2, \ldots, \phi_N$ be the underlying signal of ring-width indices for years $1, 2, \ldots, N$ respectively. This signal is common to all oak trees growing in the region. Let $x_{i,j}$ be the index for the jth tree in year i. Then the statistical model is

$$x_{i,j} = \phi_i + e_{i,j}$$

where $e_{i,j}$ represents the noise component for tree j in year i. We assume that all the ϕ_i and the $e_{i,j}$ are independent, that the signals come from a distribution having mean 0 and variance ζ^2, and that the noise component for tree j in year i has mean 0 and variance τ^2. Finally let $R = \zeta^2/\tau^2$, which is often referred to as the *signal-to-noise ratio*.

12.3.4   Consequences of the model

An immediate consequence of this model (see Litton and Zainodin, 1991) is that the correlation coefficient between indices of any two trees in the region for the same year is given by

$$\rho = \sqrt{\frac{R}{R+1}}.$$

In contrast, the correlation coefficient for two indices from different years and different trees will be $\rho = 0$.

For example, suppose that the signal-to-noise ratio, R, for a particular region was 0.4. Then the theoretical correlation between two indices in the same year but from different trees is given by

$$\rho = \sqrt{\frac{0.4}{1.0 + 0.4}} = 0.53.$$

This is very much the underlying correlation; what we would observe in practice would be a stochastic realization of this.

If we assume that both the signal and the noise components have Normal distributions, then the indices for two different trees in the same year have a Bivariate Normal distribution with means 0, equal variances and correlation equal to ρ (see Section 6.3.6). Given the sequences of indices for two trees, we would estimate ρ by calculating the sample correlation coefficient, which we denote by r, between the two sets of indices.

Consider trees 1 and 2 each with n indices represented by $x_{1,1}, x_{2,1}, \ldots, x_{n,1}$ and $x_{1,2}, x_{2,2}, \ldots, x_{n,2}$ respectively. Then the sample correlation, r, between the two sets of indices is given by

$$r = \frac{\sum_{i=1}^{n}(x_{i,1} - \bar{x}_1)(x_{i,2} - \bar{x}_2)}{\sqrt{(\sum_{i=1}^{n}(x_{i,1} - \bar{x}_1)^2)(\sum_{i=1}^{n}(x_{i,2} - \bar{x}_2)^2)}}$$

where

$$\bar{x}_1 = \frac{1}{n}\sum_{i=1}^{n} x_{i,1} \text{ and } \bar{x}_2 = \frac{1}{n}\sum_{i=1}^{n} x_{i,2}.$$

We say that two trees having the same number of indices are *contemporary* if the first index of tree 1 corresponds to the same calendar year's growth as the first index of tree 2. If the two trees are contemporary r should be a good estimate of ρ and so would be greater than 0. On the other hand, if the trees are not contemporary, the underlying correlation ρ would be 0 and so r should be close to 0.

We note that Fisher's Z-transformation (Fisher, 1921)

$$z = 0.5 \ln\left(\frac{1+r}{1-r}\right)$$

is approximately normally distributed with mean and variance given by

$$\nu(\rho) = 0.5 \ln\left(\frac{1+\rho}{1-\rho}\right) \text{ and } \frac{1}{n-3}$$

respectively. Returning to our example with $R = 0.4$ and $\rho = 0.53$, we have that the value of z calculated between two 112-year-old contemporary trees has a Normal distribution with mean given by

$$\nu(\rho) = 0.5 \ln\left(\frac{1+0.53}{1-0.53}\right) = 0.59$$

and variance given by

$$\frac{1}{112-3} = 0.0092.$$

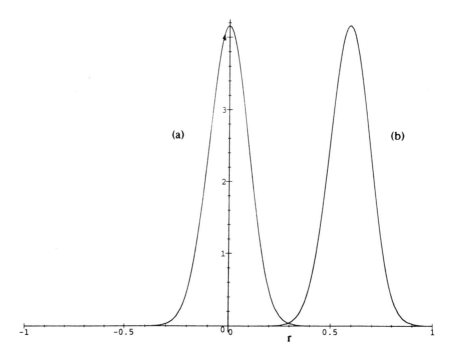

Figure 12.6   Theoretical distributions of z, Fisher's Z-transformation, between two 112-year-old trees from a small region (a) when they are not contemporary and (b) when the trees are contemporary.

Using the properties of the Normal distribution (see Section 5.10.3), there is about 95% chance that the sample correlation will lie between $0.59 - 2 \times \sqrt{0.0092}$ and $0.59 + 2 \times \sqrt{0.0092}$. That is, z will lie between 0.40 and 0.78 with probability 0.95. It will lie between 0.29 and 0.78 with probability about 0.99.

If now we consider two trees that are not contemporary, z will be an observation from a Normal distribution with mean and variance of 0 and 0.0092 respectively. In this case the 95% and 99% ranges are -0.19 to 0.19 and -0.29 to 0.29 respectively. The two distributions are given in Figure 12.6 from which we can see that there is very little overlap between them. Thus for trees of 112 or more rings in this hypothetical region there should be little danger of incorrectly identifying a match.

The above model easily extends to dating an "average" sequence composed of several samples against a master chronology for a larger and therefore less homogeneous region such as the British Isles. Litton and Zainodin (1991) show that the underlying correlation between a well-replicated master chronology and an average sequence composed of J samples is approximately

$$\rho_J = \sqrt{\frac{R}{R + S + J^{-1}}},$$

where S is a factor which allows for the fact that trees in different parts of the region respond slightly differently in the same year. In other words, the climatic signal varies slightly from one part of the region to another.

It is easy to see that the above distributional results for z apply but with ρ replaced by ρ_J. For a rather large region such as the British Isles, R would be about 0.3 and S about 0.4. Suppose that our average has 112 indices and is composed of, say, 8 samples ($J = 8$). Then the underlying correlation between the average and the master chronology at the correct date is

$$\rho = \sqrt{\frac{0.3}{0.3 + 0.4 + 8^{-1}}} = 0.36$$

which is quite a lot less than that found earlier. This is because we are looking at the correlation between trees over a much wider region and so the responses in terms of yearly growth will vary more.

With $\rho = 0.36$, we have $\nu(\rho) = 0.38$. Hence the 95% and 99% ranges for r are 0.19 to 0.57 and 0.09 to 0.67. In this case there is a more substantial overlap between the distributions of r at the correct and incorrect positions (see Figure 12.7). Thus it is possible that an incorrect position could give a higher sample correlation and therefore a higher value of z than that observed at the correct relative position. Obviously this could lead to the reporting of erroneous dates.

12.3.5   The data

The data, denoted by x, consist of the master tree-ring chronology and the tree-ring indices for the sequence under consideration.

12.3.6   The prior

Let θ denote the true but unknown date of the first index of the sample (so that $\theta + n - 1$ is the date of the last index). Then, using the traditional approach of giving each year equal weight, we are saying that the prior probability that θ takes the value j is given by

$$p(\theta = j) = \frac{1}{N - n + 1} \quad j = 1, 2, \ldots, N - n + 1.$$

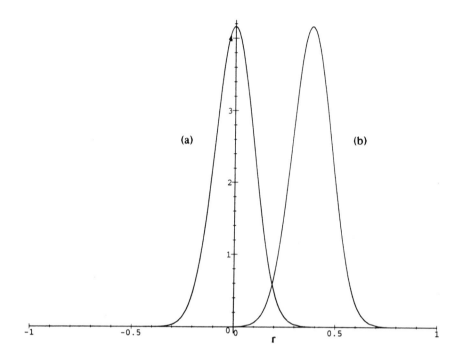

Figure 12.7   Theoretical distributions of z, Fisher's Z-transformation, between an average sequence of 112 indices composed of eight samples and a master chronology, for a large region such as the British Isles (a) when it is at an incorrect date and (b) when the average sequence is at the correct date.

12.3.7   The likelihood

For each relative position, denoted by j, of the sample against the master chronology we calculate the sample correlation coefficient, denoted by r_j. Then, using Fisher's Z-transformation, we calculate

$$z_j = 0.5 \ln \left(\frac{1 + r_j}{1 - r_j} \right) \quad \text{for } j = 1, 2, \ldots, N - n + 1.$$

Note that r_j, and hence z_j, is a function of the tree-ring data, $\boldsymbol{x}$. Thus the likelihood is given by

$$l(\theta = j; \boldsymbol{x}) \propto \exp\{-0.5(n - 3)(z_j - \nu(\rho_J))^2\}$$

since z_j has approximately a Normal distribution with mean $\nu(\rho_J)$ and variance $1/(n - 3)$.

12.3.8   The posterior

Using Bayes' Theorem, the posterior distribution of the date of the start of the average sequence is given by

$$p(\theta = j | \boldsymbol{x}) \propto p(\theta = j) \, \exp\{-0.5(n - 3)(z_j - \nu(\rho_J))^2\}$$

where $p(\theta = j)$ is the prior probability that the date is j. If we assume that, a priori, all dates are equally likely, then

$$p(\theta = j | \boldsymbol{x}) \propto \exp\{-0.5(n - 3)(z_j - \nu(\rho_J))^2\}.$$

12.3.9   Case Study II

To illustrate this approach we use an average tree-ring chronology composed of seven samples from Anne Hathaway's Cottage, Stratford-upon-Avon, England. (Anne Hathaway was William Shakespeare's wife.) This chronology has 144 indices and has been dated using conventional dendrochronological techniques against the East Midlands Master Chronology (Laxton and Litton, 1988) to span the period AD 1319–1462 (Alcock et al., 1991). Now we will attempt to "date" it using another master chronology for oak, one which spans the period AD 401–1981 (Pilcher, personal communication) and which is composed of samples from all over the British Isles.

Litton and Zainodin (1991) show that for a region such as the British Isles, R, the signal-to-noise ratio, lies in the interval 0.20 to 0.35 and S is about 0.40. Hence we expect the underlying correlation between a master chronology and an average sequence composed of seven samples, ρ_7, to lie between 0.50 and 0.65. For ease of calculation we give equal weight to values of ρ_7 in this interval in which case the posterior probabilities are given by

$$p(\theta = j | \boldsymbol{x}) \propto p(\theta = j) \sum \exp\{-0.5(n - 3)(z_j - \mu(\rho_7))^2\} \qquad (12.3)$$

where the summation is over all values of ρ_7 from 0.50 to 0.65 in steps of 0.01. In the light of no prior information about the date, we give every year from AD 401 to AD 1838 inclusive equal prior probability so that $p(\theta = j) = 1/1438 \approx 0.0007$.

After calculating the posterior probabilities, we find two positions with probability much greater than the prior probability of 0.0007. These positions correspond to the date of the last index being AD 1462 or AD 1474 with the posterior probabilities being 0.9986 and 0.0014 respectively. Thus, a posteriori, we can be fairly certain, but not 100% certain as traditional dendrochronologists demand, that the date of the last index is AD 1462.

This date of AD 1462 is in accord with the previously published result for Anne Hathaway's Cottage when it was dated against several well-replicated regional chronologies appropriate for the central part of England. Interestingly, if we shorten the average sequence by 40 indices, thus making it only 104 indices long, the posterior probability of the last index being AD 1462 is reduced to just over 0.84. In this case we have good evidence for the date but it is certainly not "beyond all reasonable doubt".

12.3.10 Discussion

The approach that we are advocating is undoubtedly controversial. Traditional dendrochronologists are strongly united against it because it would, so they claim, introduce into the literature false dates (or ones about which they are not totally confident). We believe, on the contrary, that it will introduce into the public domain dates for particular buildings that have remained undated until now. Moreover, with the date will be some assessment of the confidence that can be attached to it. This expression of confidence will be conditional on the currently available information used in the analysis. If further data become available at a later time, our confidence may increase if the later data support the original date, or it may decrease if they do.

12.4   Combining dendrochronological and radiocarbon information

12.4.1 Background

As we have said above, traditional tree-ring dating techniques do not always work. In fact dendro-laboratories are proud of the fact that they are prepared to say when they cannot date a particular sample, thereby raising the credibility of the results that are published. However, large numbers of timbers remain undated, perhaps as many as 75%, by traditional methods. Obviously the Bayesian approach can be used to produce dates together with some indication of our confidence or degree of belief. But even so, reporting that there is a 60% chance that the date is AD 1433 is not as precise as one would like. Can we do better?

One possible scenario is that we have attempted to date a sample using the process given in the previous section but no conclusive date resulted. However on the basis on the above analysis, some dates are, *a posteriori*, more likely than others. Moreover, from a historical/archaeological viewpoint the sample and its date may be of great significance, so great that additional funding is available to carry out a series of relatively expensive radiocarbon dates on the sample and wiggle-matching techniques may be used (see Sections 9.6 and 9.7).

But can we combine both the radiocarbon and the dendrochronological information? If so, how?

12.4.2   Statistical model of wiggle-matching

Wiggle-matching was described in Case Studies III and IV in Chapter 9 and we only give a brief review of the salient points here. Suppose we have a tree-ring sample that fails to date "beyond all reasonable doubt". Pearson (1986) proposed that sections of the sample taken at known relative years apart should be sent for radiocarbon analysis. Suppose m separate sections are selected and let the date of the first section relative to the date of the first ring of the sample be denoted by ϕ_1. Let the date of the kth section relative to that of the $(k-1)$st section be ϕ_k. Then the absolute date of the kth section is given by

$$\theta_k = \theta + \phi_1 + \cdots + \phi_k \quad \text{for } k = 2, 3, \ldots, m$$

where θ is the date of the first ring.

12.4.3   The data

Given a particular section, say section k, y_k will denote the estimated radiocarbon age and σ_k the corresponding standard deviation reported by the radiocarbon laboratory. Here y_k is a realization of the random variable

$$Y_k \sim N\big(\mu(\theta_k), \ \sigma_k^2 + \sigma^2(\theta_k)\big),$$

where $\mu(\theta_k)$ denotes the piece-wise linear high-precision calibration curve and $\sigma^2(\theta_k)$ models the error in the curve (see Section 9.6.4).

12.4.4   The prior

In this context we do have some a priori evidence about which dates are likely and which are unlikely. This evidence is expressed in terms of $p(\theta = j|x)$, the posterior probability that the date of the first index is j given the tree-ring data, x. We propose to use this as our prior for θ when analysing the radiocarbon data.

12.4.5   The likelihood

Assuming that conditional on the θ_k the radiocarbon determinations are independent of each other, we have

$$l(\theta; y) = \prod_{k=1}^{m} l(\theta_k; y_k),$$

where

$$\mathbf{y} = (y_1, y_2, \ldots, y_m), \boldsymbol{\theta} = (\theta_1, \theta_2, \ldots, \theta_k),$$

$$l(\theta_k; y_k) \propto \omega_k(\theta_k)^{-1} \exp\left\{-(y_k - \mu(\theta_k))^2/2\omega_k^2(\theta_k)\right\}$$

and

$$\omega_k^2(\theta_k) = \sigma_k^2 + \sigma^2(\theta_k).$$

12.4.6   The posterior

In mathematical terms, since the tree-ring information and the radiocarbon information are independent of each other, the posterior distribution of the calendar date, θ, of the first index based upon the tree-ring data, $\mathbf{x}$, and the radiocarbon data, $\mathbf{y}$, is given by

$$p(\theta|\mathbf{x}, \mathbf{y}) \propto l(\theta; \mathbf{y})p(\theta|\mathbf{x})$$

where $p(\theta|\mathbf{x})$ is the posterior of θ based upon the tree-ring data, $\mathbf{x}$, and $l(\theta; \mathbf{y})$ is the likelihood of θ based upon the radiocarbon data, $\mathbf{y}$. This is an example of the sequential updating of beliefs which we discussed in Section 7.6.

12.4.7   Case Study III

We illustrate our approach with an example first reported by Pearson (1986). The timber sample, coded Q3927, was taken from a horizontal mill at Ardnagross, Co. Westmeath, Ireland. It had 151 rings but no signs of sapwood. When measured, an unusual ring-width pattern was noticed and so, when it failed to conclusively match the relevant chronologies, the dendrochronologists were not particularly surprised (Baillie, personal communication). For example, when this single sample is compared with the British Isles master chronology, the Bayesian tree-ring dating methodology described above produces eight dates with posterior probabilities of more than 0.001. These dates, with their corresponding correlation coefficients and posterior probabilities, are given in columns (i) to (iii) of Table 12.5. Clearly, based on the tree-ring evidence alone there is no clear and unambiguous date that can be confidently reported.

However, the mill was of sufficient importance that sections of the sample were sent for radiocarbon analysis. The sections consisted of bi-decadal sets of rings taken from rings 49 to 148 inclusive so that we have $k = 5$ sections. The centre of the oldest section corresponds to rings 58/59 and the centre of the youngest section to rings 138/139. Thus we take $\phi_1 = 58$ and $\phi_2 = \phi_3 = \phi_4 = \phi_5 = 20$. The radiocarbon determinations for the five sections are, from earliest to latest, 1290, 1263, 1268, 1292 and 1270 bp, each with a quoted standard deviation of 15.

Table 12.5 Results of dating the mill at Ardnagross, Co. Westmeath, Ireland. Only those dates with a posterior probability greater than 0.01 are reported.

(i) Date of last index	(ii) Correlation coefficient	(iii) Posterior probability using tree-ring data only	(iv) Posterior probability using radiocarbon and tree-ring data
598	0.281	0.27	0.00
788	0.257	0.07	0.99
947	0.250	0.05	0.00
972	0.256	0.07	0.00
1065	0.252	0.06	0.00
1068	0.242	0.03	0.00
1455	0.275	0.19	0.00
1834	0.268	0.14	0.00

We now re-examine the data using the posterior probabilities obtained when analysing the tree-ring data as prior probabilities for the calibration of the radiocarbon. We are dealing with a single sample ($J = 1$) and so we expect ρ_1, the correlation between the sample and the master chronology for the British Isles that we used in the previous example, to lie between 0.3 and 0.5. Thus we take the summation in (12.3) to be from $\rho_1 = 0.3$ to 0.5 in steps of 0.01. The results produced are given in column (iv) of Table 12.5. Notice how the date of AD 788 has a posterior probability of just under 0.99. Does this evidence satisfy the criterion of "beyond all reasonable doubt"? The only two other dates not reported in Table 12.5 with posterior probabilities larger than 0.001 are AD 779 and AD 774 with the probabilities being 0.0041 and 0.0014 respectively. The associated correlations between sample and master chronology at these dates are 0.152 and 0.110 respectively. In others words, by using both sets of data and combining them using the Bayesian paradigm, we have very strong evidence, based upon currently available data and knowledge, that the date of the sample is AD 788. However, more evidence could come to light sometime which may or may not confirm this.

12.4.8 Sensitivity analysis

Of course, an immediate concern of the dendrochronological community is the reliability of dates based on radiocarbon dating. We have already mentioned in Section 9.8.1 that Baillie (1990) pointed out that about a third of radiocarbon determinations were in error by over 200 years and that the standard deviations reported by the radiocarbon laboratories were unduly optimistic. There are two kinds of errors that could adversely affect the validity of this novel Bayesian approach; these are related to the accuracy and the precision of the radiocarbon determinations.

To assess the sensitivity of our results to possible errors in the reported precision of the results, we reanalysed the Ardnagross data using larger standard deviations than those reported by Pearson (1986) in his original paper. The posterior probability that the date of the last index is AD 788 is still just under 0.99 if the standard deviations of the radiocarbon determinations are multiplied by a factor of 2. That is, if the standard deviations are taken as 30 instead of 15, the odds in favour of AD 788 are about the same. If the standard deviations are taken as 60 and 120 the posterior probability becomes 0.98 and 0.91 respectively. In other words, the result is fairly robust against errors in the reported precision (as represented by the laboratory standard deviations) of the radiocarbon dating process.

The sensitivity of our results due to possible errors in the accuracy of the radiocarbon determinations is much more difficult to assess. On the one hand, if only one determination is grossly out of line with the rest, the method developed by Christen (1994a, b) and illustrated in Section 9.8 may be used to detect it. On the other hand, if the majority (or all) of the determinations are affected by some systematic bias, then essentially all the determinations will be in error by roughly the same amount, in which case the wiggle-matching will take place at the wrong part of the calibration curve. One way to avoid such a systematic laboratory bias would be not to let one laboratory carry out all the analyses on the sections; instead several different laboratories should be used.

12.5   Summary

We have in this chapter tried to demonstrate the potential of the Bayesian approach to two archaeological dating problems. We are sure that the approach could benefit other areas of research — it is up to the archaeological community to apply the methodology, perhaps with some assistance from the statistical community, to their dating problems.

13

The way forward

13.1 Introduction

It has been our objective, throughout this book, to describe, explain and illustrate the Bayesian approach to the interpretation of archaeological data. We have introduced the philosophy, developed the statistical theory and examined a variety of case studies. Bayesian statistics is a rapidly developing field and we are convinced that over the next decade the number and range of applications will increase dramatically, not only in archaeology but in many other disciplines such as medicine, economics, environmental studies and so on. An awareness of the importance of Bayesian analysis is already growing — the authors of a recently published textbook on data analysis have commented (Daly *et al.*, 1995, p. 560) that their book:

> ... is based largely on current statistical practice and describes the ideas and methods underlying almost all major statistical computer packages which are currently commercially available. These, in turn, are based on a particular interpretation of probability (the *frequentist* interpretation) — a particular view of what the notion of probability means. But there are other views.

They then go on to say that:

> Of these the most important is undoubtedly the subjective view, leading to a school of statistics called the **Bayesian school**. In brief, this school asserts that probability has no objective meaning in the real world, but simply represents the degree of belief of the researcher that something is true. Data are collected and analysed and this leads to a modification of the researcher's beliefs in ways described by appropriate statistical techniques. One of the problems with applying such methods until recently has been the mathematical intractability of the analyses. Particular forms, often unrealistic, for the probability distributions involved had to be assumed in order that the analysis could be carried

out. In recent years, however, developments have been made such that these problems no longer apply. More realistic distributional forms can be adopted and the power of the computer means that we can sidestep analytic mathematical solutions. It is certainly the case that Bayesian methods are growing in popularity and, when Bayesian software becomes readily available, they may well compete with traditional methods.

In addition to an increased awareness of the existence of the Bayesian approach to data interpretation, there is an increasing number of specialist books which introduce the reader to more detailed practicalities. Gelman *et al.* (1995, p. 468–469) note in their concluding chapter that

> ... we have focused on the construction of models (especially hierarchical ones) to relate complicated data structures to scientific questions, checking the fit of such models, and investigating the sensitivity of conclusions to reasonable modeling assumptions. From this point of view the strength of the Bayesian approach lies in (1) its ability to combine information from multiple sources (thereby in fact allowing greater 'objectivity' in final conclusions), and (2) its more encompassing accounting of uncertainty about the unknowns in a statistical problem.

This statement succinctly summarizes both the philosophical and practical issues which we have attempted to convey throughout this book. The increasing number of illustrations of the Bayesian paradigm applied to real problems and the fact that Markov chain Monte Carlo methods can be implemented on readily accessible hardware mean that applications of Bayesian methods are growing very rapidly. Our aim in writing this book has been to provide archaeologists with the understanding they require to play their part.

Some readers may have been disappointed to see that we have not provided any simple recipes that can be followed in order to apply the Bayesian methodology to archaeological data. This is because of our conviction that it is impossible, and that each problem should be tackled afresh; a model needs to be developed for each particular situation, the prior information must be elicited for that problem and the posterior distributions must be interpreted in the light of the current situation. Of course, there will be some carry-over from one problem to another, but care must be taken not to apply blindly a model developed for one problem to another similar, but slightly different, one. All the same, there are some general principles which do apply. In the first part of this final chapter, we summarize these guidelines and offer some general thoughts about undertaking Bayesian analyses. In the later part, we indicate some of the challenges which face the archaeological and statistical communities.

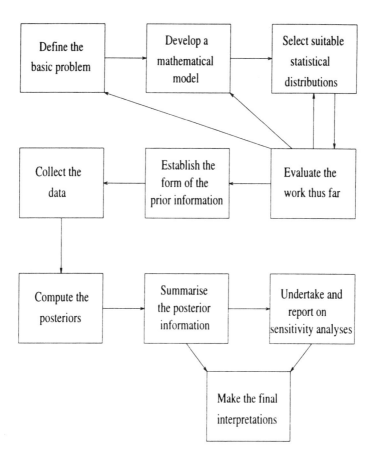

Figure 13.1   A general framework for Bayesian investigations.

13.2   Some general principles

13.2.1   The basic steps

Let us consider the basic steps necessary for working within the Bayesian framework (see Figure 13.1).

Step 1: Define the problem

Researchers on a new collaborative project need to gain a mutual understanding of the basic archaeological problem. The step is not unique to Bayesian investigation; it is a necessary part of all interdisciplinary research projects, but its importance (and the time it takes) can often be underestimated.

Step 2: Develop a mathematical model

Here the (often rather hazy) ideas developed in Step 1 are transformed to the more precise language of mathematics, by defining suitable notation and relating the parameters to each other by means of equations, inequalities and other mathematical expressions.

Step 3: Develop suitable statistical models

Having developed our basic mathematical model, we need to identify any parts of it which contain random or uncertain components and will consequently need representing by probability distributions. This step requires understanding both of the mathematical properties of the model selected and of the insights gained in Step 1. The technical aspects of this process will mean that most of the input comes from a statistical expert.

Step 4: Evaluate the work thus far

All researchers involved need another brainstorming session. Are the mathematical model and the statistical distributions selected suitable for the real problem under investigation? Have we over-simplified the archaeological problem? Have we been too ambitious in our attempts to cover all the fine points of the archaeologists' ideas? Do we have available to us the practical tools that we are going to need to work with these models? If we are not completely happy with the answers to such questions, we need to go back an appropriate number of steps and try again!

Step 5: Establish the form of the prior information

This is the point at which the investigation becomes more obviously Bayesian. For all researchers this is still a challenging step; there is relatively little guidance in the literature on how to elicit and formulate prior information. Many workers find that they develop new strategies and skills with each problem they tackle. We have already looked at this issue in each of the case studies, but we shall also discuss it in a more general way later in this chapter.

Step 6: Collect the data

Data collection is, of course, at the core of all applied statistical analyses and is the point at which the bulk of any necessary field or laboratory work takes place. Data collection undertaken as part of an investigation within the Bayesian framework need not be different from any other, but it may be. In particular we would stress here that Bayesian approaches to sample selection are often rather different from those adopted by classical statisticians and this may have a profound effect upon the practical approaches used to obtain the data (see Section 4.6).

Step 7: Compute the posterior

This step may be the most complex, technical and computer-intensive. Decisions made earlier, particularly at Step 3, about the statistical modelling will have consequences for implementation. Has a conjugate prior been used? Is numerical integration the best way forward, or should sampling methods (for example Markov chain Monte Carlo) be adopted instead? At this step it is extremely important that what is practical and convenient in technical terms is weighed against the requirements of the whole research programme.

Step 8: Summarize the posterior information

All the workers will need to meet up once again. Having computed the posteriors the work of interpretation begins. Plots, tables and summaries can be discussed and used to make archaeological interpretations of the situation under analysis.

Step 9: Sensitivity analyses

This is the stage at which we undertake a review of the model and its adequacy by considering issues such as its robustness and sensitivity.

Step 10: Interpretation

Bayesian interpretation usually involves looking both forward and backward: backward in the sense that we must ask how well the combination of data and prior can help us to further our understanding of the problem; forward in the sense that today's posterior becomes tomorrow's prior. New data may be collected or additional observations made and we must expect to update our own interpretations accordingly.

The comments made here are rather general, and the steps outlined bear a close resemblance to those of the modelling cycle as discussed in Chapter 3. These similarities arise because the Bayesian approach is model-based and so has the modelling cycle at its core. In the remainder of this chapter, we focus on certain aspects which are crucial to the success of Bayesian data investigation — namely prior elicitation, calculating posteriors and interpreting the results. Before we do so, however, we discuss the topic of problem definition.

13.2.2   Problem definition

Throughout this book (and particularly within our case studies) we have tried to emphasize the basic steps essential to all Bayesian investigations. There is, however, one vital component that we have not perhaps paid quite enough attention to, namely problem definition. Good problem definition is, of course, not uniquely important to Bayesian statistical investigations — it applies to the classical approach as well.

Excavators face a particular difficulty in that they destroy the very object of their study: once a site has been excavated it cannot be dug again. A conservative policy (and probably the right policy) is to record every possible detail for later reference, but a consequence is that immense quantities of data are accumulated. Whilst the excavator will have certain objectives in mind, the relationship between data and research is not always as direct as we might hope. Although relatively few published reports exist which testify to this there is one particularly frank description of

> an object lesson in the near disaster that comes from over-ambitious and under-resourced proposals.

This is provided by Fieller (1993), a statistician who has collaborated on many archaeological projects. The reference quoted above is to a project at Cresswell Craggs in Derbyshire, U.K. where (Fieller, 1993, p. 290)

> it was not clear at the planning stage ... just how much data would accumulate. The exceptional level of detail of the excavation resulted in enormous numbers of finds (tens of thousands) and consequently colossal amounts of data. Apart from the unparalleled opportunities for mistakes in data recording and transcription errors, many of the data were incomplete because of inadequate resources being available for the more specialist tasks
> ...

Certainly archaeological field work, and excavation in particular, is a complex operation, frequently with a number of competing objectives, which will tend to compromise each other. Moreover the outcomes are unpredictable: at any stage of the work unexpected finds can alter the priorities of a project. This complexity may lead some workers to shy away from attempting to make categorical statements about their research objectives and the routes they plan to use to meet them. From a Bayesian statistical perspective we would urge against this. Defining the problem clearly at the outset does not mean that one cannot change one's mind in the light of new data; it simply means that one has a firm base upon which to build.

In particular, we would stress the importance of dialogue during the planning and definition of a project. For example, statisticians often expect to define "what is reasonable" in terms of data quality and quantity. Suggestions about "what is statistically sensible" can be helpful to archaeologists who are thinking about problem definition for the first time, but often "what is reasonable" can only be dictated by the conditions that prevail in the field.

Given such practical considerations, the need in any archaeological project for unambiguous problem definition is clear. Time spent making sure that the statisticians are really tackling the question(s) originally posed is always rewarded by less frustration in the field or laboratory. In our experience, there is no substitute for patience: an afternoon of introductory presentations in which those involved explain their questions and their suggested solutions, or perhaps a round-table discussion with drawings, plans and sketches forming the basis of a detailed exposition of ideas.

13.2.3   Prior elicitation

Once experts in a team have reached a stage where they share vocabulary, understand each other's ideas, and have developed a model, it is necessary to take the important and challenging step of eliciting the prior information and converting it to prior distributions. Naturally, most of the technical work will be undertaken by a statistician. Nonetheless, specification of prior distributions in a form suitable for incorporation into the analysis is something in which archaeologists must be expected to share. Consequently, they will need to learn new skills.

Many of the issues related to specifying the prior have been summarized by Hogarth (1975) who notes that it is often difficult to get non-statisticians to provide their information in the form of distributions. Sometimes examples, like the "bar" and "restaurant" example given in Chapters 4 and 5, can be used to demonstrate the important features of prior definition without focussing on a specific real problem, the details of which might distract from the more theoretical issues which need to be conveyed. Sketches and graphs can allow both statisticians and archaeologists to explain their ideas in a pictorial manner, before any notation or statistical distributions are introduced. Once the distributions are formalized, the statistician can demonstrate to the archaeologists just what effect their assessments will have on the final results and, if necessary, provide some kind of sensitivity analysis to support this.

It is also important to remember that there are some situations in which very little prior knowledge is available. This is particularly common in new research areas where not much previous research has been carried out. In such cases a non-informative prior may be used (see Section 7.7).

In addition, statisticians eliciting prior information will need to bear in mind that all individuals have a different range of abilities both in their own field and in statistics. If attempting to elicit information from a number of experts it is particularly important to allow for this. Indeed, Hogarth (1975, p. 282–283) suggests a number of formal methods for doing so. He also points out that most non-statisticians do not naturally think of chance in terms of probability and may, even if given probabilities, translate them into frequencies.

For those particularly interested in knowledge elicitation we should draw attention to Cooke (1991) in which a successful attempt is made to demonstrate the strong similarities between the knowledge elicitation issues in a wide range of fields. Its title is *Experts in Uncertainty: Opinion and Subjective Probability in Science* and it offers an impressively wide coverage of the elicitation, application and interpretation of subjectively stated expert knowledge. The book is, by no means, written exclusively from a Bayesian perspective and does not use archaeological examples, but nevertheless the work addresses many theoretical and practical issues that readers of the current volume might find useful.

13.2.4   Evaluation of posterior distributions

As noted in Chapter 8, there are three main routes to evaluating posterior distributions:

(i) by obtaining an analytical solution often achieved by using conjugate priors,
(ii) by using quadrature methods, and
(iii) by adopting Markov chain Monte Carlo methods.

Given the detailed explanation of these methods in Chapter 8, all that need concern us here is how one selects the most appropriate approach in any given practical situation.

Although it may be possible to model prior uncertainty by means of conjugate priors (see Section 8.3) so that the form of the posterior is the same as that of the prior, such distributions are not always suited to archaeological problems. As a result, in order to represent our prior knowledge realistically, we are forced to use prior distributions which may make computation of the posterior difficult. In the past this might have been seen as a real practical barrier, today this is no longer necessarily so.

In general quadrature methods are only really viable for evaluating posteriors in situations when the number of parameters is relatively small. Commonly, as we have seen in some of our case studies, archaeological problems require models with high dimensionality. In such cases, we really have very little practical choice but to use an MCMC approach. Such techniques are now widely accepted not just for their ability to tackle problems found to be intractable using other methods, but because of their ease and speed of use. In recent years an enormous amount of research effort has been invested in developing and improving the efficiency of MCMC methods for obtaining Bayesian posteriors and it is certain that they will play a vital role in future applications in a wide range of subject areas including archaeology.

13.2.5   Joint posteriors and marginal posteriors

Having constructed a suitable model, formulated the prior knowledge and collected the data, what do we actually need in the way of posterior information? At the most basic level the answer to this question is interpretable representations of the posterior densities of the parameters of the model. But what does this mean in practice? For most Bayesian analyses of real problems, the model will be multi-parameter and as a consequence the posterior distribution of one parameter may very well affect the values taken by all the other parameters of interest. This interaction means that we will obtain most detailed information about the posterior densities by considering a number of parameters at the same time and interpreting their inter-relations.

13.2.6   Interpretable summaries

Generally speaking the various techniques used for descriptive statistics have also proved effective for summarizing posterior information: posterior means and posterior standard deviations (as in reporting the on-site and off-site phosphate contents in Chapter 10), highest posterior density regions (in the case of the multi-modal posteriors from radiocarbon dating, see Chapter 9), dot-density plots (in the case of the analysis of field survey in Chapter 10) and dendrograms (as in the investigation of pottery provenance in Chapter 11). Thus display methods such as these are to be recommended for ease of interpretation. However the very familiarity which is their strength may also have its dangers. A dot-density plot which merely summarizes the raw data will appear to the reader very similar to a dot-density plot which is the result of painstaking data analysis. Statistically the two are light years apart, but a casual reading might trap the reader into supposing them equivalent.

Although statisticians are accustomed to using summaries of posterior distributions such as posterior means and the like, we have noted throughout the book that there are cases in which this is not appropriate. Radiocarbon calibration, for example, due to the "wiggly" nature of the curve, results in marginal posterior distributions which are multi-modal. Such distributions cannot realistically be reported using standard summary measures, so histograms and/or highest posterior density regions are to be preferred (see Chapter 9 for examples of both of these). Highest posterior density regions may need a little more explanation to the non-expert reader than, say, histograms, but are already broadly accepted in the literature.

13.3   Realizing the potential

The case studies in earlier chapters have focussed on a variety of archaeological problems, in particular

  (i) a range of problems of archaeological chronology,
 (ii) the analysis of spatial data from surveys, and
(iii) questions of provenance and source.

The selection is fortuitous and was constrained not by theoretical considerations, but by the fact that these are the areas to which the Bayesian paradigm has actually been applied. Plainly there are many areas of archaeological research where adopting the Bayesian approach could be fruitful: environmental questions, the agricultural economy, demography, climate, technology, trade and so forth.

It must be said that the problems considered so far have been at the scientific end of the spectrum of archaeological investigation. Again this bias is coincidental; there is no reason why questions of social significance, for example, should not be addressed in similar ways. Statistics play a vital part in the social sciences, and we can think of no reason why questions reflecting on social structure and social hierarchy, for example, should not be open to Bayesian analysis.

Are there limits to the archaeological problems which can be broached using the Bayesian paradigm? This question is not easily answered. Plainly there is a requirement for a formal mathematical framework: the models we operate with are mathematical and statistical. There is no theoretical reason why any province of human experience should not be open to quantification. Certainly there is general debate, for example, about the extent to which belief systems can be the subject of archaeological enquiry at all, that is, the debate over cognitive archaeology. For those who deny the possibility, inference is impossible; for those who accept it, quantification is both possible and useful. And consequently the Bayesian approach to data investigation is likely to provide a way forward.

13.4   How to realize the potential

13.4.1   Challenges for the archaeological community

If the advantages of Bayesian archaeology are to be realized then certain emphases, requirements and skills need to be developed by archaeologists, and will need to feature in the archaeological curriculum. Interdisciplinary collaboration between archaeologists and statisticians is at the heart of the process. This demands that archaeologists have sufficient numeracy to be comfortable with playing their part in the modelling cycle. They require the ability to analyse a situation, to define the archaeological problem explicitly — in such a way that it is realistic but not so over-refined as to lack all generality. In addition, they need to be prepared to communicate their ideas to statistical experts and to explain in simple terms the importance of their own work within the wider archaeological framework.

With the possible exception of numeracy, these skills are already recognized as being of great importance by the archaeological profession. Indeed, in the past, the most successful archaeologists have been willing to work with experts from a multiplicity of backgrounds and to draw on many different skills — not least communication with their colleagues and the general public. Since archaeologists are also generally agreed about the importance of background knowledge and prior information, we believe that they will soon learn to view the Bayesian approach to data analysis as a very important interpretational tool.

In summary, what is required is an ability to think in terms of uncertainty, probabilities and distributions, familiarity with the language and notation of probability, and a readiness to interpret results summarized through density plots and various statistical estimators. This will come through experience and training: how to develop and handle mathematical models, how to think in statistical terms, how to frame prior probabilities and interpret posteriors. Such skills, like most in archaeology, are probably best learned through real applications.

13.4.2   Challenges for the Bayesian statistical community

The major challenge facing the Bayesian community is to achieve general acceptance that their methodology provides a logical and coherent framework within which to update (scientific) beliefs. This acceptance will not occur overnight, but will be won by Bayesians demonstrating their wares in the market-place, that is, by collaborating with experts in many fields of research and showing that Bayesian methods do work in practice as well as in theory. Of course this will involve the Bayesian statistical community in a major "advertising campaign" in order to win support.

The advent, in the early 1990's, of Markov chain Monte Carlo methods provided Bayesians with the computation power only dreamed of a decade or so earlier. But it would be a very foolish statistician who believed that all the computational problems had been solved. There is still the need to develop more efficient algorithms in response to the ever more complex problems which need to be analysed.

Moreover, there are other challenges that Bayesians need to address. For example, how best to elicit and quantify prior beliefs remains very much an open question that will be with us for the foreseeable future. And the complementary problem of how best to summarize posterior beliefs in a form readily understandable by non-statisticians is a potentially fruitful area of academic research. Both areas require carefully formulated research if non-statisticians are to be convinced of the general applicability of the Bayesian approach to data analysis.

13.4.3   Challenges at the interface

Clearly, we cannot discuss challenges for archaeology and challenges for statistics as if they were completely divorced from one another. Throughout this book we have emphasized the importance of collaboration and interdisciplinary support and understanding. In this final chapter we wish to reiterate this.

Firstly, we note that the proportion of Bayesians amongst the applied statistical community is still relatively small. This situation is changing fast and we hope that the growth in their numbers will keep pace with the demand for their skills both from archaeologists and from specialists in many other disciplines.

Secondly, a change of paradigm in statistics will not be sufficient. There are also implications for the resourcing of practical archaeology. In the past (in Britain at least) most funding has been concentrated on the initial field work. Funding for later interpretative work, such as statistical analyses, has tended to be rather neglected. Throughout this book, we have attempted to highlight the importance we place on all workers playing their part throughout the planning, field work and interpretation stages. Projects will need to be organized and resourced with these requirements in mind. Increasingly, we find ourselves being consulted before field work begins; this is a positive step in the right direction — statisticians are experts (just like ceramicists, bones experts and the like), and need to be included in the early stages of projects.

Finally, we note that using currently available sampling routines, MCMC methods still require the use of powerful computers. This means that archaeologists may struggle to undertake complex computations of posteriors for themselves. Provided that collaboration with statisticians is seen as normal, this should pose no problem; powerful workstations are standard tools amongst the statistical community. If archaeologists wish to begin to undertake Bayesian computations for themselves (and some clearly already do, for example using the software of Ramsey, 1994), then it will be necessary to set aside portions of excavation (or post-excavation) budgets for the purchase of computer hardware and programming skills. We are sure that readers of this book will agree, however, that "going it alone" in this manner is not, in most cases, the ideal way forward since, more so even than with classical statistical methods, the dangers of "black box" Bayesian statistics are immense. The culture of interdisciplinary collaboration needs to be fostered and encouraged.

13.5   Epilogue

In many ways the Bayesian approach, with its emphasis on prior understanding and modelling, distils the essence of what, already, is accepted as good archaeological practice, and even in that limited sense its widespread adoption will bring benefits. But there are even greater gains, for in addition to the Bayesian insistence on clarity and explicitness, the approach permits, indeed insists on, the different domains of expertise, so characteristic of the bringing together of archaeological research, into a coherent whole. Finally and above all else, Bayesian statistics is a framework for interpretation, and we hope to have indicated in this book what we strongly believe to be the case: that our understanding of the past will be greatly enriched by the adoption of the Bayesian approach to interpreting archaeological data.

References

M. J. Aitken. *Science-based Dating in Archaeology.* Longman, London, 1990.

N. W. Alcock, R. E. Howard, R. R. Laxton, C. D. Litton, and D. H. Miles. Leverhulme Cruck Project (Warwick University and Nottingham University Tree-ring Dating Laboratory) results: 1990, *Vernacular Architecture*, 22:45–47, 1991.

M. S. Aldenderfer, editor. *Quantitative Research in Archaeology: Progress and Prospects.* Sage Publications, California, 1987.

R. O. Allen, A. H. Luckenbach, and C. G. Holland. The application of instrumental neutron activation analysis to a study of prehistoric steatite artefacts and source material. *Archaeometry*, 17(1):69–83, 1975.

M. G. L. Baillie. *Tree-ring Dating and Archaeology.* Croom Helm, London, 1982.

M. G. L. Baillie. Checking back on an assemblage of published radiocarbon dates. *Radiocarbon*, 32(3):361–366, 1990.

M. G. L. Baillie and J. R. Pilcher. A simple cross-dating program for tree-ring research. *Tree-Ring Bulletin*, 33:7–14, 1973.

M. G. L. Baillie and J. R. Pilcher. Make a date with a tree. *New Scientist*, 117:48–51, 1988.

N. D. Balaam, K. Smith, and G. J. Wainwright. The Shaugh Moor project: fourth report, environment, context and conclusion. *Proceedings of the Prehistoric Society*, 48:203–278, 1982.

V. Barnett. *Comparative Statistical Inference.* Wiley, New York, second edition, 1982.

M. J. Baxter. *Exploratory Multivariate Analysis in Archaeology.* Edinburgh University Press, Edinburgh, 1994.

T. R. Bayes. An essay towards solving a problem in the doctrine of chances. *Philosophical Transactions of the Royal Society*, 53:370–418, 1763.

J. M. Bernardo and A. F. M. Smith. *Bayesian Theory.* Wiley, Chichester, 1994.

J. Bernoulli. *Ars Conjectandi.* Thurnisiorum, Basel, 1713.

D. A. Berry. *Statistics: A Bayesian Perspective.* Duxbury Press, Belmont, California, 1995.

J. Besag. On the statistical analysis of dirty pictures. *Journal of the Royal Statistical Society*, 48(3):259–302, 1986.

J. Besag and P. J. Green. Spatial statistics and Bayesian computation. *Journal of the Royal Statistical Society Series B*, 55(1):25–37, 1993.

A. M. Bieber, D. W. Brooks, G. Harbottle, and E. V. Sayre. Applications of multivariate techniques to analytical data on Aegean ceramics. *Archaeometry*, 18(1):59–74, 1976.

L. I. Boneva. A new approach to a problem of chronological seriation associated with the works of Plato. *Mathematics in the Archaeological and Historical Sciences*, eds., F. R. Hodson, D. G. Kendall and P. Tautu, 173–185, Edinburgh University Press, Edinburgh, 1971.

S. Bowman. *Interpreting the Past: Radiocarbon Dating.* British Museum Publications, London, 1990.

S. G. E. Bowman and M. N. Leese. Radiocarbon calibration - current issues. *American Journal of Archaeology*, 99 (1):102–105, 1995.

G. E. P. Box and G. C. Tiao. *Bayesian Inference in Statistical Analysis.* Addison-Wesley, Reading, Mass., 1973.

L. D. Broemeling and H. Tsurumi. *Econometrics and Structural Change.* Marcel Dekker, New York, 1987.

C. E. Buck. *Towards Bayesian Archaeology.* PhD thesis, University of Nottingham, Nottingham, 1994.

C. E. Buck, W. G. Cavanagh, and C. D. Litton. The spatial analysis of site phosphate data. *Computer Applications and Quantitative Methods in Archaeology, 1988*, pages 151–160, British Archaeological Reports, Oxford, 1988.

C. E. Buck, W. G. Cavanagh, and C. D. Litton. Image segmentation methods for archaeological field survey data. *New Tools from Mathematical Archaeology*, pages 55–68, Scientific Information Center of the Polish Academy of Sciences, Warsaw, Poland, 1990.

C. E. Buck, J. A. Christen, J. B. Kenworthy, and C. D. Litton. Estimating the duration of archaeological activity using ^{14}C determinations. *Oxford Journal of Archaeology*, 13(2):229–240, 1994b.

C. E. Buck and C. D. Litton. A computational Bayes approach to some common archaeological problems. *Computer Applications and Quantitative Methods in Archaeology, 1990*, eds., K. Lockyear and S. P. Q. Rahtz, 93–99, Tempus Reparatum, British Archaeological Reports, Oxford, 1991.

C. E. Buck and C. D. Litton. Mixtures, Bayes and archaeology. In A. P. Dawid, J. M. Bernardo, J. Berger and A. F. M. Smith, editors, *Bayesian Statistics 5.* Oxford University Press, Oxford, (forthcoming).

C. E. Buck, C. D. Litton, and S. J. Shennan. A case study in combining radiocarbon and archaeological information: the Early Bronze Age settlement of St. Veit-klinglberg, Land Salzburg, Austria. *Germania*, 2:427–447, 1994a.

C. E. Buck, C. D. Litton, and A. F. M. Smith. Calibration of radiocarbon results pertaining to related archaeological events. *Journal of Archaeological Science*, 19:497–512, 1992.

P. Budd, A. M. Pollard, B. Scaife, and R. G. Thomas. Evaluating lead isotope data: further observations. *Archaeometry*, 35(2):252–259, 1993.

P. Budd, A. M. Pollard, B. Scaife, and R. G. Thomas. The possible fractionation of lead isotopes in ancient metallurgical processes. *Archaeometry*, 37(1):143–150, 1995.

D. N. Burghes and M. S. Barrie. *Modelling with Differential Equations.* Ellis Horwood, Chichester, 1981.

A. Caiger-Smith. *Tin Glaze Pottery in Europe and the Islamic World.* Faber and Faber, London, 1973.

M. O. H. Carver. Theory and practice in urban pottery seriation. *Journal of Archaeological Science*, 12(5):353, 1985.

G. Casella and E. I. George. Explaining the Gibbs sampler. *The American Statistician*, 46(3):167–174, 1992.

H. W. Catling, J. F. Cherry, R. E. Jones, and J. T. Killen. The Linear B inscribed stirrup jars and West Crete. *Annual of the British School of Archaeology at Athens*, 75:49–113, 1980.

W. G. Cavanagh, S. Hirst, and C. D. Litton. Soil phosphate, site boundaries and change-point analysis. *Journal of Field Archaeology*, 15(1):67–83, 1988.

W. G. Cavanagh and R. R. Laxton. The structural mechanics of the Mycenaean Tholos tombs. *Annual of the British School of Archaeology at Athens*, 76:109–140, 1981.

J. M. Chambers, W. S. Cleveland, B. Kleiner, and P. A. Tukey. *Graphical Methods for Data Analysis*. Wadsworth, Belmont, California, 1983.

J. A. Christen. *Bayesian Interpretation of ^{14}C Results*. Ph.D. dissertation, University of Nottingham, Nottingham, 1994a.

J. A. Christen. Summarising a set of radiocarbon determinations: a robust approach. *Applied Statistics*, 43(3):489–503, 1994b.

J. A. Christen and C. D. Litton. A Bayesian approach to wiggle-matching. *Journal of Archaeological Science*, 22:719–725, 1995.

O. H. J. Christie, J. A. Brenna, and E. Straume. Multivariate classification of Roman glasses found in Norway. *Archaeometry*, 21:233–241, 1979.

A. J. Clark. Geophysical and chemical assessment of air photographic sites. *Archaeological Journal*, 134:189–193, 1977.

D. L. Clarke, editor. *Models in Archaeology*. Methuen, London, 1972a.

D. L. Clarke. Models and paradigms in contemporary archaeology. *Models in Archaeology*, ed., D. L. Clarke, 1–60, Methuen, London, 1972b.

R. M. Clymo, F. Oldfield, P. G. Appleby, G. W. Pearson, P. Ratnesar, and N. Richardson. The record of atmospheric deposition on a rainwater-dependent peatland. *Philosophical Transactions of the Royal Society of London Series B*, 327:331–338, 1990.

R. M. Cooke. *Experts in Uncertainty: Opinion and Subjective Probability in Science*. Oxford University Press, Oxford, 1991.

G. L. Cowgill. Distinguished lecture in archaeology: beyond criticizing new archaeology. *American Anthropologist*, 95(3):551–573, 1993.

P. T. Craddock, M. R. Cowell, M. N. Leese, and M. J. Hughes. The trace element composition of polished flint axes as indicator of source. *Archaeometry*, 25(2):135–163, 1983.

P. T. Craddock, D. Gurney, F. Pryor, and M. J. Hughes. The application of phosphate analysis to the location and interpretation of archaeological sites. *Archaeological Journal*, 142:361–376, 1985.

F. Daly, D. J. Hand, M. C. Jones, A. D. Lunn, and K. J. McConway. *Elements of Statistics*. Addison Wesley, Wokingham, 1995.

P. E. Damon, D. J. Donahue, B. H. Gore, A. L. Hatheway, T. W. Jull, A. J. T. Linick, P. J. Sercel, L. J. Toolin, C. R. Bronk, E. T. Hall, R. E. M. Hedges, R. Housley, I. A. Law, C. Perry, G. Bonani, S. Trumbore, W. Woelfli, J. C. Ambers, S. G. E. Bowman, M. N. Leese, and M. S. Tite. Radiocarbon dating of the shroud of Turin. *Nature*, 337(611):611–615, 1989.

B. de Finetti. *Theory of Probability: A Critical Introductory Treatment*. volume 1, Wiley, New York, 1974.

B. de Finetti. *Theory of Probability: A Critical Introductory Treatment*. volume 2, Wiley, New York, 1975.

M. H. DeGroot. *Optimal Statistical Decisions*. McGraw-Hill, New York, 1970.

M. H. DeGroot. *Probability and Statistics*. Addison-Wesley, Reading, Mass., 1986.

A. P. Dempster, Bayesian inference in applied statistics. In J. M. Bernardo, M. H. De Groot, D. V. Lindley and A. F. M. Smith, editor, *Bayesian Statistics*, 266–281, Oxford University Press, Oxford, 1980.

L. Devroye. *Non-uniform Random Variate Generation*. Springer, New York, 1986.

F. Djindjian. Fifteen years contributions of the French school to quantitative archaeology. *Computer Applications and Quantitative Methods in Archaeology, 1989*, pages 207–218, British Archaeological Reports, Oxford, 1989.

J. E. Doran and F. R. Hodson. *Mathematics and Computers in Archaeology.* Edinburgh University Press, Edinburgh, 1975.

J. Earman. *Bayes or Bust? A Critical Examination of Bayesian Confirmation Theory.* MIT, Cambridge, Mass., 1992.

B. English and P. R. Freeman.Prehistoric Scottish fish bones. *Bulletin in Applied Statistics*, 8:120–133, 1981.

M. D. Escobar and M. West. Bayesian density estimation and inference using mixtures. *Journal of the American Statistical Association*, 90:577–585, 1995.

W. Feller. *An Introduction to Probability Theory and its Applications.* volume I. Wiley, New York, third edition, 1968.

N. R. J. Fieller. Archaeostatistics: old statistics in ancient contexts. *The Statistician*, 42:279–295, 1993.

N. R. J. Fieller and E. Flenley. Statistical analysis of particle size and sediments In *Computer Applications and Quantitative Methods in Archaeology, 1987*, eds., C. L. N. Ruggles and S. P. Q. Rahtz, pages 79–94, British Archaeological Reports, Oxford, (1988).

R. A. Fisher. On the probable error of a coefficient of correlation deduced from a small sample. *Metron*, 1(1), 1921.

M. Fletcher and G. R. Lock. *Digging Numbers: Elementary Statistics for Archaeologists.* Oxford University Committee for Archaeology, Oxford, 1991.

H. C. Fritts. *Tree Rings and Climate.* Academic Press, New York, 1976.

N. H. Gale and Z. A. Stos-Gale. Evaluating lead isotope data: comments on E. V. Sayre, K. A. Yener, E. C. Joel and I. L. Barnes. *Archaeometry*, 34(2):311–317, 1992.

E. Geake, Bronze-age computer dating. *New Scientist*. 142(1924):17, 1994.

A. E. Gelfand and A. F. M. Smith. *Bayesian Practice.* Wiley, London, forthcoming.

A. Gelman and D. B. Rubin. A single series from the Gibbs sampler provides a false sense of security. *Bayesian Statistics 4*, eds., A. P. Dawid, J. M. Bernardo, J. Berger and A. F. M. Smith, 627–633, Oxford University Press, Oxford, 1992a.

A. Gelman and D. B. Rubin. Inference from iterative simulation using multiple sequences. *Statistical Science*, 7(4):457–511, 1992b.

A. G. Gelman, J. B. Carlin, H. S. Stern, and D. B. Rubin. *Bayesian Data Analysis.* Chapman and Hall, London, 1995.

S. Geman and D. Geman. Stochastic relaxation, Gibbs distributions and the Bayesian restoration of images. *I.E.E.E. Transactions in Pattern Analysis and Machine Intelligence*, 6:721–741, 1984.

W. Gilks, S. Richardson, and D. J. Spiegelhalter, editors. *Markov Chain Monte Carlo in Practice*, Chapman and Hall, London, 1995.

B. R. Hallam, S. E. Warren, and C. Renfrew. Obsidian in the Western Mediterranean: characterisation by neutron activation analysis and optical emission spectroscopy. *Proceedings of the Prehistoric Society*, 42:85–110, 1976.

N. Hammond, A. Aspinall, S. Feather, J. Hazelden, T. Gazard, and S. Agrell. Maya jade: source location and analysis. *Exchange Systems in Prehistory*, eds., T. K. Earle and J. E. Ericson, pages 35–67, Academic Press, New York 1977.

W. K. Hastings. Monte Carlo sampling methods using Markov chains and their applications. *Biometrika*, 57:97–109, 1970.

J. Heyman. The stone skeleton. *International Journal of Solid Structures*, 2:249–299, 1966.

J. Heyman. *The Masonry Arch*. Ellis and Horwood, Chichester, 1982.

J. Hillam, C. M. Groves, D. M. Brown, M. G. L. Baillie, J. M. Coles, and B. J. Coles. Dendrochronology of the English Neolithic. *Antiquity*, 64:210–220, 1990.

D. V. Hinkley. Inference about the change-point in a sequence of random variables. *Biometrika*, 57:1–17, 1970.

I. Hodder and C. Orton. *Spatial Analysis in Archaeology*. Cambridge University Press, Cambridge, 1976.

R. M. Hogarth. Cognitive processes and the assessment of subjective probability distributions. *Journal of the American Statistical Association*, 70:271–294, 1975.

C. Howson and P. Urbach. *Scientific Reasoning: The Bayesian Approach*. Open Court, Illinois, second edition, 1993.

R. E. Hughes. Intrasource chemical variability of artifact-quality obsidians from the Casa-Diablo area, California. *Journal of Archaeological Science*, 21(2):263–271, 1994.

J. G. Hurst, D. S. Neal, and H. J. van Beuningen. Pottery produced and traded in North-West Europe 1350-1650. *Rotterdam Papers*, VI, 1986.

H. S. Jeffreys. *Theory of Probability*. University Press, Oxford, third edition, 1961.

I. Johnson, ed. *Methods in the Mountains: Proceedings of UISPP Commission IV Meeting, Mount Victoria, Australia. August 1993*. Archaeological Computing Laboratory, University of Sydney, Sydney, 1994.

N. L. Johnson and S. Kotz. *Distributions in Statistics: Continuous Univariate Distributions I*. Wiley, Chichester, 1970.

N. L. Johnson and S. Kotz. *Distributions in Statistics: Continuous Univariate Distributions II*. Wiley, Chichester, 1970.

N. L. Johnson and S. Kotz. *Distributions in Statistics: Continuous Multivariate Distributions*. Wiley, Chichester, 1972.

R. E. Jones. *Greek and Cypriot Pottery: A Review of Scientific Studies*, volume 1 of *Fitch Laboratory Occasional Paper*, The British School at Athens, Athens, 1986.

W. J. Judge and L. Sebastian, editors. *Quantifying the present and predicting the past: theory, method and application of archaeological predictive modeling*. U.S. Department of the Interior Bureau of Land Management, Denver, Colorado, 1988.

D. G. Kendall. Seriation from abundant matrices. *Mathematics in the Archaeological and Historical Sciences*, eds., F. R. Hodson, D. G. Kendall and P. Tauta, 215–252, Edinburgh University Press, Edinburgh, 1971a.

D. G. Kendall. Abundance matrices and seriation in archaeology. *Zeit, Wahrscheinlichkeitstheorie und Verwandte Gebiete*, 17:104–112, 1971b.

K. Kintigh. Intra-site spatial analysis in archaeology. *Mathematics and Information Science in Archaeology: A Flexible Framework*, ed., A. Voorrips, 165–200, Holos, Bonn, 1990.

N. Köckler. *Numerical Methods and Scientific Computing using Software Libraries for Problem Solving*. Oxford University Press, Oxford, 1994.

A.N. Kolmogorov. *Grundbegriffe der Wahrscheinlichkeitsrechnung*. Berlin, 1933.

D. Kronberg. Distribution of cross-correlations in two-dimensional time series with application to dendrochronology. Technical Report 72, Department of Theoretical Statistics, University of Aarhus, Aarhus, Denmark, 1981.

M. Lavine and M. West. A bayesian method for classification and discrimination. *The Canadian Journal of Statistics*, 20(4): 451–461, 1992.

R. R. Laxton. A measure of pre-q-ness with applications to archaeology. *Journal of Archaeological Science*, 3:43–54, 1976.

R. R. Laxton and C. D. Litton. Information theory and dendrochronology. *Science and Archaeology*, 24:9–24, 1982.

R. R. Laxton and C. D. Litton. *An East Midlands master tree-ring chronology and its use for dating vernacular buildings.* University of Nottingham: Department of Classical and Archaeological Studies, Nottingham, 1988. Monograph Series III.

R. R. Laxton and C. D. Litton. Construction of a Kent master dendrochronological sequence for oak, A.D. 1158 to 1540. *Medieval Archaeology,* XXXIII:90–98, 1989.

R. R. Laxton and J. Restorick. Seriation by similarity and consistency. In *Computer Applications and Quantitative Methods in Archaeology, 1989,* 229–240, British Archaeological Reports, Oxford, (1989).

W. F. Libby, E. C. Anderson, and J. R. Arnold. Age determination by radiocarbon content: worldwide assay of natural radiocarbons. *Science,* 109:227–228, 1949.

D. V. Lindley. *Introduction to Probability and Statistics from a Bayesian Viewpoint.* University Press, Cambridge, 1965.

D. V. Lindley. *Making Decisions.* Wiley, London, second edition, 1985.

C. D. Litton. Methods and problems of dendrochronology as exemplified by Bede House Farm, North Luffenham, Leicestershire. *Historic Buildings and Dating by Dendrochronology,* ed., G. I. Meirion-Jones, Oxbow Press, Oxford, (forthcoming).

C. D. Litton and M. N. Leese. Some statistical problems arising in radiocarbon calibration. *Computer Applications and Quantitative Methods in Archaeology, 1990,* eds., K. Lockyear and S. P. Q. Rahtz, pages 101–109, Tempus Reparatum, Oxford, 1991.

C. D. Litton and H. J. Zainodin. Statistical models of dendrochronology. *Journal of Archaeological Science,* 18:429–440, 1991.

T. Madsen. Multivariate statistics and archaeology. *Multivariate Archaeology: Numerical Approaches in Scandinavian Archaeology,* ed., T. Madsen, pages 7–27, Jutland Archaeological Society Publications, Moesgard, 1988.

S. W. Manning and B. Weninger. A light in the dark: archaeological wiggle matching and the absolute chronology of the close of the Aegean Late Bronze Age. *Antiquity,* 66:636–663, 1992.

K. V. Mardia, J. T. Kent, and J. M. Bibby. *Multivariate Analysis.* Academic Press, London, 1982.

N. Metropolis, A. W. Rosenbluth, M. N. Rosenbluth, A. H. Teller, and E. Teller. Equation of state calculations by fast computing machines. *Journal of Chemical Physics,* 21:1087–1092, 1953.

W. J. Meyer. *Concepts of Mathematical Modelling.* McGraw-Hill, New York, 1984.

A. M. Mood, F. R. Graybill, and D. C. Boes. *Introduction to the Theory of Statistics.* McGraw-Hill, New York, 1974.

B. J. T. Morgan. *Elements of Simulation.* Chapman and Hall, London, 1984.

N. K. Namboodiri. *Matrix Algebra: An Introduction.* Sage, Beverly Hills, CA,.1984.

J. C. Naylor and A. F. M. Smith. An archaeological inference problem. *Journal of the American Statistical Association,* 83(403):588–595, 1988.

A. O'Hagan. *Probability: Methods and Measurements.* Chapman and Hall, London, 1988.

A. O'Hagan. *Kendall's Advanced Theory for Statisticians: Bayesian Inference.* volume 2B, Edward Arnold, London, 1994.

A. C. H. Olivier. Excavation of a Bronze Age funerary cairn at Manor Farm, near Borwick, North Lancashire. *Proceedings of the Prehistoric Society,* 53:129–186, 1987.

C. R. Orton. *Mathematics in Archaeology.* William Collins, Glasgow, 1980.

C. R. Orton. Quantitative methods in the 1990s. *Computer Applications and Quantitative Methods in Archaeology, 1991,* eds., G. Lock and J. Moffett, pages 137–140, Tempus Reparatum, British Archaeological Reports, Oxford, 1992.

P. A. Parkes. *Current Scientific Techniques in Archaeology.* Croom Helm, Beckenham, Kent, 1986.

J. Pavuk, ed. *Actes du XIIe Congrès International des Sciences Préhistoriques et Protohistoriques: Bratislava, 1-7 Septembre 1991,* volume 1–4, Bratislava, 1993. Union Internationale des Sciences Préhistoriques et protohistoriques.

M. F. Pazdur and A. Krzanowski. Fechados radiocarbónicos para los sitios de la cultura Chancay. *Estudios Sobre la Cultura Chancay, Perú,* ed., A. Krzanowski, 115–132, Universidad Jaguelona, Kraków, Poland, 1991.

G. W. Pearson. Precise calendrical dating of known growth-period samples using a 'curve fitting' technique. *Radiocarbon,* 28(2A):292–299, 1986.

G. W. Pearson, B. Becker, and F. Qua. High-precision C-14 measurement of German and Irish oaks to show the natural C-14 variations from 7890 to 5000 BC. *Radiocarbon,* 35(1):93–104, 1993.

G. W. Pearson, J. R. Pilcher, M. G. L. Baillie, D. M. Corbett, and F. Qua. High-precision ^{14}C measurements of Irish oaks to show the natural ^{14}C variations from AD 1840 - 5210 BC. *Radiocarbon,* 28(2B):911–34, 1986.

G. W. Pearson and M. Stuiver High-precision calibration of the radiocarbon time scale, 500-2500 BC. *Radiocarbon,* 28(2B):839–862, 1986.

G. W. Pearson and M. Stuiver. High-precision bi-decadal calibration of the radiocarbon time scale, 500–2500 BC. *Radiocarbon,* 35(1):25–33, 1993.

E. Pernicka. Evaluating lead isotope data: comments on E. V. Sayre, K. A. Yener, E. C. Joel and I. L. Barnes. *Archaeometry,* 34(2):322–326, 1992.

E. Pernicka. Crisis or catharsis in lead isotope analysis? *Journal of Mediterranean Archaeology,* 8(1):59–64, 1995.

E. Pernicka, F. Begemann, S. Schmitt-Strecker, and A. P. Grimanis. On the composition and provenance of metal artefacts from Poliochni on Lemnos. *Oxford Journal of Archaeology,* 9(3):263–298, 1990.

S. Piggott. *Ancient Europe from the Beginnings of Agriculture to Classical Antiquity.* Edinburgh University Press, Edinburgh, 1965.

S. J. Press *Bayesian Statistics: Principles, Models and Applications.* Wiley, New York, 1989.

B. Proudfoot. The analysis and interpretation of soil phosphorus in archaeological contexts. In D. A. Davidson and M. L. Shackley, editors, *Geoarchaeology,* 93–113, Duckworth, London, 1976.

A. Raftery and S. Lewis. How many iterations in the Gibbs sampler? In *Bayesian Statistics 4,* eds., A. P. Dawid, J. M. Bernardo, J. Berger and A. F. M. Smith, 765–776, Oxford University Press, Oxford, 1992.

C. B. Ramsey. Analysis of chronological information and radiocarbon calibration: the program. *Archaeological Computing Newsletter,* 41:11–16, 1994.

D. W. Read. Some comments on the use of mathematical models in anthropology. *American Antiquity,* 39(1):3–15, 1974.

D. W. Read. Archaeological theory and statistical methods: discordance, resolution, and new directions. *Quantitative Research in Archaeology: Progress and Prospects,* ed., M. S. Aldenderfer, 151–184, Sage Publications, California, 1987.

R. Reece. Are Bayesian statistics useful to archaeological reasoning? *Antiquity,* 68:848–850, 1995.

C. Renfrew and P. Bahn. *Archaeology: Theories, Methods and Practice.* Thames and Hudson, London, 1991.

C. Renfrew, J. E. Dixon, and J. R. Cann. Obsidian and early cultural contact in the Near East. *Proceedings of the Prehistoric Society,* 32: 30–72, 1966.

C. Renfrew, J. E. Dixon, and J. R. Cann. Further analysis of Near Eastern obsidians.

Proceedings of the Prehistoric Society, 34: 319–331, 1968.

P. M. Rice. *Pottery Analysis: A Source Book*. University of Chicago Press, Chicago, Ill., 1987.

T. W. Richards. The composition of Athenian pottery. *American Chemical Journal*, 17:152–154, 1895.

B. D. Ripley. *Statistical Inference for Spatial Processes*. Cambridge University Press, Cambridge, 1988.

G. O. Roberts. Convergence diagnostics of the Gibbs sampler. *Bayesian Statistics 4*, eds., A. P. Dawid, J. M. Bernardo, J. Berger and A. F. M. Smith, pages 777–784, Oxford University Press, Oxford, 1992.

W. S. Robinson. A method for chronologically ordering archaeological deposits. *American Antiquity*, 16:293–301, 1951.

S. M. Ross. *A First Course in Probability*. Macmillan, New York, third edition, 1988.

J. H. Rowe. Stratigraphy and seriation. *American Antiquity*, 26(3):324–330, 1961.

C. Ruggles. You can't have one without the other? I.T. and Bayesian statistics and their possible impact within archaeology. *Science and Archaeology*, 28:8–15, 1986.

E. V. Sayre, K. A. Yener, E. C. Joel, and I. L. Barnes. Statistical evaluation of the presently accumulated lead isotope data from Anatolia and surrounding regions. *Archaeometry*, 34(1):73–106, 1992.

S. Shennan. *Quantifying Archaeology*. Edinburgh University Press, Edinburgh, 1988.

S. J. Shennan. Ausgrabungen in einer frühbronzezeitlichen siedlung auf dem Klinglberg, St. Veit im Pongau, Salzburg (1985–1988). *Archaeologia Austriaca*, 73:35–48, 1989.

E. A. Slater and J. O. Tate, editors. *Science and Archaeology Glasgow 1987*. British Archaeological Reports, Oxford, 1988.

A. F. M. Smith. A Bayesian approach to inference about a change-point in a sequence of random variables. *Biometrika*, 62:407–416, 1975.

A. F. M. Smith. Bayesian computational methods. *Philosophical Transactions of the Royal Society London Series A*, 337:369–386, 1991.

A. F. M. Smith and G. O. Roberts. Bayesian computation via the Gibbs sampler and related Markov chain Monte Carlo methods. *Journal of the Royal Statistical Society Series B*, 55(1):3–23, 1993.

J. Q. Smith. *Decision Analysis: A Bayesian Approach*. Chapman and Hall, London, 1988.

J. K. St Joseph and D. R. Wilson. Air reconnaissance: recent results, 41. *Antiquity*, 50:237–239, 1976.

A. H. Stroud *Approximate Calculation of Multiple Integrals*. Prentice-Hall, Englewood Cliffs, New Jersey, 1971.

A. Stuart and J. K. Ord. *Kendall's Advanced Theory of Statistics: Classical Inference and Relationships*, volume 2. Edward Arnold, London, fifth edition, 1991.

M. Stuiver and G. W. Pearson. High-precision calibration of the radiocarbon time scale, AD 1950 - 500 BC. *Radiocarbon*, 28(2B):805–838, 1986.

M. Stuiver and G. W. Pearson. High-precision bi-decadal calibration of the radiocarbon time scale, AD 1950–500 BC and 2500–6000 BC. *Radiocarbon*, 35(1):1–23, 1993.

M. Stuiver and P. Reimer. Extended ^{14}C data base and revised Calib 3.0 ^{14}C age calibration program. *Radiocarbon*, 35(1):215–230, 1993.

A. Thom. A statistical examination of the megalithic sites in Britain. *Journal of the Royal Statistical Society Series A*, 118:275–295, 1955.

A. Thom. The megalithic unit of length. *Journal of the Royal Statistical Society Series A*, 125:243–251, 1962.

A. Thom. The larger units of length of megalithic man. *Journal of the Royal Statistical Society Series A*, 127:527–533, 1964.

A. Thom. *Megalithic Sites in Britain*. Clarendon Press, Oxford, 1967.

H. A. Thompson and R. E. Wycherley. *The Athenian Agora XIV*. American School of Classical Studies, Princeton, 1972.

L. Tierney. Markov chains for exploring posterior distributions. *The Annals of Statistics*, 22(4):1701–1728, 1994.

M. S. Tite. *Methods of Physical Examination in Archaeology*. Seminar Press, London, 1972.

P. F. Velleman and D. C. Hoaghlin. *Applications, Basics, and Computing of Exploratory Data Analysis*. Duxbury Press, Boston, Mass., 1981.

A. Voorrips. *Mathematics and Information Science in Archaeology: A Flexible Framework*. Holos, Bonn, 1990.

A. J. B. Wace. *Chamber Tombs at Mycenae*. Oxford, Society of Antiquaries, 1932.

B. Weninger. High-precision calibration of archaeological radiocarbon dates. *Acta Interdisciplinaria Archaeologica*, IV:11–53, 1986.

T. M. L. Wigley, P. J. Jones, and K. R. Briffa. On the average value of correlated time series, with applications in dendrochronology and hydrometeorology. *Journal of Climatic and Applied Meteorology*, 23:201–213, 1984.

D. R. Wilson. *Air Photo Interpretation for Archaeologists*. Batsford, London, 1982.

H. J. Zainodin. *Statistical Models and Techniques for Dendrochronology*. Ph.D. dissertation, University of Nottingham, Nottingham, 1988.

J. A. Zeidler, C. E. Buck, and C. D. Litton. The integration of archaeological phase information and radiocarbon results from the Jama River valley, Ecuador: a Bayesian approach. *Latin American Antiquity*, forthcoming.

Index

Index compiled by Geoffrey C. Jones